Lecture Notes Biochemistry

J. K. CANDLISH BSc, PhD

Associate Professor, Biochemistry Department,
Faculty of Medicine,
National University of Singapore

BLACKWELL SCIENTIFIC PUBLICATIONS

OXFORD LONDON

EDINBURGH BOSTON MELBOURNE

© 1984 by
Blackwell Scientific Publications
Editorial offices:
Osney Mead, Oxford, OX2 0EL
8 John Street, London, WC1N 2ES
9 Forrest Road, Edinburgh, EH1 2QH
52 Beacon Street, Boston
    Massachusetts 02108, USA
99 Barry Street, Carlton
    Victoria 3053, Australia

First published 1984

Typeset by Oxprint Ltd, Oxford and printed and
bound at the Alden Press, Oxford

DISTRIBUTORS

USA
    Blackwell Mosby Book Distributors
    11830 Westline Industrial Drive
    St Louis, Missouri 63141

Canada
    Blackwell Mosby Book Distributors
    120 Melford Drive, Scarborough
    Ontario, M1B 2X4

Australia
    Blackwell Scientific Book Distributors
    31 Advantage Road, Highett
    Victoria 3190

Lecture Notes on Biochemistry

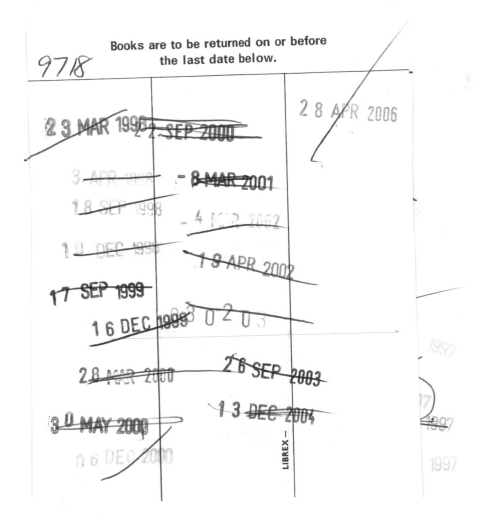

Preface

A biochemistry course of one sort or another is now obligatory for students in a large number of subjects—one can think immediately of pharmacy, genetics, botany, zoology, bacteriology, nutrition and dietetics, dentistry, veterinary medicine, and of course human medicine itself. There is no reason to suppose that biochemistry is intrinsically uninteresting or difficult to understand, but there is every reason to think that the majority of students nevertheless see biochemistry merely as an imposed part of their vocational course, a hurdle to be overcome during the progress to a professional qualification. Under these circumstances the desire for a basic understanding of the subject is lacking; it is the assumption that understanding is sought and can be imparted which, however, directs the endeavours of the biochemistry teacher and of the writer of texts. Encountering hordes of students who merely want to amass sufficient facts to reproduce in an examination and gain promotion in the course, such teachers and writers are apt to be disappointed. This, however, is not a realistic response. The teacher is in a sense trying to impart an appreciation of his life's work, but the student can hardly be expected to set an equivalent evaluation on that, groping as he is towards an MBBS or a BDS or a BVM; he or she is subject to the competing claims of other disciplines, and moreover wants some spare time to see life, on the realistic assumption that youth does not last forever.

Having assimilated a minimum of chemistry and of basic biology, there is no reason why the student should not be introduced to a concise compendium of metabolic pathways which comprise the main part of biochemistry examinations. That is the rationale of this book. 'Understanding' is not neglected, but is here subordinated to the presentation of readily reproducible facts. 'Understanding' is a questionable concept in any case. As a professional biochemist I may feel that I 'understand' protein synthesis, but I have a mental picture either of some formulae connected by arrows on a page, or perhaps of molecules interacting at enzyme surfaces. What relationship these imaginings have to events inside the cell is a moot point and in any case a student can quickly bring me to the limits of my understanding, or knowledge, by putting a few consecutive 'why's.

Thus although this text forms a part of the well-known Blackwell 'Lecture Notes' series, in a real sense it is a 'crammer'. Crammers were formerly Latin and Greek crammers, and there is a fair analogy here. Many groaning youths who had their noses ground into the classics seemingly learned very little at the time, yet were proud, in later years, to be able to translate Latin tags or derive the meaning of a word from its Greek roots. The mature mind

'Mlle de Saint-Yves . . . pretended for some time not to understand him; and he was obliged to explain himself more clearly. One word, used with some reserve, brought on another less delicate, which was succeeded by one still more expressive.'

Voltaire: *L'Ingenue*

CRAMMER 1635 1. One who or that which crams poultry, etc. 2. *Colloq.* One who crams pupils for an examination, etc. rarely a student who crams his subject 1813 3. *slang* a lie 1862.

Shorter Oxford English Dictionary 1973

later realizes that the classics are an integral part of Western consciousness and culture. In the same way biochemistry has become a background subject for the great family of life sciences. Even a little knowledge is valuable, and later on, if it is necessary, a return can be made to the subject and a deeper insight gained.

In this text, perhaps unusually, there is a certain amount of repetition to avoid too much cross-referencing. Certain topics tend to arise in various areas of biochemistry, perhaps in slightly different guises, and they have sometimes been repeated with a different emphasis rather than cross-referenced. This should help students to work through specific topics without going backwards and forwards to pick up threads. (I have not, however, been able to omit cross-references entirely.)

It is in trying to achieve a logical and coherent sequence, within the mass of facts and concepts which (if we may use the word 'designed' for a moment) were certainly not designed to fit into a book with a beginning, an end, and a middle, that writers of textbooks find themselves particularly tied up in knots. The difficulty of course resides in deciding what knowledge is a prerequisite for some other topic. The plan adopted here is as follows. Two chapters, one on chemistry, the other on biological terminology, are presented prior to structural biochemistry and the elements of metabolism respectively. They can be skipped by anyone familiar with the material. The elements of metabolism are given in 7–10 as the catabolic pathways, and 11 deals with the various biosyntheses. This might seem to some to separate catabolism and anabolism to an unfortunate degree, but it can be justified on the grounds that:

1 many substances such as cholesterol, proteins, and nucleic acids are synthesized in animals by complex pathways which have no analogies in catabolism; and

2 it is one of the tenets of biochemistry that a substance is always synthesized by a pathway which is not the reverse of its catabolism.

After the anabolic pathways, the student could pass on either to 12, the section on regulation, or to 13, which incorporates a miscellany of topics which come into examination questions.

Finally, there is a section labelled 'Checklists' (14.1) which should be self-explanatory on inspection, as should 'Examinations in biochemistry' (15).

Thus a student who knows no chemistry or biology might glance through 1 and 6 before going on to 2–5, 7–10, 11, and 13, missing out the regulation section (12) as being too tricky on a first approach. Somebody with chemistry and biology might work through 2–5, 7–10, 11, and 13, and only go on to 12, 14, and 15 should time permit. It is evident that the topics in red on the contents page represent the basics among basics, biochemistry pared to the bone.

The sections within 11 and 13, and indeed those within the sequence 7–10 can be taken in any order, whichever seems most pressing.

The diagrams are all reproducible by the student. None consist of photographs of famous biochemists and none are merely ornamental. Most have a short mnemonic or summary diagram attached, further to emphasize the framework.

At various points, short glossaries are included where it is felt that the terminology may seem tortuous. These are also referenced in the contents page.

Finally, the limitations associated with any attempt to set down biochemistry in book form, indeed, to get it over in any way at all, have been set down in 6.2 for those who might be interested.

Acknowledgements

I am indebted, for many corrections and helpful discussions, to Dr. Carol Brownson (City of London Polytechnic), Dr. Carol Lovelidge (Guy's Hospital Medical School), to Dr. Rosanne Mullings and Dr. Teo Tian Seng (National University of Singapore) as well as to the staff of Messrs Blackwells, especially Mr Robert Campbell, for their patience and for giving me the opportunity to express some ideas about biochemistry teaching.

Contents

Abbreviations Employed

ACP	Acyl carrier protein
ACTH	Adrenocorticotrophic hormone
ADP	Adenosine diphosphate
AMP	Adenosine monophosphate
ATP	Adenosine triphosphate
ATPase	Adenosine triphosphatase
B_6	Coenzymes derived from pyridoxal phosphate and related compounds
B_{12}	Cobamide coenzymes (and precursor vitamins)
C	Cytosine
cAMP	Cyclic AMP
cDNA	Complementary DNA
CDP	Cytidine diphosphate
CMP	Cytidine monophosphate
CoA-SH	Coenzyme A
CoQ	Coenzyme Q or abiquinone
C-terminal	Amino acid residue with free α-carboxyl
CTP	Cytidine triphosphate
cyt	Cytochrome
d	Deoxyribose
DNA	Deoxyribonucleic acid
DOPA	Dihydroxyphenylalanine
ϵ	Electron
FAD	Flavin adenine dinucleotide (oxidized)
FH_4	Tetrahydrofolic acid
FMN	Flavin mononucleotide (oxidized)
FSH	Follicle-stimulating hormone
G	Gibb's free energy
G	Guanine
GDP	Guanosine diphosphate
GMP	Guanosine monophosphate
GSH	Glutathione (reduced)
GTP	Guanosine triphosphate
GTT	Glucose tolerance test
H	Enthalpy
Hb	Haemoglobin
HDL	High density lipoprotein
IgG	Immunoglobulin 'G'
ITP	Inosine triphosphate
J	Joules
K_m	Michaelis constant
LDH	Lactate dehydrogenase
LDL	Low density lipoprotein
LH	Luteinizing hormone
mRNA	Messenger RNA
NAD^+	Nicotinamide adenine dinucleotide (oxidized)
$NADP^+$	Nicotinamide adenine dinucleotide phosphate (oxidized)
N-terminal	Amino acid residue with free α-amino
Pi	Inorganic phosphate
P/O (ratio)	Atoms phosphate esterified to atoms oxygen consumed
PPi	Pyrophosphate
PRPP	Phosphoribosyl pyrophosphate
RNA	Ribonucleic acid
rRNA	Ribosomal RNA
SAM	S-adenosyl methionine
S	Entropy
STH	Somatotrophin (growth hormone)
T	Thymine
tRNA	Transfer RNA
TPP	Thiamine pyrophosphate
TSH	Thyroid-stimulating hormone
TTP	Thymidine triphosphate
U	Uracil
UDP	Uridine diphosphate
UDPG	Uridine diphosphoglucose
UMP	Uridine monophosphate
UTP	Uridine triphosphate
VLDL	Very low density lipoproteins

1 The Language of Chemistry

1.0 Preamble

To learn biochemistry you do not have to know as much chemistry as is often supposed, for there is a certain specialization about the structures and reactions to be encountered. However, some chemistry is certainly a prerequisite, if only because the language of chemistry is used; moreover, in no case does a living organism obviate the laws of chemistry. There follows a brief refresher, or alternatively, introduction to biochemistry. It can be skipped if, running your eyes over the italicized items, you are fairly sure of their significance.

1.1 Atoms and molecules

An *element* is composed of atoms, all of the same type. However, within an element the precise composition of the atoms may vary and the variations are called *isotopes* or *nuclides*. Each atom consists of a *nucleus* containing *neutrons* (differences in the number of which, within the same element, give rise to the isotope phenomenon) having zero charge and a fixed number of *protons*, each of which has a unit positive charge. The nucleus is surrounded by *electrons*, each having unit negative charge equal to that of the proton, and arranged in layers or shells. Atoms of different elements combine with each other, usually in fixed ratios, to give *compounds*. Observation of these ratios suggests that each element has a definite combining power, that is, its *valency* or valencies. Thus, one chlorine atom (Cl) can exist in combination with one hydrogen atom (H) in hydrogen chloride, HCl. Four H can exist in combination with one carbon (C) as methane, CH_4. Four Cl can be combined with one C in carbon tetrachloride, CCl_4. HCl, CH_4, and CCl_4 represent *molecules* and are therefore *molecular formulae*. H and Cl appear to have a combining power of one, C of four. Oxygen (O) seems to have a combining power of two so that water, or hydrogen oxide, can be represented by the molecular formula H_2O.

Valency is related to the nature of the electrons in the outer shell of an element. Valencies of the elements important in biochemistry are given in Table 1.1.1, which also shows *atomic weights*. Atomic weights have been worked out for all the elements in terms of the relation of their mass to the mass of the carbon atom, taken as 12. Biochemists are often not as progressive as chemists and still tend to use the term 'atomic weight' when 'atomic mass unit' and 'dalton' are now preferred, although the numerical values of all these are identical. *Molecular weight* (also used when 'relative molecular

Table 1.1.1 Important elements in biochemistry

Element	Symbol	Valency	Atomic weight
Hydrogen	H	1	1
Carbon	C	4	12
Nitrogen	N	3	14
Oxygen	O	2	16
Sodium	Na	1	23
Magnesium	Mg	2	24
Phosphorus	P	5 or 3	31
Sulphur	S	2	32
Chlorine	Cl	1	35.5
Potassium	K	1	39
Iron	Fe	2 or 3	56
Iodine	I	1	127

mass' or 'molar mass' are more correct) is found by adding the atomic weights in a molecular formula. Thus water, H_2O, has a molecular weight of $(2 \times 1) + 16 = 18$ and glucose $C_6H_{12}O_6$ has a molecular weight of $(6 \times 12) + (12 \times 1) + (6 \times 16) = 180$. Expressed in grams the molecular weight is the *gram molecular weight*. A gram molecular weight always contains the same number of molecules, and this number is the *mole*, abbreviation *mol*.

The outer shell of electrons tends to gain or lose electrons such that it will achieve the same number of electrons as the outer shell of one of the rare gases: helium, neon, argon, krypton, or xenon. The gases, which are un-reactive or 'inert', evidently have outer shell configurations which are stable. (There are two electrons in the outer shell of helium, and eight in the others.) Since, by the second law of thermodynamics, systems tend to move to the most stable state, there is a tendency for atoms to gain or lose electrons from the outer shell to achieve the condition of an inert gas.

In most of the organic compounds to be encountered electrons are not completely gained or lost but are shared, in *covalent bonds*. The hydrogen atom has one electron, and we can represent this for the time being as H . Carbon has four electrons in the outer shell, represented

$$\times \overset{\times}{\underset{\times}{C}} \times$$

If they interact such that four hydrogens gain a share of one electron to give them the outer shell of helium, and carbon gains a share of four electrons to give it the outer shell of neon, then methane, CH_4, may be represented

$$H \overset{H}{\underset{H}{\overset{\times \bullet}{\underset{\bullet \times}{C}}}} H$$

Oxygen has six electrons in the outer shell, and so needs two more to give it the same number as neon. Water may then be represented as

$$H \overset{\times \times}{\underset{\times \times}{O}} H \quad or \quad H—O—H \quad or \quad H_2O$$

Two shared electrons are represented as . or——. Ethane, in which one carbon shares electrons with another carbon, is written

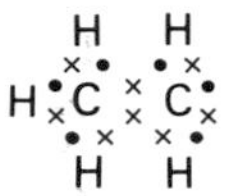

$$H \overset{\times}{\underset{\times}{\overset{\bullet}{\underset{\bullet}{C}}}} \overset{H}{\underset{H}{}} \overset{\bullet}{\underset{\bullet}{\overset{\times}{\underset{\times}{C}}}} H \quad or \quad C_2H_6 \quad or \quad H_3C \cdot CH_3 \quad or \quad H_3C—CH_3$$

In a *saturated* compound (such as ethane), any two atoms are linked by no more than one shared pair of electrons. In an *unsaturated* compound, four electrons may be shared by two carbon atoms. Four shared electrons (i.e. two covalent bonds) are denoted by $=$. Ethene is

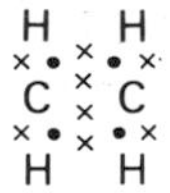

$$\overset{H}{\underset{H}{}} \overset{\times}{\underset{\times}{\overset{\bullet}{\underset{\bullet}{C}}}} \overset{\times}{\underset{\times}{\overset{\bullet}{\underset{\bullet}{C}}}} \overset{H}{\underset{H}{}} \quad or \quad C_2H_4 \quad or \quad H_2C{=}CH_2$$

It is because carbon atoms so readily share electrons with each other that very long chains, branched or unbranched, may exist, giving rise to the diversity of organic compounds. Chains may be fused into ring structures like cyclohexane (Fig. 1.1.1).

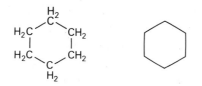

Fig. 1.1.1 Cyclohexane.

Rings may also be aromatic, usually represented as alternate single and double bonds in Fig. 1.1.2 (a)–(d), but sometimes as in Fig. 1.1.2 (e).

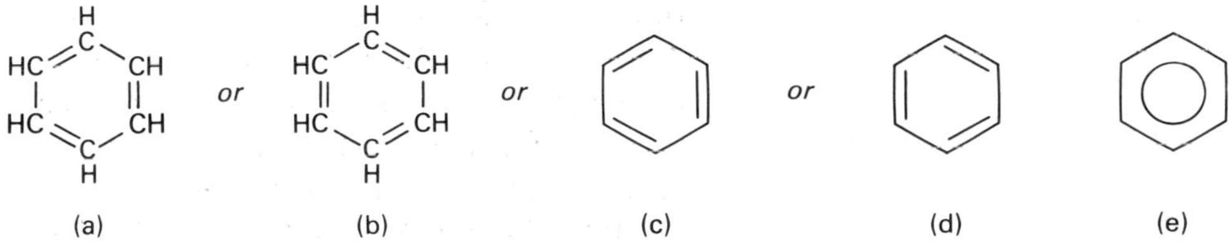

Fig. 1.1.2 Benzene.

(a)   (b)   (c)   (d)   (e)

Fig. 1.1.3 π-orbitals.

As for all formulae, these are an attempt to represent, on the surface of a piece of paper, the arrangement of nuclei and electrons in space. According to *orbital theory*, which describes the area in space where an electron is most likely to exist, alternate single and double bonds have no existence in benzene; rather, orbitals overlap above and below the plane of the ring to give regions of evenly distributed electron density, called π-bonds, as in Fig. 1.1.3. (Of course, all overlapping electron orbitals are bonds.)

The cyclohexane and benzene rings are homocycles, that is the rings contain only one type of atom: carbon. Heterocycles contain some other atom. Prominent heterocycles are shown in Fig. 1.1.4.

Fig. 1.1.4

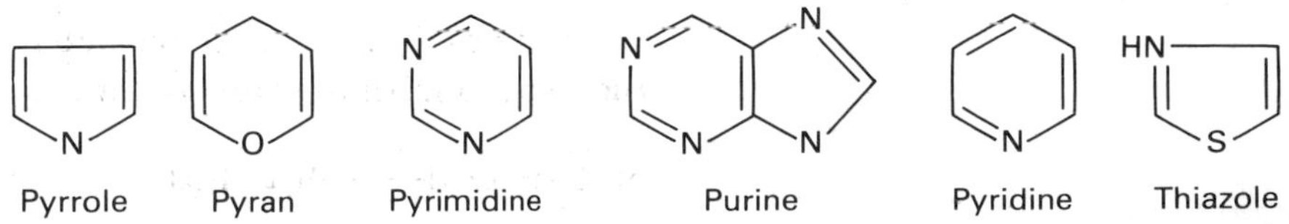

Pyrrole   Pyran   Pyrimidine   Purine   Pyridine   Thiazole

There are other types of bonding besides covalent bonding. *Electro-valent* or *ionic bonding* occurs when entities of opposite charge are held together by electrostatic attraction. This occurs in crystals and in large molecules where they contribute to cohesion between chains. At its simplest, we can represent it as

$$\vdash\!\text{NH}_3{}^+ \ \ {}^-\text{OOC}\!\dashv$$

Coordinate linkages are a type of covalency but both electrons in the bond are donated by one atom. For example, in certain compounds nitrogen and iron are linked by coordination:

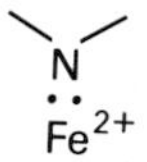

$$\underset{\overset{..}{\text{Fe}^{2+}}}{\text{N}}$$

Hydrogen bonds occur when hydrogen is covalently bonded to a more electro-negative element (that is one which has the ability to attract shared electrons and thus increase the negative charge in its own vicinity). If electrons are drawn away from the hydrogen atom, it acquires a small positive charge and so can attract another electronegative atom. Prominent electronegative elements are O, F, and N. So an amine and a carboxyl function can be mutually attracted because the H of the amine has the small charge repre-sented as $\delta+$

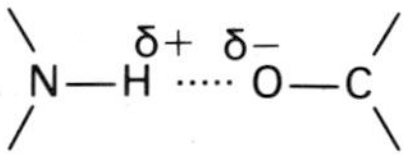

$$\overset{\backslash}{\underset{/}{\text{N}}}\text{—H}\ \overset{\delta+}{}\ \overset{\delta-}{\cdots\cdots\text{O}}\text{—}\overset{/}{\underset{\backslash}{\text{C}}}$$

In a hydrogen bond, a hydrogen atom serves as a bridge between two electronegative atoms, holding one by a covalent bond and the other by electrostatic forces. This electrostatic bond is about one-tenth of the strength of a normal covalent bond.

van der Waals Forces Two atoms approaching each other become subject to an attractive force at about 0.3–0.4 nm distance. Such interactions are known as van der Waals bonds or forces. At even shorter distances there is a repulsive effect. A single van der Waals bond is weak in itself but a cluster of them in a restricted area can produce significant cohesion between molecules.

Hydrophobic bonds Water tends to form the maximum number of hydrogen bonds with itself (Fig. 1.1.5).

Fig. 1.1.5

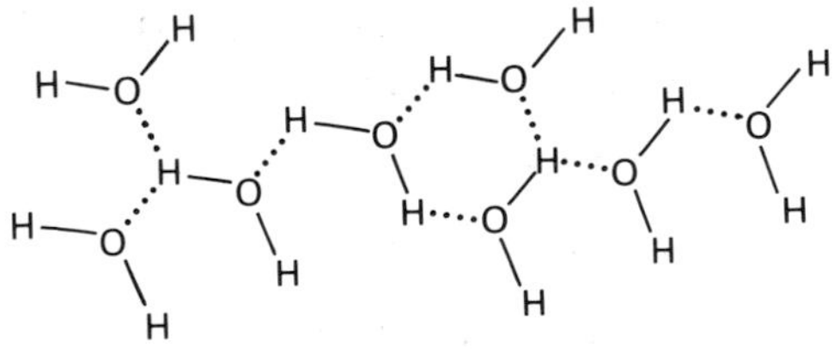

Fig. 1.1.6

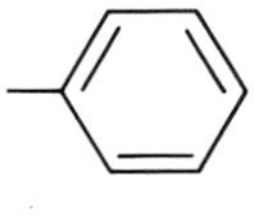

Benzyl

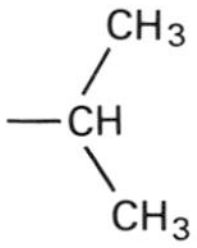

Isopropyl

If apolar functional groups such as benzyl and isopropyl (Fig. 1.1.6) are introduced into an aqueous environment, they disrupt this ordered structure, causing lacunae or cavities. If two cavities occupied by apolar functions come together, then the number of disrupted water-water hydrogen bonds can be partially restored, and the apolar groups appear to have the facility to bind to each other. This is the basis of the hydrophobic bond. As for hydrogen bonds and van der Waals forces, they are significant only when they occur cumulatively.

1.2 Functional groups

Chains and rings bear *functional groups*, that is arrangements of atoms which have definite properties. They can be written in various ways, but are normally given as compactly as possible in formulae; some of the functional groups prominent in biological materials are given in Table 1.2.1.

1.3 Ionization

Whereas covalent bonding involves the sharing of electrons, they can also be transferred completely from one atom to another such that both achieve the outer shell of an inert gas. An atom which has either lost or gained one or more electrons is called an *ion*. Thus sodium (Na) has one electron in its outer shell, and chlorine (Cl) has seven. If Na were to lose one, and Cl gain one, then each would have the outer shell of an inert gas (neon and argon respectively), but Na would then have a positive charge (a *cation*) and chlorine a negative charge (the *anion chloride*).

$$Na^{\bullet} \quad {}_{\times}^{\times\times}Cl_{\times}^{\times} \;\; or \; Na^+ \; Cl^-$$

In a solid crystalline lattice the cation and anion are held together in this way by electrostatic attraction, that is they are bound by an ionic or electrovalent bond, but in solution they become separately *solvated* (surrounded by water molecules). The term *ionization* refers to the tendency of a functional group or atom to gain or lose electrons and thus bear a charge. The carboxyl function ionizes thus to give carboxylate and a hydrogen ion.

$$-COOH \rightarrow -COO^- + H^+$$

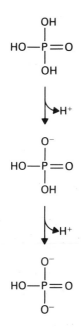

Fig. 1.3.1

Some substances have two or more ionizing functions, such as phosphoric acid (Fig. 1.3.1). If in this way the ionization produces a hydrogen ion (which you can also call a proton), the ionizing substance is termed an *acid*. Pure water ionizes thus

$$H_2O \rightleftharpoons H^+ + OH^-$$

or more accurately

$$2H_2O \rightleftharpoons H_3O^+ + OH^-$$

so H_2O is an acid.

The rate of formation of H^+ and OH^-, which can be designated v_1, is

Table 1.2.1 Some functional groups

Structural representation	Name	Example	
—OH	Alcohol, hydroxyl	C_2H_5OH	Ethyl alcohol
$-\overset{\overset{O}{\parallel}}{C}-H$ or —CHO	Aldehyde	$CH_3—CHO$	Acetaldehyde
$>C{=}O$ or $>C{=}O$	Ketone, carbonyl	$CH_3—CO—CH_3$	Acetone
—C—O—C—	Ether	$CH_3—O—CH_3$	Diethyl ether
—C—S—C—	Thioether	$H_3C—S—CH_2—CH_2—\overset{\overset{NH_3{}^+}{\mid}}{C}H\,COO^-$	Methionine
⟨benzene ring⟩—OH	Phenol	$^-OOC—\overset{\overset{NH_3{}^+}{\mid}}{C}H—CH_2$—⟨benzene ring⟩—OH	Tyrosine
$—NH_2$ or $—NH_3{}^+$	Amino	$^-OOC—CH_2—NH_3{}^+$	Glycine
—COOH or $—COO^-$	Carboxyl	$CH_3—COOH$	Acetic acid
$-\overset{\overset{O}{\parallel}}{C}-NH_2$ or $—CONH_2$	Amide	⟨pyridine ring⟩$CONH_2$	Nicotinamide
—SH	Thiol, sulphydryl	$HS—H_2C—\overset{\overset{NH_3{}^+}{\mid}}{C}H—COO^-$	Cysteine
—S—S—	Disulphide	$^-OOC—\overset{\overset{NH_3{}^+}{\mid}}{C}H—CH_2—S—S—H_2C—\overset{\overset{NH_3{}^+}{\mid}}{C}H—COO^-$	Cystine
$-\overset{\overset{O}{\parallel}}{C}-O—$ or —CO—O—	Ester	$R—O—CO—R'$	Waxes*
$-\overset{\overset{O}{\parallel}}{C}-S—$ or —CO—S—	Thioester	$CH_3—CO—S—CoA$	Acetyl CoA*

*Representations of complex formulae

proportional to $[H_2O]$ (the concentration of H_2O in mol/l). The rate of the back reaction v_2 is proportional to the product $[H^+][OH^-]$. It follows that

$$v_1 = k_1[H_2O] \text{ and } v_2 = k_2[H^+][OH^-]$$

where k_1 and k_2 are constants.

At equilibrium the rates of the forward and back reactions are equal, so

$$k_1 [H_2O] = k_2 [H^+][OH^-]$$

or $k_1/k_2 = [H^+][OH^-]/[H_2O]$

k_1/k_2 can be replaced by a new constant, the equilibrium or dissociation constant K, so

$$K = [H^+][OH^-]/[H_2O]$$

Water is a weak acid, and $[H_2O]$ is so much larger than $[H^+][OH^-]$ in an aqueous system that it may be regarded as a constant. This yields yet another constant K_w, the ion product of water, and

$$K_w = [H^+][OH^-]$$

Experimentally K_w is found to be 10^{-14} at 25 °C. In an aqueous system at 25 °C this is invariable, and so if $[H^+]$ is 10^{-3} mol/l then $[OH^-]$ must be 10^{-11} mol/l. If $[H^+]$ is 10^{-7} mol/l then $[OH^-]$ is also 10^{-7} mol/l. This is evidently the point of neutrality. If $[H^+]$ is greater than 10^{-7} then H^+ ions predominate and the solution is acidic. If OH^- predominates then the solution is alkaline. The concentration of hydrogen ions is still usually expressed on the pH scale, in which the concentration of H^+ in mol/l is transformed to the negative logarithm. Thus at the neutral point

$$[H^+] = 10^{-7} \text{ mol/l}$$
$$- \log [H^+] = - \log 10^{-7}$$
$$= 7$$

Acid solutions have a pH below 7, alkaline solutions above 7.

Just as a functional group with a tendency to donate a hydrogen ion or proton is an acid, so a group with a tendency to accept a proton is a *base*. Therefore the hydroxyl ion OH^- is a base. A prominent basic function in biological compounds is the amino group.

$$-NH_2 + H^+ \rightarrow -NH_3^+$$

Conversely again, the $-NH_3^+$ is an acid.

In proteins there is an important basic function in the side chain derived from imidazole, which ionizes as shown in Fig. 1.3.2. At the pH of the body fluids the imidazole group is partly in the ionized (acidic) form and partly in the unionized (basic) form. If H^+ becomes available in a solution containing imidazole (due to the introduction of acid), then the form on the left in Fig. 1.3.2 (base) will tend to take up some of it to give the form on the right (acid) so that the equilibrium is displaced towards the right. In other words, the

Fig. 1.3.2

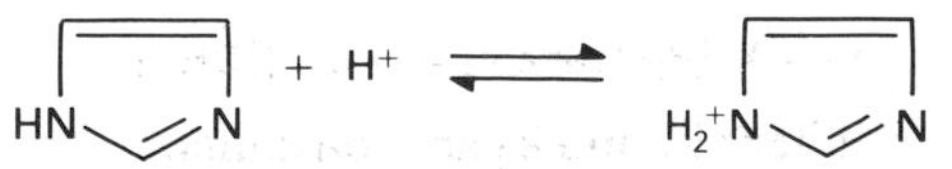

addition of acid is to some extent cancelled, so that we call imidazole a *buffer*. A buffer tends to cancel the increase in acidity caused by the addition of (small) amounts of H^+ or the rise in alkalinity due to the addition of (small) amounts of OH^-. The bicarbonate ion is a buffer since it tends to accept a proton to yield carbonic acid.

$$H_2CO_3 \rightleftharpoons HCO_3^- + H^+$$

Acid strength is expressed as pK or pK_a, which is the pH at which half the total dissociation of H^+ takes place. The lower the pK_a, the greater is the tendency to shed H^+. The dissociation constant for the reaction above, K, can be expressed as

$$K = [H^+][HCO_3^-]/[H_2CO_3]$$

Taking negative logarithms

$$-\log K = -\log ([HCO_3^-]/[H_2CO_3])$$

By definition $-\log[H^+]$ is pH, and we define $-\log K$ as pK. Rearranging

$$pH = pK + \log[HCO_3^-]/[H_2CO_3]$$

This is the Henderson-Hasselbalch equation, predictive of pH in buffering media. pK strictly refers to the 'activities' of ionizing species, not their concentrations; where concentrations are used, however, one may refer to the apparent dissociation constant K' yielding pK'. Applying the Henderson-Hasselbalch equation to imidazole, above, would give Fig. 1.3.3.

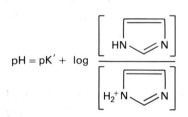

Fig. 1.3.3

1.4 Reactions and catalysis

A chemical reaction is usually written to show the proportions of reacting atoms or molecules. For the oxidation of glucose

$$C_6H_{12}O_6 + 6O_2 \rightarrow 6CO_2 + 6H_2O$$

One molecule of glucose reacts with six molecules of oxygen to yield six molecules each of carbon dioxide and water. Taking account of such proportions is called *stoichiometry*. The arrow here is of the 'one-way' type, to show that the equilibrium is far to the right, or that the reaction, for practical purposes, is irreversible. Many reactions are habitually written with a double arrow $\longleftrightarrow$ or $\rightleftharpoons$ which in biochemistry implies that the equilibrium mixture has considerable amounts of both reactants and products, for example

glucose 6-phosphate $\rightleftharpoons$ fructose 6-phosphate

When glucose is oxidized, as above, it is observed that the reaction is accompanied by a release of energy, detectable as heat. Energy considerations are the province of *thermodynamics*. When a reaction proceeds in such a way that energy is released, it is a matter of observation that not all the energy can be used to perform useful work. The part of it which can, however, be effectively utilized is called the free energy change, ΔG. In the oxidation of one mole of glucose ΔG is about 688 kcal (2.7 MJ). By convention, if energy is evolved it is accorded a negative sign. If energy is taken up ΔG is positive. The total energy change in a reaction is termed the enthalpy change, ΔH. This must be made up from ΔG plus that amount of energy which is lost and cannot be used to perform useful work, or

$\Delta G = \Delta H -$ fraction of energy lost

The part which is lost is found to be greater the higher the temperature, although temperature is not the only factor involved. There is also a change of state, referred to as a change in *entropy*, ΔS. We can quantify the energy lost as the product of the temperature and this measure of the change in state, ΔS. Thus

$\Delta G = \Delta H - T \Delta S$

The units of ΔG and ΔH are both joules. The temperature unit is the kelvin. Thus ΔS must have the units $J. K^{-1}$ to allow dimensional equality in the equation,

$J = J - K (J/K)$

There is nothing mysterious about entropy. It is a factor introduced to account for the impossibility of harnessing all the energy in a changing system and it must necessarily have the units J/K. Since the above equation in effect accounts for the enthalpy change in a reaction, it is another expression of the first law of thermodynamics, that energy cannot be created or destroyed. Moreover it incorporates the second law, which states that in a system moving towards an equilibrium there is an obligatory increase in entropy. Or, systems cannot move away from equilibrium.

However much information we have about the energy change in a reaction we are not in a position to predict its velocity. If glucose is allowed to stand in oxygen, in thermodynamic terms the reaction to give CO_2 and H_2O (ΔG being negative) is spontaneous, but as a matter of common observation it is very slow. The rate of the reaction can be speeded up by various means: raising the temperature perhaps, or using one or more *catalysts*. A catalyst cannot alter the free energy change of a reaction, but it speeds up the rate of the reaction by lowering the activation energy. In a solution molecules are in random thermal motion but do not react unless they have some extra energy, that is the *activation energy* at the point of collision. A catalyst decreases the activation energy necessary for reaction by orientating the molecules such that they can react one with another, or undergo a reaction pathway different from that in the absence of the enzyme (Fig. 1.4.1).

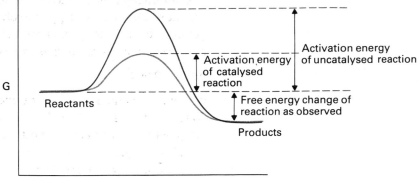

Fig. 1.4.1 The activation energy of a catalysed and uncatalysed reaction.
Much greater energy of activation is required by the uncatalysed mechanism. A catalyst does not, however, alter the observed free energy change.

Activation energy of uncatalysed reaction

Activation energy of catalysed reaction

Free energy change of reaction as observed

G

Reactants

Products

Progress of reaction in time

The initial and final energy states are not altered by a catalyst; only the activation energy is changed. This means that ΔG for a reaction cannot be altered by a catalyst. In any reaction ΔG obviously has some relationship to the state of equilibrium; one can imagine that if the equilibrium is pushed in a certain direction then more or less energy becomes available. If ΔG is not altered by a catalyst, then we may conclude that the final equilibrium, quantified by the equilibrium constant, will not be altered either. A catalyst does not alter the equilibrium mixture in a reaction, only the rate at which that equilibrium is attained.

A reaction proceeds at a rate proportional to the concentration of the reactants. If A and B are mixed, yielding C and D, as in

$$A + B \rightarrow C + D$$

then usually the rate of formation of C and D (the 'forward' reaction), v_1, is proportional to $[A][B]$; or,

$$v_1 = k_1[A][B]$$

Conversely the rate of formation, v_2, of A and B (the 'back' reaction) is

$$v_2 = k_2[C][D]$$

At equilibrium the rate of the forward reaction is equal to the rate of the back reaction

$$v_1 = v_2 \quad or \quad k_1/k_2 = [C][D]/[A][B]$$

k_1/k_2 yields a new constant K, the equilibrium constant. If K is greater than unity, the equilibrium evidently favours the forward reaction. (Again K applies strictly to 'activities'; K′ is the apparent equilibrium constant, when concentrations are used.)

Free energy released in a movement towards equilibrium obviously has some relationship to the magnitude of the change, that is how much of C and

D is produced from A and B. The greater the flux, purely on commonsense grounds, the more energy is likely to be available. Put another way, the greater the K, the greater the ΔG. The relationship, derived from the thermodynamic reasoning, is expressed as

$$\Delta G° = -RT\ln K$$

where $\Delta G°$ is the standard free energy change and the reactants are all at concentration 1 mol/l. R is the universal gas constant, T is temperature (degrees kelvin) and ln is logarithm to the base e.

Three types of reactions are particularly important in biochemistry. *Oxidation/reduction* involves loss or gain of electrons. The removal of electrons is oxidation; addition of electrons is reduction. In an oxidation/reduction (redox) reaction the electrons lost by the reducing agent are gained by the oxidizing agent. Oxidation was originally defined as combination with oxygen,

$$C + O_2 \rightarrow CO_2$$

By extension the definition means removal of hydrogen, for example see Fig. 1.4.2, which would be coupled to oxygen utilization, as

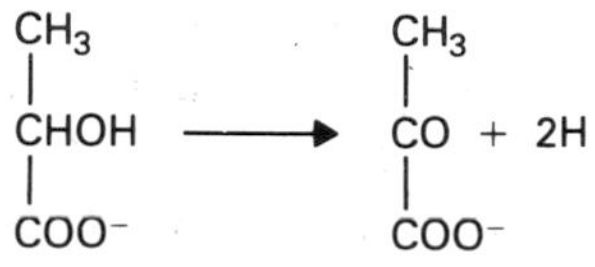

$$2H + \tfrac{1}{2}O_2 \rightarrow H_2O$$

and by further extension, to loss of electrons, since H consists of one proton and one electron. Thus in the reaction

$$Fe^{2+} \rightarrow Fe^{3+} + \text{electron}$$

Fe^{2+} is oxidized because it loses an electron. Written to include oxygen the reaction would be

$$2Fe^{2+} + \tfrac{1}{2}O_2 + 2H^+ \rightarrow 2Fe^{3+} + H_2O$$

The *redox potential* is the quantitative expression of reducing potency: it is calculated by reference to a hydrogen electrode, where hydrogen gas at standard pressure and temperature yields electrons thus

$$H_2 \rightarrow 2H^+ + 2\,\text{electrons}$$

The potential of this half-cell (similar to one electrode of a battery) is arbitrarily assigned a redox potential of 0 volts at pH0. At pH 7.0 it is -0.42 V. The stronger a reducing agent, the more negative is its redox potential. Conversely, oxidizing agents have a positive redox potential: for the O_2/H_2O half-cell it is $+0.82$ V.

Hydrolysis is the splitting of a covalent bond by introduction of the elements of water. Examples in biochemistry are extremely numerous and the hydrolysis of an ester is one:

$$R—CO—OR + H_2O \rightarrow R—COOH + HOR$$

$$
\begin{array}{ccc}
CH_3 & & CH_3 \\
| & & | \\
CHOH & \longrightarrow & CO + 2H \\
| & & | \\
COO^- & & COO^-
\end{array}
$$

Fig. 1.4.2

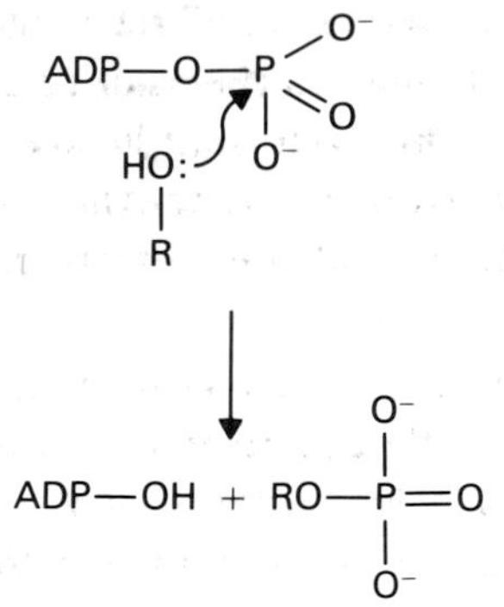

Fig. 1.4.3 Nucleophilic attack. ADP represents a nucleoside, R a sugar.

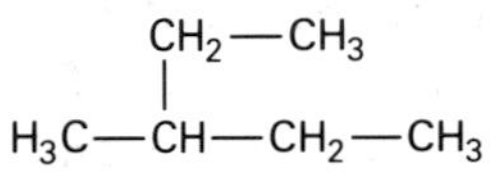

(a) 3-Methyl pentane

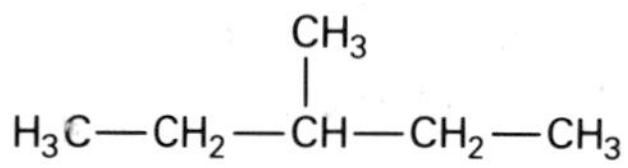

(b) 3-Methyl pentane

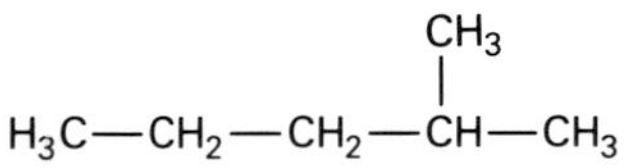

(c) 2-Methyl pentane

Fig. 1.5.1

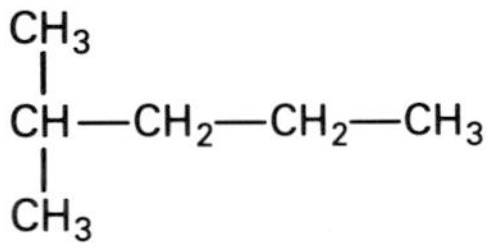

(The reverse reaction, by which the elements of water are eliminated, is known as 'condensation'.)

Nucleophilic attack is the reaction of an electron-rich centre with an electron-poor centre (the electron-rich atom or group is the *nucleophile*). Sugars with a nucleophilic hydroxyl may attack nucleoside phosphate esters to give sugar phosphate esters (Fig. 1.4.3).

1.5 Isomers

The *molecular formula* of a compound is a simple way of indicating the proportions of atoms combined to make up the molecule. Thus C_2H_6O is the molecular formula of ethyl alcohol or ethanol. However, C_2H_6O is also the molecular formula of dimethyl ether or 'ether'. If these are, however, written as C_2H_5OH and $CH_3.O.CH_3$ respectively, as *structural formulae*, it is clear that one is an alcohol, the other an ether. Two such compounds, having the same kind and number of atoms, but arranged differently, are termed isomers. The exact manners in which structural isomers are written is not important as long as the functional groups are correctly positioned relative to each other. Thus 3-methyl pentane written as in Fig. 1.5.1 (a) is the same as Fig. 1.5.1 (b) but different from its isomer 2-methyl pentane written as in Fig. 1.5.1 (c), which is the same as

Similarly a compound such as nicotinamide may be written as in Fig. 1.5.2.

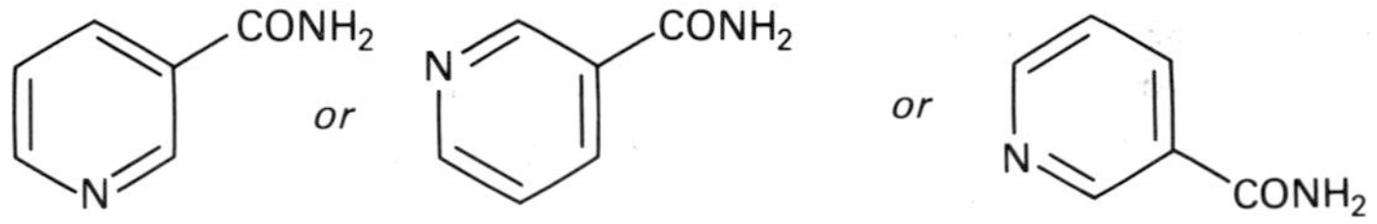

Fig. 1.5.2 Nicotinamide.

In the aromatic ring of benzene or pyrimidine the alternative representations, shown in Fig. 1.5.3, do not represent isomerism. Either of the forms may be written since the electrons are considered to be delocalized, above and below the plane of the rings.

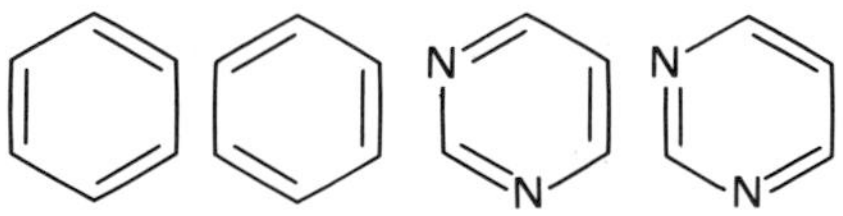

Fig. 1.5.3   (a) Benzene   (b) Pyrimidine

Isomerism may be conceptually divided into *structural isomerism*, as above, and *stereoisomerism*. The latter refers to compounds which are identical in their atoms and functional groups, and moreover all functional groups occupy the same positions relative to each other in chains or rings, but have some difference in the way the groups are orientated in space. A problem immediately becomes apparent in that such orientations have to be represented in the plane of a sheet of paper, and of course there are precise rules for doing this.

Of the important types of stereoisomerism, *optical isomerism* is frequently encountered in biochemistry. It arises from the nature of the carbon atom, which can be considered, when it bears four substituents or atoms, to be at the centre of a tetrahedron. If all the substituents are different, then the molecule is asymmetric, having a mirror image molecule which cannot be superimposed on it. The relationship is conveniently represented by the ball-and-stick type of model in Fig. 1.5.4. The prefixes D- and L- refer to the orientation about a reference carbon, which is arbitrarily chosen. All the amino acids in mammalian tissues bear the prefix L- because they have the same configuration about the reference carbon (the α-carbon). In no sense are they 'opposite' to the common sugar designation D- of which the reference carbon is the one penultimate one in the open-chain form (see 3.2). The D- and L- (Fischer) convention is not the only one for representing optical isomerism but is still the most widely used in biochemistry.

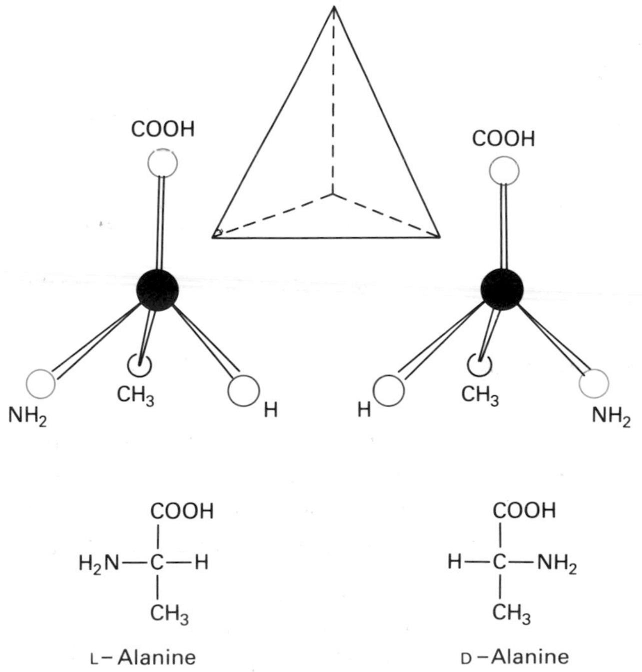

Fig. 1.5.4 Ball-and-stick 'models' to represent optical isomerism.
The (black) carbon atom has four different atoms or substituents, represented by different colours, and these can be imagined to mark off the corners of a tetrahedron with the carbon at the centre. The two forms shown are mirror images of each other, that is there is no way that one can be superimposed on the other such that the substituents are adjacent. The corners of the tetrahedron can further be imagined to be projected onto the plane of a piece of paper. If the compound is the amino acid alanine then the substituents are —NH$_2$,— COOH, —CH$_3$ and—H. The form on the left is projected as shown and is designated L-alanine while the form on the right is D-alanine. The D-isomer is not synthesized by mammalian tissues.

Geometric isomerism is the other important type of stereoisomerism, and is found in compounds with one or more double bonds. The overlapping

13

orbitals in a single bond are such that free rotation can, in general, occur round the bond

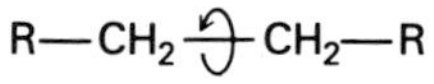

$$R—CH_2 \text{—} CH_2—R$$

However, the four electrons in a double bond have orbitals such that free rotation is not possible. A double-bonded structure has two possible orientations for substituent groups either on the same side (*cis*) or on opposite sides (*trans*), as in Fig. 1.5.5. Restricted orientation is also a feature of fused ring systems (Fig. 1.5.6). In decalin, substituents on 9 and 10 may be *cis* or *trans* to each other. Such isomerism is exhibited, among biological substances, by the steroids (3.1 and 13.1).

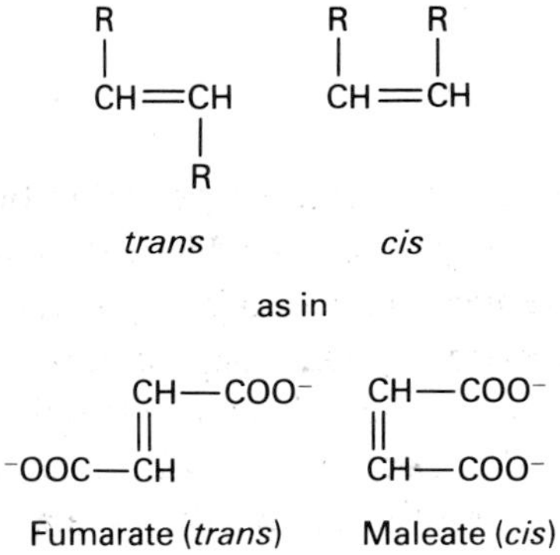

R
|
CH=CH
|
R

trans

R   R
| |
CH=CH

cis

as in

CH—COO⁻
||
⁻OOC—CH

Fumarate (*trans*)

CH—COO⁻
||
CH—COO⁻

Maleate (*cis*)

Fig. 1.5.5

Fig. 1.5.6   *trans*-Decalin derivative   *cis*-Decalin derivative

4 Biopolymers

4.0 Preamble

The bonds exhibited by the intermediate compounds are repeated in the *biopolymers*, that is high molecular weight compounds found in tissues. In all cases the syntheses of these large molecules are specialized processes which we have to study separately. One class of compounds appearing in 2.1 and 3.1 are missing below, namely the lipids. Although in many cases they combine with polymers to give complex molecules, for example the lipoproteins, they do not occur in long chains as do the carbohydrates, amino acids, and nucleotides. However, lipids, due to their property of excluding water, tend to associate together in globules or (if stabilized) *micelles*.

Once again, it is as well to learn these structures thoroughly, as a firm foundation for any discussion of their syntheses and biological functions. Before proceeding to metabolism you should therefore know, as though by ancient habit, the formulae of twenty or so key substances. This is only the first of many exercises purely in memorizing.

4.1 Four formulae

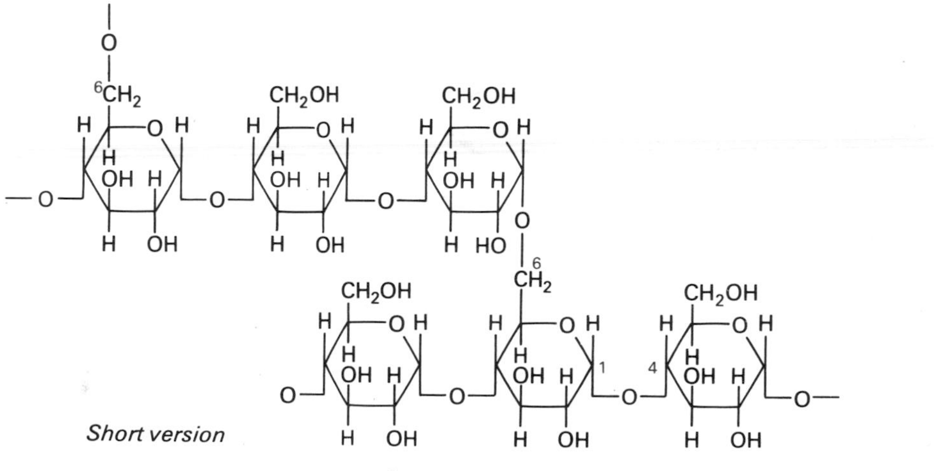

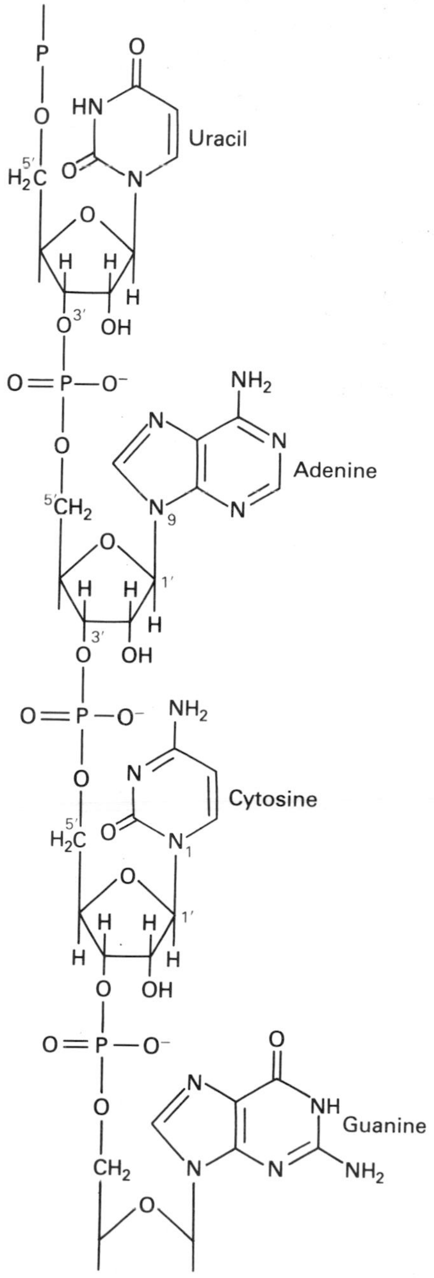

Uracil
Adenine
Cytosine
Guanine

2 RNA

Short version

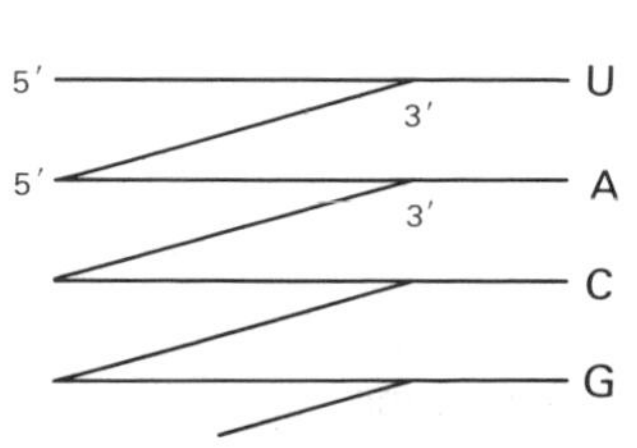

1 Glycogen

(Cont. overpage.)

Guanine

Cytosine

Adenine

Thymine

Short version

5' end ——C····G—— 3' end

——T····A——

3' end   5' end

or even

5'   3'

C····G
T····A

3'   5'

···· Hydrogen bonds

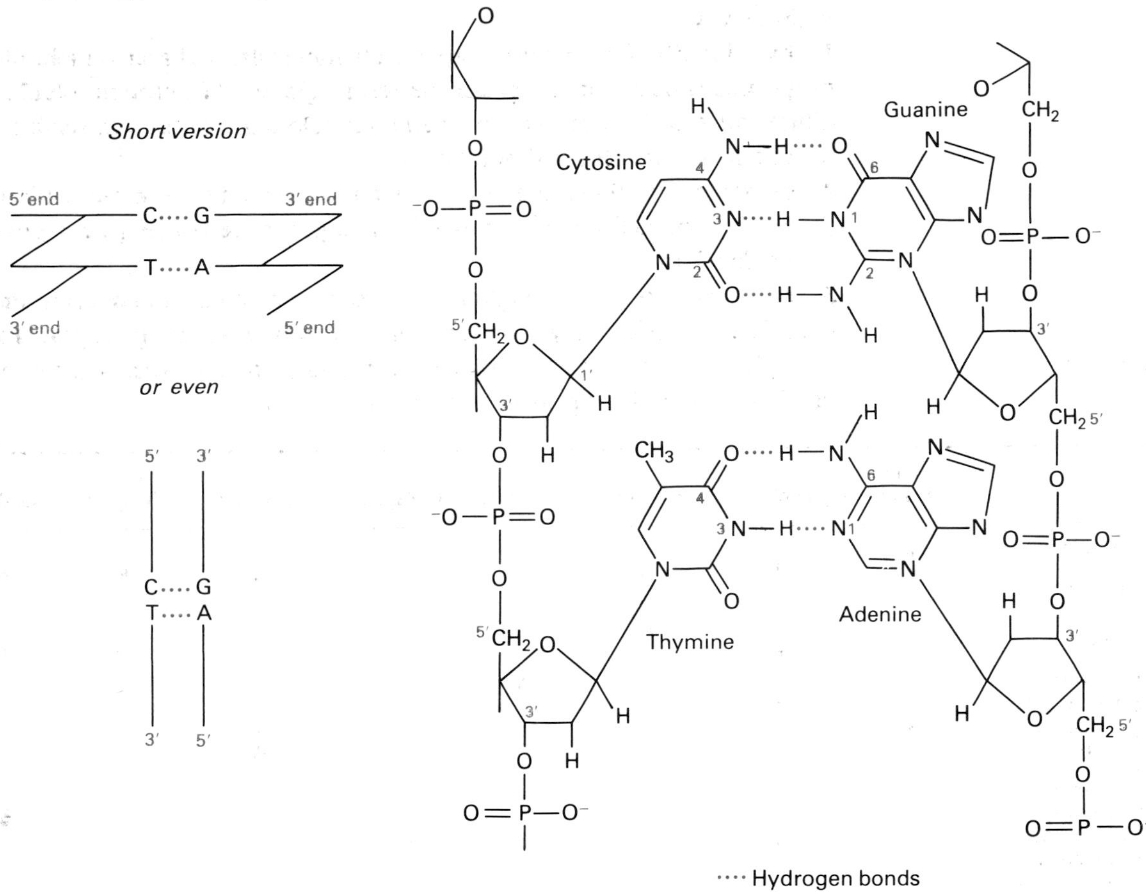

4 α-CHAIN OF HAEMOGLOBIN

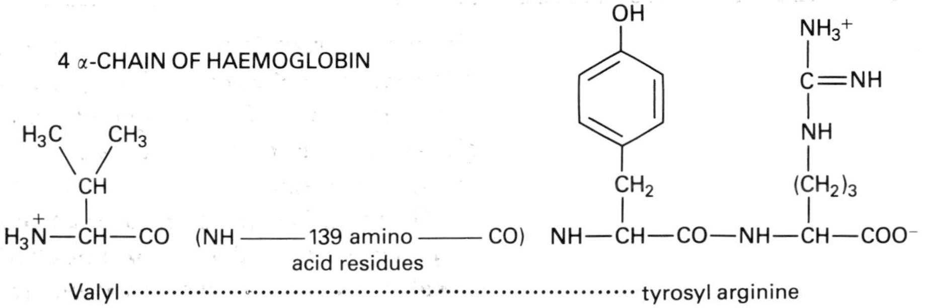

$$H_3C \quad CH_3$$

$$\underset{\underset{H_3\overset{+}{N}-CH-CO}{\quad}}{CH}$$ (NH —— 139 amino —— CO) NH—CH—CO—NH—CH—COO⁻
acid residues

Valyl ·· tyrosyl arginine

4.2 Naming proteins

There are not many polysaccharides to be encountered in elementary bio-chemistry, and the large range of nucleic acids has sustained the coining of only two general names: RNA and DNA. However, a considerable vocabulary attaches to the proteins. These have been traditionally classified in three ways:

1 by solubility: for practical purposes all nomenclature based on solubility properties is obsolete except for the terms 'globulin' (a protein soluble in dilute salt solutions but not in water) and 'albumin' (a protein soluble in water and also in dilute salt solutions);

2 by structure: there are only two distinctions, between the globular (meaning compact) and the fibrous (meaning extended or lengthy) types of molecular shape;

3 by composition: although many proteins contain amino acids only (simple proteins), many more are complexed with sugar (glycoproteins), fats (lipoproteins), or haem (haemoproteins). Many other ligands are possible and similarly provide descriptive prefixes.

Table 4.2.1 Checklist for protein nomenclature

		Serum albumin	γ-Globulin	Fibrinogen	Myosin	Collagen	Haemoglobin	Trypsin	Insulin
By solubility	Albumins	●							
	Globulins		●	●	●		●	●	●
	Histones								
By structure	Fibrous			●	●	●			
	Globular	●	●				●	●	●
By function	Enzymes				●			●	
	Antibodies		●						
	Transport	●					●		
	Respiratory						●		
	Structural				●	●			
	Contractile								
	Effector								●
By conjugation	Glycoproteins	●	●	●		●			
	Lipoproteins	●							
	Nucleoproteins								
	Haemoproteins						●		
	Metalloproteins						●		
	Non-conjugated				●			●	●

Of course a protein may have an extended shape (fibrous), be soluble in dilute salt but not water (globulin) and contain carbohydrate (glycoprotein) so that the nomenclature systems are not collectively exclusive. Table 4.2.1 gives a guide to this concurrency for some important proteins which will be encountered later.

The word 'haemoglobin' does not imply a single, unique substance in the way that 'sucrose' does. 'Haemoglobin' refers to a series of proteins of closely-related occurrence, function, and properties. There are variations within each named protein (better termed 'protein system') which may demand further nomenclature. For example in adult humans there are haemoglobins termed HbA, HbA_2, and HbA_{1c}. 'Collagen' is the name for fibrous, non-extensible proteins found in connective tissues, but individual collagens vary significantly one from another; in this case, instead of numbers, the terms 'dermis collagen' or 'tendon collagen' or 'bone collagen' are used.

5 Enzymes

5.0 Enzymes as catalysts

Chemical reactions in cells, collectively known as *metabolism*, are catalysed by the class of proteins called *enzymes*. The separate reactions of metabolism are feasible in the absence of enzymic catalysis, but the final equilibria would be obtained, in most cases, only after very extended periods of time. Enzymes, like inorganic catalysts, accelerate progress towards equilibrium, but do not affect the final equilibrium in itself. As protein catalysts, enzymes bear similarities to inorganic catalysts, but there are important differences (Table 5.0.1). Notably, enzymes generally exhibit marked specificity. Sometimes specificity is absolute, when the transformation of a single unique reacting substance, or *substrate*, is catalysed. A few enzymes are of low specificity, and there are intermediate descriptions (Table 5.0.2).

Table 5.0.1 Enzymes and inorganic catalysts compared

Enzymes	Inorganic catalysts
All chemically similar (proteins)	Chemically diverse (metals, salts, acids)
Easily inactivated (labile)	Stable to heat but can be 'poisoned'
Usually specific	Often non-specific
Specifically inhibited	Usually not inhibited by the same substances which inhibit enzymes
(By definition) unchanged at the end of a reaction; thus do not alter final equilibrium, only rate	(By definition) unchanged at the end of a reaction; thus do not alter final equilibrium, only rate

Table 5.0.2 Range of enzyme specificity

Decreasing specificity	Examples
Absolute, one unique substrate	$H_2N-CO-NH_2 \xrightarrow{\text{urease}} 2NH_3 + CO_2$
Stereochemical, one enantiomer as substrate	$H_2N-CHR-COO^- \xrightarrow[\text{acid oxidase}]{\text{D-amino}} OC-COO^-$ with R
Group or function specificity	$R-CH_2OH \xrightarrow[\text{dehydrogenase}]{\text{alcohol}} R-CHO$
'Low'	$X-Y + GHS \xrightarrow[\text{S-transferase}]{\text{glutathione}} X-SG + YH$

Note: even when a group of substances serve as substrates, the enzyme is usually absolutely specific in the stereochemical sense. For example L-amino acid oxidases catalyse the oxidation of many L-amino acids, but are without activity towards D-amino acids.

5.1 Enzymes as proteins

Proteins are *labile* structures, easily disrupted. If disruption goes as far as to cause loss of biological activity, the change is termed *denaturation*. Denaturation causes loss of catalytic activity in enzymes, of capacity to combine with oxygen in the case of haemoglobin, and of effector activity in polypeptide hormones. *Denaturants* destroy the weak types of interaction (hydrogen bonds, ionic linkages, hydrophobic bonds, van der Waals forces) which maintain the unique three-dimensional shape of each protein molecule.

For the purposes of description, protein structure is divided into four hierarchical levels (Table 5.1.1). Denaturation is disruption of all but the primary level of structure; in other words, all or some of the weak interactions are destroyed but the covalent bonds specifying amino acid sequence

Table 5.1.1 Hierarchy of protein structures Denatured proteins exhibit only the first level, as shown by the box

	Level	Description	Types of bonding
	Primary	The specific sequence of amino acids in a single protein chain	Peptide bonds (covalent) denatured protein
or	Secondary	Manner in which a segment of a protein chain may be folded into a helical or sheet structure	Hydrogen bonds
	Tertiary	Total shape of a single chain	Hydrogen bonds, disulphide bonds ionic bonds, van der Waals forces, hydrophobic bonds
	Quaternary	Association of separate chains into polymeric structure	Ionic bonds, van der Waals forces, hydrophobic bonds

remain intact. The best known denaturant is heat, but in contrast to this some enzymes are cold-labile. Strong acids and alkalis, as well as very concentrated salts and amines, also denature proteins. Except in some special cases, denaturation is considered to be irreversible. Sensitivity to heat and a denaturant results in the phenomenon known as the optimum temperature (Fig. 5.1.1). All reactions are speeded up by heating, and enzymic reactions are no exception, but at a certain point, depending on the nature of the enzyme molecule, disruption of the delicate interactions maintaining the higher levels of structure is initiated, and the rate of reaction begins to decline. The optimum is not a sharply definable temperature, but rather a range. Even at normal body temperature, many enzymes are unstable, so that continuous synthesis is necessary if their biological activities are to be maintained.

The *optimum pH* can also be viewed as a consequence of the protein nature of enzymes. Side chains such as those in Fig. 5.1.2 (here all written in

Fig. 5.1.1 The optimum temperature for an enzyme-catalysed reaction.

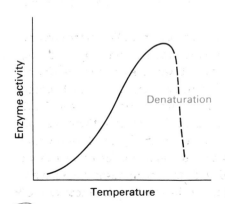

Fig. 5.1.2

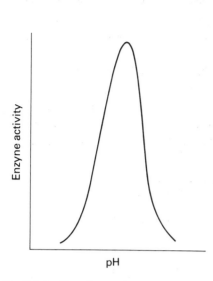

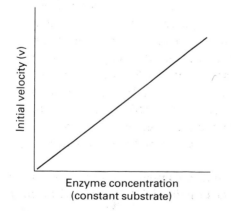

$$\underset{\text{Aspartyl}}{\begin{array}{c} COOH \\ | \\ CH_2 \\ | \\ -CH- \end{array}} \quad \underset{\text{Glutamyl}}{\begin{array}{c} COOH \\ | \\ CH_2 \\ | \\ CH_2 \\ | \\ -CH- \end{array}} \quad \underset{\text{Arginyl}}{\begin{array}{c} NH_3^+ \\ | \\ C=NH \\ | \\ NH \\ | \\ (CH_2)_3 \\ | \\ -CH- \end{array}} \quad \underset{\text{Lysyl}}{\begin{array}{c} NH_3^+ \\ | \\ (CH_2)_4 \\ | \\ -CH- \end{array}} \quad \underset{\text{Histidyl}}{\begin{array}{c} \\ CH_2 \\ | \\ -CH- \end{array}} \quad \underset{\text{Cysteinyl}}{\begin{array}{c} SH \\ | \\ CH_2 \\ | \\ -CH- \end{array}}$$

Fig. 5.1.3 pH-optimum of an enzyme-catalysed reaction.
You would be expected to depict the curve as being much narrower and sharper than the one for optimum temperature, although this would depend entirely on the scale for the horizontal axis, of course.

Fig. 5.2.1 The effect of enzyme concentration on the velocity of an enzymic reaction.
The enzyme is assumed to be in very much smaller concentration than the substrate; it is by definition present in catalytic quantities only. Evidently the velocity is directly proportional to the enzyme concentration.

their protonated forms) lose or gain protons with rise or fall (respectively) of pH. If particular forms are necessary for enzyme activity (e.g. the ionized or positively charged histidyl side chain) then activity falls as the pH is removed from that necessary to maintain maximum protonation. This pH is the one at which enzyme activity is optimal (Fig. 5.1.3).

Most bell-shaped pH curves are interpreted to suggest two or more ionizable groups involved in catalytic activity. Generally the optima are in the range 6–9, but this is not invariable. Pepsin, for example, has a pH optimum of about 2, apparently highly adapted to the acid conditions of the stomach.

Often, of course, the substrate as well as the enzyme is ionizable. This obviously complicates the effect of pH on the reaction; the enzyme cannot be active if the substrate is not of suitable ionic form to bind to it.

5.2 Rate of an enzymic reaction

Enzymes do not alter the final equilibrium of a reaction but accelerate the attainment of that equilibrium. If they are to change the final equilibrium, this would necessarily alter K in the equation

$$\Delta G^\circ = -RT \ln K$$

and in turn, this would mean a change in the value of ΔG°; that is to say, the enzyme would add or subtract energy. This is impossible if the enzyme is to remain unchanged at the end of the reaction, which by observation is the case.

Consider an enzyme reaction with a single substrate S. When the enzyme E is added to a solution of S it speeds up the reaction, forming products. Both S and E can be varied in concentration. If we measure the velocity or rate of the reaction first while keeping [E] constant, and then while keeping [S] constant, we usually obtain the graphs shown in Figs 5.2.1 and 5.2.2 respectively. From consideration of Fig. 5.2.2 it is clear that there is some maximum velocity (V_{max} which is approached at high [S]) at which we may presume that the enzyme is saturated with substrate. It is also evident that the lower the [S] at which V_{max} is achieved, the more easily the enzyme becomes saturated with substrate, and this is in some sense a measure of

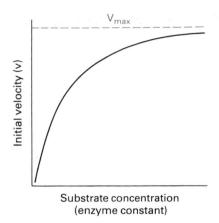

Initial velocity (v)

V_{max}

Substrate concentration
(enzyme constant)

Fig. 5.2.2 Effect of substrate concentration on the velocity of an enzyme-catalysed reaction. The curve is asymptotic to the maximum velocity V_{max}. Evidently there is a stage at which all the enzyme present is saturated with reacting substrate. This does not apply to an allosteric enzyme (see Fig. 5.2.4).

catalytic power or activity. It is desirable to quantify this activity. The approach is to visualize a combination between enzyme and substrate to form a complex, ES, which can dissociate to give E and S, or E plus the products of the reaction. The result is the Michaelis-Menten equation, which is worked out on the basis of five assumptions.

1 The enzymic reaction can be represented thus:

$$S + E \underset{v_2}{\overset{v_1}{\rightleftharpoons}} ES \overset{v_3}{\rightarrow} E + products$$

There is a combination between E and S to give a complex ES which breaks down to E and products. We are considering an equilibrium moving from a preponderance of S to a preponderance of products. We need not consider the theoretical back reaction because we wish only to consider the initial rates of reaction, before large amounts of product can accumulate.

2 E being a catalyst, [E] is very much smaller than [S] at all times. Free enzyme at any one time is [E]−[ES], but [S] is so much bigger than [ES] that at all times free substrate is taken to be identical to [S].

3 The velocity of each component of the reaction is proportional to a constant (*velocity constant*) and the concentration of reactants (this is standard physical chemistry), so

$v_1 = k_1 [E - ES][S]$

$v_2 = k_2 [ES]$

$v_3 = k_3 [ES]$

4 The overall velocity v is determined by the breakdown of ES

$v = v_3 = k_3 [ES]$

5 At the maximum velocity when [S] is very high, E is totally saturated with S so [E] = [ES];
since at that point

$V_{max} = k_3 [ES]$

then

$V_{max} = k_3 [E]$

At a specific velocity the rate of breakdown of ES, $(v_2 + v_3)$ must be equal to the rate of formation of ES (v_1) so

$v_1 = v_2 + v_3$

We can substitute in this according to item 3.

$k_1 [E - ES][S] = k_2 [ES] + k_3 [ES]$

Divide by k_1 and rearrange

$[E][S] - [S][ES] = \left\{(k_2 + k_3)/k_1\right\} [ES]$

Call $(k_2 + k_3)/k_1$ a new constant, K_m. From item 5 above, $[E] = V_{max}/k_3$
Substituting for E,

$$(V_{max}/k_3)[S] - [S][ES] = K_m[ES]$$

Rearrange

$$V_{max}[S] = k_3(K_m[ES] + [S][ES])$$
$$= K_m[ES]k_3 + [S][ES]k_3$$

From item 4 above

$$v = k_3[ES]$$

Substituting for $k_3[ES]$ we have

$$V_{max}[S] = K_m v + [S]v$$

Rearranging,

$$v = V_{max}[S]/(K_m + [S])$$

This is the Michaelis-Menten equation. You can memorize it, but with a recall of the five assumptions it is not difficult to derive. More important to grasp is that the form of the equation is that of a rectangular hyperbola, such as in Fig. 5.2.2. This is in agreement with experimental measurements on many enzymes, suggesting the correctness of the assumptions used to derive the equation.

Consider the situation when the substrate concentration is such that half the maximum velocity has been achieved. Then

$$v = \frac{1}{2}V_{max}$$

substitute V_{max} for v in the Michaelis-Menten equation and solve

$$\frac{1}{2}V_{max} = V_{max}[S]/([S] + K_m)$$
$$2[S] = [S] + K_m$$
$$K_m = [S]$$

Or, the Michaelis constant is the substrate concentration at which there is half the maximum velocity. It is a measure of the affinity of enzyme for substrate and has the units mol/l. A low K_m means in general a high affinity of enzyme for substrate (saturation achieved at low $[S]$); a high K_m means low affinity. It is helpful to remember that it is derived from $(k_2 + k_3)/k_1$ so the larger the k_1, the smaller the K_m.

The asymptotic nature of the curve of v against $[S]$ (Fig. 5.2.2) makes it experimentally difficult to determine V_{max}. To avoid this difficulty various modifications to the Michaelis-Menten equation have been developed. The best known is that of Lineweaver and Burk, in which reciprocals are taken

$$1/v = 1/\left\{ V_{max}[S]/([S] + V_{max}) \right\}$$

which rearranges to

$$1/v = K_m / V_{max} \cdot [S] + 1/V_{max}$$

This is an equation of the type y = mx + c, so that if l/v is plotted against l/[S], a straight line is obtained (Fig. 5.2.3) of which the slope is K_m/V_{max} and the intercept of the y-axis is $1/V_{max}$. K_m can then be readily determined by measurements of the graph. Michaelis-Menten kinetics are observed for many enzymes, where the assumptions made in deriving the equation seem to hold.. However, in some enzymic reactions, other kinetics (non-Michaelis-Menten) may be observed, suggesting that one or more of the assumptions is not true. For certain enzymes if, as before, [E] is kept constant and v is measured for increasing [S], an S-shaped or sigmoid curve results (Fig. 5.2.4). Examination of the curve shows that v increases slowly at first with increasing [S], then increases very rapidly, and finally increases more slowly again. The interpretation is that the enzyme has multiple binding sites for the substrate and when [S] is low little of it is bound, so that the velocity is low. With increasing [S] more is bound, but binding at certain sites increases the affinity of other sites on the same molecule, and the velocity rises rapidly. At length, of course, all sites approach saturation and the velocity increases more slowly towards the inevitable V_{max}. Michaelis-Menten kinetics cannot cope with this situation, that is multiple and *cooperative* binding altering the properties of a molecule towards a ligand (*allostery*). There are, however, graphical transformations to allow determination of [S] at half maximum velocity.

5.3 The mechanism of enzyme action

Each enzyme functions as a catalyst by binding substrate or substrates at a specific locale, termed the *active site* or *catalytic site*. It is firmly established that the active site represents a small proportion of the surface of the enzyme molecule. The forces initially binding the substrate to the active site are described as 'weak', that is non-covalent, consisting of salt, hydrophobic and hydrogen bonds, and van der Waals forces. However these are seen as sufficient to bring the substrate molecule into a position where the *activation energy* necessary for the reaction is lowered. Any reaction proceeds because molecules acquire a transition state of higher energy than the sum of the energies of the reactants; an enzyme lowers the activation energy required to achieve a transition state (Fig. 1.4.1). It is easy to imagine, in a qualitative way, how this might happen: binding of the substrate to the enzyme surface could modify bond strength within the substrate. Binding could cause orientation of the substrate such that the probability of reaction with a second substrate is increased; such changes in the condition of the substrate might lessen the energy required to achieve a transition state.

The concept that the enzyme and substrate may react with each other in modified form is described as 'induced fit'. This is an extension of the

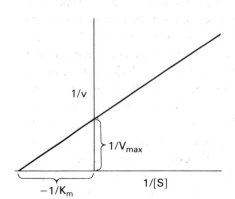

Fig. 5.2.3 The Lineweaver-Burk plot for the determination of K_m and V_{max}.
Plotting l/v against l/[S] allows the unknowns to be determined from the intercepts on the two axes. An additional check is that, from the Lineweaver-Burk equation, the slope of the line is K_m/V_{max}.

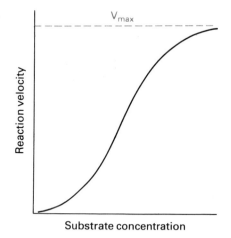

Fig. 5.2.4 Graph of velocity against substrate concentration for an enzymic reaction in which there is cooperative binding of the substrate.

COO⁻
|
CHOH   fumarase
|
CH₂   ⇌
|
COO⁻

CH—COO⁻
‖
⁻OOC—CH

Malate   Fumarate

Fig. 5.3.1

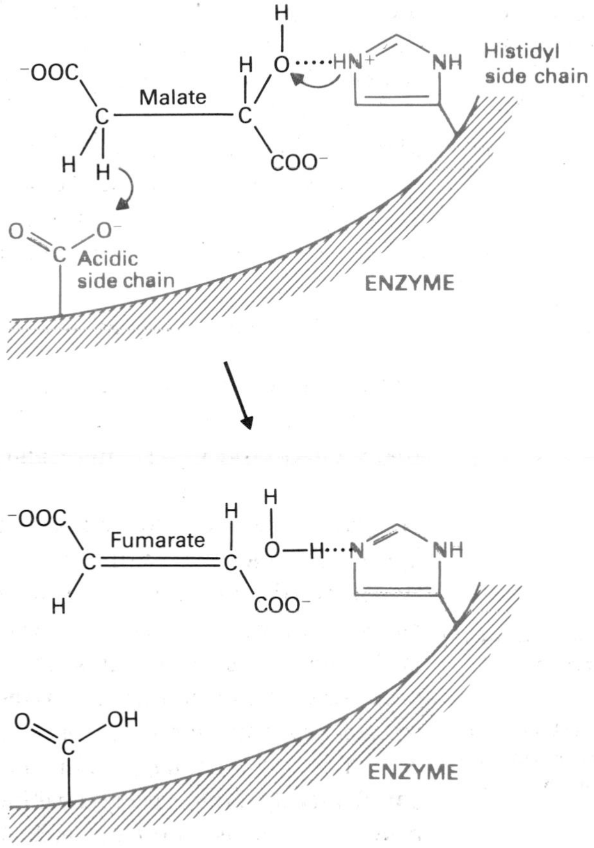

'lock-and-key' theory of enzyme activity: both lock and key are now thought to be deformable.

If you are asked to describe an 'enzyme mechanism', what is required is an exact description of the shifts of electrons in a substrate, when bound to an enzyme surface, such that transformation into product takes place. Various mechanisms have been proposed for various enzymes. Here we are not concerned with a survey, as it were, but with identifying a mechanism which can be understood and if possible reproduced from memory. The mechanism which then recommends itself is that of fumarase. Fumarase catalyses the reaction shown in Fig. 5.3.1. The reaction is readily reversible under physiological conditions. Its mechanism is envisaged as a translocation of protons (Fig. 5.3.2) and one or the other hydrogen atoms ringed (it is not known which at the time of writing) is transferred, setting up a series of electron shifts. Since all enzymes catalyse the forward and back reactions equally to accelerate attainment of the final equilibrium, the fumarate to malate transfer takes place by a reversal of the mechanism in the figure.

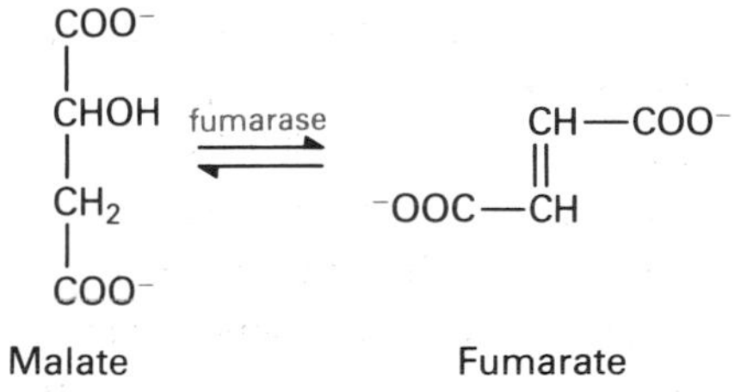

Fig. 5.3.2 Proposed mechanism of action of fumarase.
It is suggested that the substrate is bound by two ionizable side chains (these constituting the active site). When malate reacts, it loses both a hydrogen atom and the −OH radical, and it is easy to imagine the two unoccupied bonds flipping back to give the double bond in fumarate.

Although the fumarase reaction mechanism as depicted here is of the simplest, it is more usual in elementary texts to describe the action of the serine proteases and this is indicated for the sake of completeness (Fig. 5.3.3).

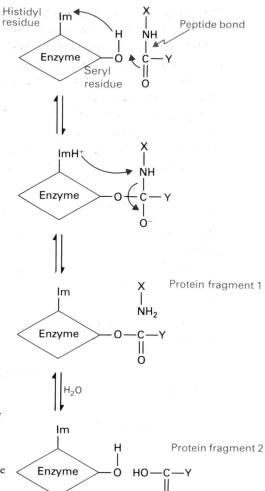

Fig. 5.3.3 The mechanism of the serine proteases.
The serine proteases are a group of enzymes whose function in protein hydrolysis depends upon the hydroxyl group of a seryl residue at the active centre. In the simplified sequence shown, the protein substrate X—NH.CO—Y is bound at the active site such that a nearby histidyl residue (marked 'Im' in the diagram) attracts a proton from the hydroxyl. A series of electron shifts yields an acylated intermediate which can be hydrolysed to regenerate the free enzyme. The sequence is obviously reversible; physiologically, enzymes such as trypsin catalyse reactions in the direction shown, yielding protein fragments in the small intestine.

Also falling within the description of 'mechanism' is the 'ping-pong' type. The analogy is not a very exact one, but imagines an enzyme being knocked back and forth between two substrates. Consider an example. Certain amino acids react with an enzyme (transaminase) to yield the corresponding keto acid. A second keto acid then reacts with the modified enzyme (bearing the nitrogen from the first amino acid) to allow synthesis of a second amino acid. In general terms this is as follows, where En′ represents the modified enzyme:

$$En + amino\ acid_1 \rightarrow En'\text{——}amino\ acid_1$$
$$En'\text{——}amino\ acid_1 \rightarrow En' + \alpha\text{-keto acid}_1$$
$$En' + \alpha\text{-keto acid}_2 \rightarrow En'\text{——}\alpha\text{-keto acid}_2$$
$$En'\text{——}\alpha\text{-keto acid}_2 \rightarrow En + amino\ acid_2$$

The apparent total reaction is:

$$Amino\ acid_1 + \alpha\text{-keto acid}_2 \rightarrow \alpha\text{-keto acid}_1 + amino\ acid_2$$

non-competitive. Non-competitive inhibition may thus mean non-competitive reversible plus irreversible or only non-competitive reversible. Since the kinetics of non-competitive reversible and irreversible are similar, the distinction is fudged. 'Uncompetitive' inhibition is distinguished from non-competitive inhibition but belongs in a more advanced treatise. There are also intermediate types.

Figure 5.5.1 also includes a rather different type of activation not hitherto discussed. Proenzymes or *zymogens*, usually digestive proteases, are activated to the active forms by peptide bond hydrolysis. Examples are the pepsinogen to pepsin and trypsinogen to trypsin activations.

5.8   Glossary of enzyme activity terms

The nomenclature surrounding the phenomena of enzyme inhibition and activation has become quite tortuous and a reference guide is included here. The whole phrase is referenced: look under 'covalent modification' rather than 'modification, covalent'. Some of the terms are not discussed in the text, but are included for (relative) completeness.

Activation An increase in the catalytic efficiency of an enzyme.

Allosteric effector See *effector*.

Allosteric inhibitor See *effector*.

Allostery With respect to an enzyme, refers to the interaction between different sites such that one can influence the catalytic properties of another. There are two types: *homotropic allostery* refers to the binding of two identical molecules, for example two molecules of substrate in a *cooperative* manner; *heterotropic allostery* is due to the cooperative binding of some other substance as well as substrate.

Catalytic subunit A subunit of an *oligomeric* enzyme bearing an active site.

Coenzyme An organic molecule functioning as a *cofactor*. There appears to be no precise definition but the term excludes metal ions, firmly bound *prosthetic groups*, and inhibitors.

Cofactor A non-protein substance participating in an enzymic reaction. This is the widest term, embracing *coenzymes*, metal-ion *activators* and *effectors*, but not inhibitors.

Cooperativity The interaction between sites in an enzyme molecule such that binding of a substrate at one site affects the extent of binding at another site.

Covalent modification Alteration of the activity of an enzyme by the attachment of some group by a covalent bond.

Effector A substance which when bound to an allosteric enzyme alters its catalytic activity. May be positive or negative (increasing or decreasing activity respectively). The word itself in 'enzymology' seems to be applied only to the phenomenon of *allostery*.

Induction Initiation or ehancement of enzyme synthesis when a substrate (*inducer*) is presented to a cell or tissue.

Inhibition Decrease in the catalytic activity or efficiency of an enzyme.

Ligand Any atom or molecule bound to a protein or enzyme.

Modifier Any substance modifying enzyme activity. Classed as positive or negative.

Modulator Equivalent to *effector* or *modifier*.

Non-competitive inhibition Inhibition other than that due to binding reversibly at the active site by a substrate analogue.

Oligomeric enzyme An enzyme composed of two or more subunits. In general, allosteric enzymes are oligomeric.

Product inhibition Decrease in the rate of an enzyme reaction due to tight binding of product to enzyme.

Prosthetic group A non-protein entity bound relatively firmly to a protein or enzyme and participating in its activity.

Regulatory subunit The subunit of an *oligomeric enzyme* which does not bind substrate, but rather binds inhibitors or activators such that the activity of the *catalytic subunit* is altered.

Substrate analogue inhibition Same as *competitive inhibition*.

Uncompetitive inhibition Inhibition due to a substance binding to the enzyme-substrate complex but not free enzyme.

Zymogen A proenzyme such as chymotrypsinogen which is activated to chymotrypsin, the catalytic form.

5.9 Coupled reactions

Implicit in the term 'metabolic pathway', there is the idea that reactions do not take place in isolation, but occur in distinct sequences. Each reaction is coupled to the one before and the one after. In chemistry it is an elementary principle that reactions with common intermediates may be added, to accomplish a total reaction. For example:

$$\text{Ascorbic acid} + 2Hg^{2+} \rightarrow \text{dehydroascorbic acid} + 2Hg^{+}$$
$$2Hg^{+} + I_2 \rightarrow 2Hg^{2+} + 2I^{-}$$
$$\text{SUM: Ascorbic acid} + I_2 \rightarrow \text{dehydroascorbic acid} + 2I^{-}$$

In essence, the mercury ions are transporting electrons to iodine. (Incidentally these reactions are a basis for the estimation of ascorbic acid in biological fluids.)

In biochemistry, a common intermediate is the coenzyme nicotinamide adenine dinucleotide, NAD^+, transferring two hydrogen atoms (in effect, H $+ H^+ + \epsilon$) thus:

$$\text{Ethanol} + NAD^+ \rightarrow \text{acetaldehyde} + NADH + H^+$$
$$NADH + H^+ + \text{pyruvate} \rightarrow NAD^+ + \text{lactate}$$
$$\text{SUM: Ethanol} + \text{pyruvate} \rightarrow \text{acetaldehyde} + \text{lactate}$$

(These reactions illustrate why alcoholics sometimes suffer from lactic acidosis.)

Another example is the phosphorylation of glucose; here we put in the free energy changes of the reactions, which involve the coenzymes adenosine triphosphate (ATP) and adenosine diphosphate (ADP).

$$
\begin{aligned}
\text{ATP} + H_2O &\rightarrow \text{ADP} + \text{phosphate} - 35\,\text{kJ} \\
\text{Glucose} + \text{phosphate} &\rightarrow \text{glucose 6-phosphate} + 16\,\text{kJ} \\
\text{SUM: Glucose} + \text{ATP} &\rightarrow \text{glucose 6-phosphate} + \text{ADP} - 19\,\text{kJ}
\end{aligned}
$$

In this case the energetics are unfavourable for a spontaneous reaction involving the production of glucose 6-phosphate from glucose and inorganic phosphate. But coupling the very exergonic hydrolysis of ATP provides the necessary free energy for the accomplishment of the total reaction.

In biochemistry, the main interest in coupled reactions pertains to this type and this is what to emphasize if asked to explain the term. It must also be emphasized that the concept of coupled reactions does not apply specifically to enzyme reactions. It cannot be repeated too often that enzymes cannot alter the energetics of a reaction; they can only enhance its progress to the final equilibrium.

5.10   Isoenzymes

Initially an enzyme is described, or at least detected, by some observable catalytic activity. Thus lactate dehydrogenase (LDH) is the protein which catalyses the removal of hydrogen atoms from lactate (Fig. 5.10.1) (and restores hydrogen atoms in the reverse reaction). It has become apparent that such a catalytic activity is inherent in a variety of molecular species, that is LDH is more than one specific protein. The separate forms are termed *isoenzymes* (or *isozymes*). The phenomenon arises since the enzyme is composed of subunits, the pattern of which is organ-dependent; thus, turning again to LDH, each molecule of it, in the human at any rate, is composed of four subunits, which are of two different types. The 'L-subunit' is typical of the liver, and the 'M-subunit' typical of skeletal muscle. Now there are five ways of combining two such subunits in groups of four, namely

4M   3M1H   2M2H   1M3H   4H

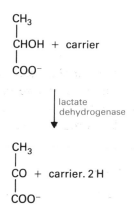

Fig. 5.10.1

These separate species, all of which are found in the serum, can be separated and quantified by various means, but the method of choice is generally electrophoresis, that is differential migration in an electric field.

When LDH is released by damaged tissue, the isoenzyme pattern of that tissue is reflected in the isoenzyme pattern of LDH in the serum. Thus the identification of the organ responsible for pathological increase in serum enzyme activity is possible. LDH may be raised in the course of myocardial infarction, or in various types of liver disease, or in muscle conditions such as muscular dystrophy. If it is the first of these, then the 4H form predominates in the serum; if the last, then it is 4M. Liver damage gives a similar pattern to skeletal muscle damage but these two can rather easily be distinguished on clinical grounds.

Other enzymes whose isoenzyme patterns are routinely examined in the course of clinical diagnosis include creatine phosphokinase and alkaline phosphatase.

Isoenzymes have been found throughout life, including plants and insects, but their biological significance remains obscure.

6 The Language of Biology and Medicine

6.0 The cell

We can formally regard the structural part of biochemistry as a sort of microanatomy, a level in the hierarchy: atom, molecule, macromolecule, organelle, cell, tissue, organism, society. Biochemistry operates largely, but not exclusively, in the molecule to cell segment. Some knowledge of cell structures and types is essential for biochemistry.

All cells have certain features in common, such as an enclosing membrane and genetic material of some sort. Other than that there is great diversity. There is a fundamental division into the *eukaryotic* (having a nucleus surrounded by a definite membrane in which meiosis takes place) and the *prokaryotic* (not so equipped). The former generally have, in addition to the nucleus, various organelles such as mitochondria and lysosomes. Some representative cell types are diagrammed in Fig. 6.0.1.

A brief glossary of the names ascribed to cell structures follows.

Chloroplast An *organelle* in the cytoplasm of photosynthesizing cells, containing chlorophyll.

Cytoplasm The cell sap or protoplasm exterior to the *nucleus*.

Cytosol The soluble part of the *cytoplasm*, that is the part outside the *organelles*.

Endoplasmic reticulum A series of tubules and vesicles in eukaryotic cells. There are two types: 'smooth' which is not studded with *ribosomes*; and 'rough' which possesses them.

Golgi apparatus Cisternae and vesicles concerned with the secretion of products synthesized by cells.

Lysosome A vesicle-like body in the cytoplasm containing hydrolytic enzymes.

Microsomes A preparation (not an *in vivo organelle*) obtained on centrifuging broken cells and consisting of fragments of *Golgi*, *vesicles* and *ribosomes*, also sometimes pieces of *plasma membrane*.

Mitochondrion (pl. mitochondria) An *organelle* enclosed by a double membrane in the cell. The inner membrane has folds which compartmentalize the interior.

Nucleus Spheroidal body containing nucleoprotein (*chromatin*) surrounded by the nucleus envelope.

Nucleolus A spherical body in the *nucleus* with the genes coding for ribosomal RNA.

Organelle A structure inside a cell with some characteristic feature.

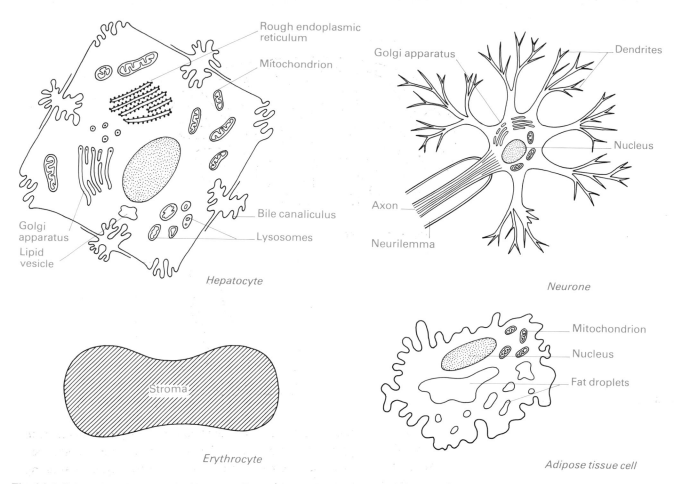

Fig. 6.0.1 Some cell types, structures as seen with the electron microscope. Not drawn to scale.

Peroxisome Body containing catalase and oxidases, which catalyse reactions of which hydrogen peroxide is a product.

Plasma membrane The surface membrane of a cell.

Plastid A general class of plant cell *organelles* of which the *chloroplast* is one.

Ribosome A protein- and RNA-containing *organelle* universally occurring in cells, consisting of two subunits. In mammalian cells these are designated 40S and 60S on the basis of their sedimentation rates, corresponding to their relative masses.

Subsequent pages will seem to pay most attention, among all the cell organelles, to the mitochondrion. Whether or not it is more important (for life) than the other cell organelles is a moot point but it does contain two of the systems on which very great stress is placed within elementary biochemistry courses, namely oxidative phosphorylation and the tricarboxylic acid cycle.

One can see mitochondria with the light microscope, for they are of the same order of size as bacteria. They appear to have various shapes but are most often thought of as 'sausage shaped'. They possess an inner and an outer membrane, and the extensions of the latter into the interior yield the

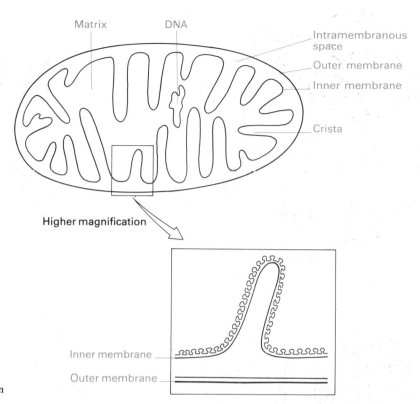

Fig. 6.0.2 Some features of the mitochondrion

folds known as *cristae* (Fig. 6.0.2). Thus the body has two spaces, namely the intermembrane space, and the matrix, which is inside the inner membrane.

When the electron microscope is employed with certain types of staining (spraying with heavy metals to provide contrast) then the inner membrane is seen to be studded with numerous small globules, attached to the membrane by stalks.

One of the most interesting aspects of mitochondrion biology relates to the theory that they originated as free-living prokaryotes (that is, bacteria of some sort) which were engulfed by the ancestors of the modern eukaryotic cells. (The same theory, in essence, also applies to the chloroplast.) The early forms of eukaryotic cell were probably anaerobic and, in taking up the aerobic prokaryotes, themselves became aerobic, greatly gaining in metabolic efficiency. Consistent with the idea is the fact that mitochondria have their own replicating DNA (usually, as in bacteria, in the form of circular duplexes) and their own ribosomes (on which protein synthesis resembles protein synthesis in bacteria). Also, mitochondrial DNA generally lacks introns (non-coding sequences) unlike eukaryotic nuclear DNA, and it has codons with the meanings of bacterial rather than nuclear DNA.

6.1 Roots

The language of biochemistry, which is derived from that of chemistry, biology, and medicine, is heavily dependent on Greek roots and a passing knowledge of these will speed up the learning process considerably, entailing much less groping after meaning. The following short glossary gives the most frequently used roots; indeed many of them are common enough in a lay vocabulary. A run through them should render unnecessary the search for the meaning of such words as avitaminosis, cytotoxic, and hepatitis. Where the root is not Greek it is marked 'L' (for Latin).

Root	Meaning	Example
ADIPO-	Fat	Adipose tissue
A-, AN-	Not	Anaerobic
–AEMIA	Blood	Anaemia
AMPHI-	On both sides	Amphibolic
ANA-	Up	Anabolism
APO-	Incompleteness	Apoprotein
-ASE	Enzyme (modern coinage)	Lipase
AQUA (L)	Water	Aqueous phase
-BOLIC	Throw	Amphibolic
CATA-	Downward	Catabolism
CHOLE-	Bile	Cholesterol (substance from gallstones)
CHROMATO-	Coloured	Chromatography
CYTO-	Cell	Erythrocyte, leucocyte
DYS-	Ill, bad, abnormal	Dystrophy
-EN, -IN	A protein (modern coinage)	Myosin, haemoglobin, collagen
ENDO-	Inside	Endopeptidase
-ERGIC	Generative	Adrenergic
ERYTHRO-	Red	Erythrocyte
EU-	Well, normal	Eukaryotic
FLAVO-(L)	Yellow	Flavoprotein
-GEN	Producing	Porphyrinogen
GLOB- (L)	A round body	Globulin
GLUCO-	Sweet	Glucose
GLYCERO-	Sweet	Glycerol
GLYCO-	Sweet	Glycogen
GONADO-	Generation	Gonadotrophin
HAEMO-	Blood	Haemoglobin
HEPATO-	Liver	Hepatitis
HYPER-	Over	Hyperglycaemia
HYPO-	Below	Hypoglycaemia
-ITIS	Disease, now inflammation	Hepatitis
KERATO-	Hard	Keratin
LACT- (L)	Of milk	Lactose
LEUCO-	White	Leucocyte
LIPO-	Fat	Lipoprotein
LYSO-, -LYSIS	Breaking up	Lipolysis
MEGA-	Great	Hepatomegaly
META-	Change	Metabolism
MITO-	Fibre, thread	Mitosis
MUTAT-(L)	To change	Mutation
MYO-	Muscle	Myofibril
NEURO-	Nerves	Neurotransmitter
OLIGO-	Some	Oligosaccharide
-ORRHAGIA	Discharge	Haemorrhage
-OSE	Sugar (modern coinage)	Sucrose
-OSIS	Condition	Acidosis, thyrotoxicosis
OSTEO-	Bone	Osteocyte
-PATHY	Grief, feeling	Myopathy, amphipathic

Root	Meaning	Example
-PHIL	Loving	Hydrophilic
SACCHROS-(L)	Sugar	Monosaccharide
SARCO-	Flesh	Sarcoplasm
SOMATO-	Body	Somatotrophin
THIO-	Sulphur	Thiolysis
TOX	Poison	Cytotoxic
TROPHO-	Changing	Adrenocorticotrophin
-URIA, URO-	Urine	Urobilinogen, glycosuria

6.2   Some thorny problems

The presentation of biochemistry in a text involves some almost intractable problems. These are aired at this point in the interests of honesty but are hereafter ignored.

Ionization states and phosphate groups

The vast majority of molecules in living tissues are ionizable. If the pK' of the dissociation of a specific group is sufficiently far removed from the pH of the medium (usually 7.4) then it is possible to state with certainty its predominating ionization state. Thus the pK' value (the pH of half-ionization) of the α-carboxyl group of the amino acids is around 2. At pH 7.4 then these carboxyls are almost completely ionized and can be written as $-COO^-$. The oft-encountered metabolite pyruvic acid is also virtually completely ionized at physiological pH. Since the predominant species is the anion, it should rather be referred to as 'pyruvate' (thus also 'oxaloacetate', 'malate', 'citrate', 'lactate', etc.). This is not always adhered to, and most biochemists have the unfortunate habit of oscillating from '. . . acid' to '-ate'. Here we try to be consistent, sticking to '-ate', as far as possible. (It is not always possible; nobody writes 'the citrate cycle', but always the 'citric acid cycle'.)

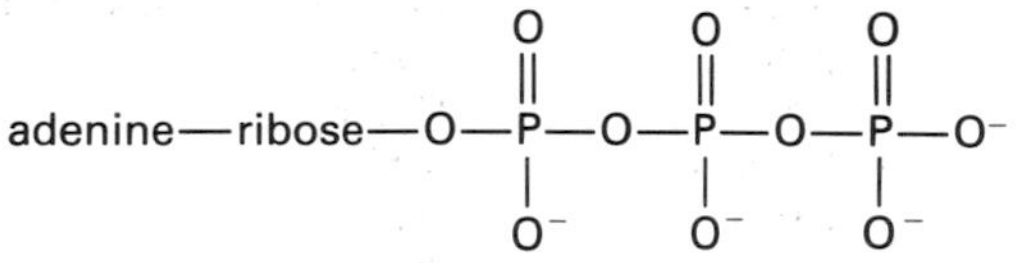

Fig. 6.2.1 ATP.

The pK's of the phosphoric acid ionizations of ATP are about 6.8. Thus at pH 7.4 they are not completely ionized. The degree of ionization can be calculated by means of the Henderson-Hasselbalch equation (1.3) and this indicates that the ratio of ionized to unionized form is four to one. Thus ATP may be written (in shorthand) as in Fig. 6.2.1 but since it is not all in this form, this is but an approximation. Moreover the phosphate is generally thought to bind magnesium, which diminishes the net negative charge. The whole difficulty can be averted by using even more shorthand, such as

Adenine-ribose-℗ — ℗ — ℗

and there is much to be said for this form. Both ways of writing ATP are used in modern textbooks and one can see advantages of using them both.

A further point regarding ionization of di- and triphosphates is the production of a proton on hydrolysis. Thus, in 'short' form

$$ATP \rightarrow ADP + P_i$$

but, in our slightly expanded form this may be written as in Fig. 6.2.2.

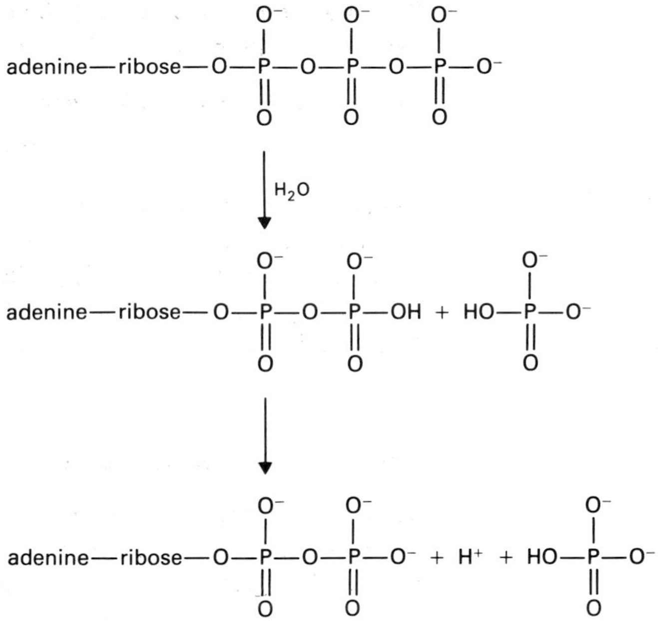

Fig. 6.2.2

So the abbreviated form should really be

$$ATP \rightarrow ADP + P_i + H^+$$

There is, however, a cyclic process whereby a large part of the ADP is re-phosphorylated to regenerate ATP. The proton is taken up again in the reaction.

$$H^+ + ADP + P_i \rightarrow ATP$$

Thus when the hydrolysis of ATP to ADP is shown (perhaps yielding energy to drive a reaction in a specific direction) this proton is usually not written. It does not affect the reaction stoichiometry.

Balancing reactions

A balanced reaction (or 'chemical equation') has the same number of atoms on the right and the left hand sides. It should also balance with respect to charges. This can usually be achieved, on paper, by the introduction of H^+ or H_2O. Thus the glutamine synthetase reaction is commonly written as in

48

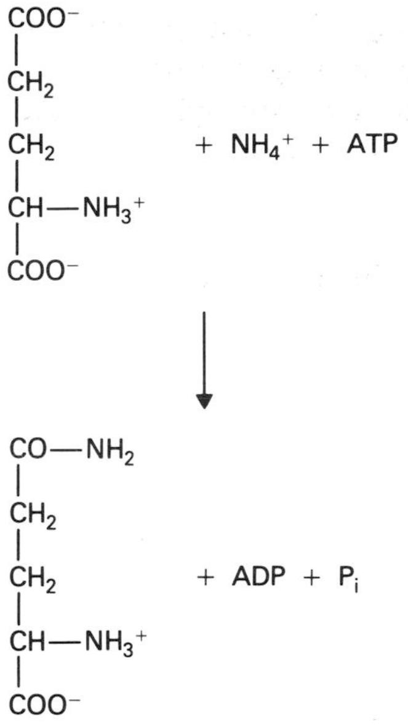

$$
\begin{array}{l}
\text{COO}^- \\
|\\
\text{CH}_2 \\
|\\
\text{CH}_2 \qquad + \text{NH}_4^+ + \text{ATP} \\
|\\
\text{CH}-\text{NH}_3^+ \\
|\\
\text{COO}^-
\end{array}
$$

$$\downarrow$$

$$
\begin{array}{l}
\text{CO}-\text{NH}_2 \\
|\\
\text{CH}_2 \\
|\\
\text{CH}_2 \qquad + \text{ADP} + \text{P}_i \\
|\\
\text{CH}-\text{NH}_3^+ \\
|\\
\text{COO}^-
\end{array}
$$

Fig. 6.2.3

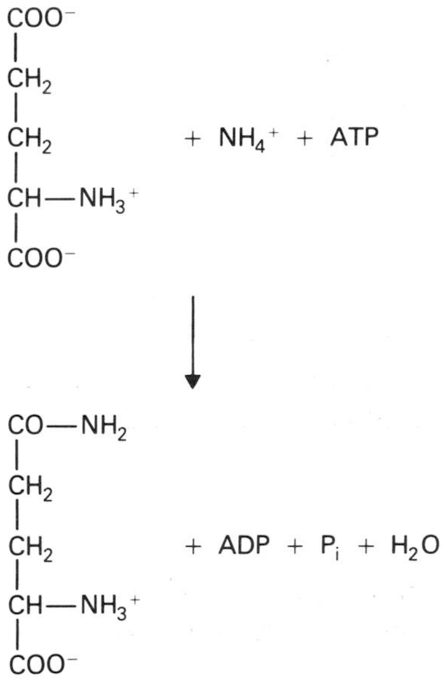

$$
\begin{array}{l}
\text{COO}^- \\
|\\
\text{CH}_2 \\
|\\
\text{CH}_2 \qquad + \text{NH}_4^+ + \text{ATP} \\
|\\
\text{CH}-\text{NH}_3^+ \\
|\\
\text{COO}^-
\end{array}
$$

$$\downarrow$$

$$
\begin{array}{l}
\text{CO}-\text{NH}_2 \\
|\\
\text{CH}_2 \\
|\\
\text{CH}_2 \qquad + \text{ADP} + \text{P}_i + \text{H}_2\text{O} \\
|\\
\text{CH}-\text{NH}_3^+ \\
|\\
\text{COO}^-
\end{array}
$$

Fig. 6.2.4

Fig. 6.2.3. It can be seen that there is insufficient oxygen and hydrogen in the products, although the charges are balanced. This can be put right by imagining the elimination of water, so it can be rewritten as in Fig. 6.2.4. Similarly the carboxylation of acetyl CoA is often written

$$\text{CO}_2 + \text{CH}_3-\text{CO}-\text{S}-\text{CoA} + \text{ATP} \rightarrow$$
$$^-\text{OOC}-\text{CH}_2-\text{CO}-\text{S}-\text{CoA} + \text{ADP} + \text{P}_i$$

This is obviously not balanced with respect to charges and to hydrogen, so an H^+ can be written on the right hand side, making the reaction

$$\text{CO}_2 + \text{CH}_3-\text{CO}-\text{S}-\text{CoA} + \text{ATP} \rightarrow$$
$$^-\text{OOC}-\text{CH}_2-\text{CO}-\text{S}-\text{CoA} + \text{ADP} + \text{P}_i + \text{H}^+$$

Many authors ignore these balancing adjustments, either for simplicity or because the addition of atoms here and there implies a knowledge about mechanism which is not really justified. However, it seems neater and more satisfactory in the first instance to balance up unless some special point is being made, and this is generally done in these pages.

Reversibility of reactions and the use of arrows

There is a rigorous thermodynamic definition of a reversible reaction, which need not concern us. In biochemistry a reversible reaction is often thought of as one which has a low ΔG (either positive or negative). In other words, it has an equilibrium constant near unity. An example is

Glucose 6-phosphate $\rightleftharpoons$ glucose 1-phosphate

for which ΔG^{01} is about 1.2 kcal/mol (5 kJ/mol). The arrow is then usually drawn as $\rightleftharpoons$ or $\longleftrightarrow$ to denote reversibility. However a reaction may be driven in an unfavourable direction by the manipulation of concentrations (ΔG^{01} applies to molar concentrations only). An example is

Dihydroxyacetone phosphate $\rightleftharpoons$ glyceraldehyde 3-phosphate

for which the equilibrium lies in favour of dihydroxyacetone phosphate ($\Delta G^{01} = +7.4$ kJ/mol). However it is glyceraldehyde 3-phosphate which is consumed in the glycolysis pathway so that the reaction is driven to the right hand side because the next reaction in the pathway is proceeding rapidly. This is the same as saying that the reaction can be driven by coupling of reactions.

Therefore one might use the arrow to indicate the direction of flow of carbon atoms in a metabolic pathway. When the carbon atoms flow in either direction, under conditions which ideally should be set out, then the double arrows could be used. This has obviously nothing to do with the magnitude of the equilibrium constant or of free energy change: it is a more 'physio-logical' convention if one is describing a pathway proceeding in a specific direction, but if this flow can be reversed under certain circumstances, how does one write the arrows? Probably as the 'two-way' type. The rigorously logical method would be to discuss the nature of each reaction and its chemical and physiological reversibility as one goes along. This is a space- and time-consuming business; compromise is necessary, and in these pages reactions are generally written to indicate flow of carbon atoms, that is the equilibrium constant and free energy change are unheeded.

Identification of cell, tissue and organism

A glance through any advanced textbook will indicate immediately that the amount of biochemical information now at hand is vast. However not all tissues in all species have been comprehensively examined. Except where biochemists are specifically retained to investigate plants or farm animals or insects by their employers, research normally devolves upon rats or mice or certain types of bacteria. Thus any reaction or pathway may not have been demonstrated in all tissues in all organisms. However it is generally out of the question to qualify every statement to this effect. The location of a reaction or pathway in a specific tissue or cell organelle is important and should be mentioned where possible. But one cannot fill an elementary text with verbiage such as '. . . it has been shown conclusively in rats and pigeons, and may also apply to man, that . . .' or 'in some animals (mice and hamsters) but not in others (man and other primates) . . .', nor is this necessary. Pathways and reactions must be taken to apply typically to mammals unless otherwise noted. Most of the basic ones apply to the vast majority of life forms anyway. Again, compromise alone is possible.

7   Respiration in the Cell

7.0   Organization of coenzymes

In cells energy is made available by the dehydrogenation of substrates, and conserved (stored) as phosphate esters. The hydrogen atoms, in most cases, are initially taken up by the coenzyme nicotinamide adenine dinucleotide or NAD^+. The molecule has a positive charge but is sometimes represented simply as NAD. The product of its reaction with 2H can be represented as $NADH + H^+$ or NAD2H. If $NADH + H^+$ or NAD2H is oxidized by oxygen

$$NADH + H^+ + \tfrac{1}{2}O_2 \rightarrow NAD^+ + O_2$$

the standard free energy at pH7, ΔG^{01}, is about 50 kcal/mol (45 kJ/mol). Thermodynamic theory dictates that this is constant whether oxidation takes place in one step as shown above, or occurs in a series of steps. In fact the energy available is released in a graded manner and generally results in the synthesis of three moles of ATP per mole NADH oxidized. Since ΔG^{01} for the reaction

$$ATP + H_2O \rightarrow ADP + P_i$$

is 7.3 kcal/mol (30 kJ/mol) there seems to be the possibility of the synthesis of about five or six moles of ATP, but a system operating at this efficiency has not evolved. The phosphorylation of three moles occurs in association with an ordered series of coenzymes located in the inner membrane of the mitochondrial membrane (Fig. 7.0.1). There is no alternative to working through the series of coenzymes, but fortunately they can be represented in abbreviated forms. Their importance and role in metabolism is generally of more interest to examiners than are their structures. NAD^+ is synthesized from the B-vitamins nicotinamide and nicotinic acid (Fig. 7.0.2). Reversible

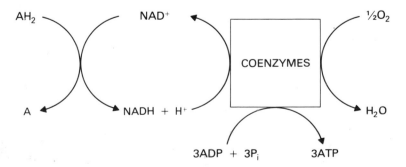

Fig. 7.0.1

oxidation can take place (Fig. 7.0.3). The quaternary nitrogen atom

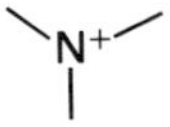

$\overset{\diagdown}{\underset{|}{N^+}}\diagup$ is the four-valent state of nitrogen.

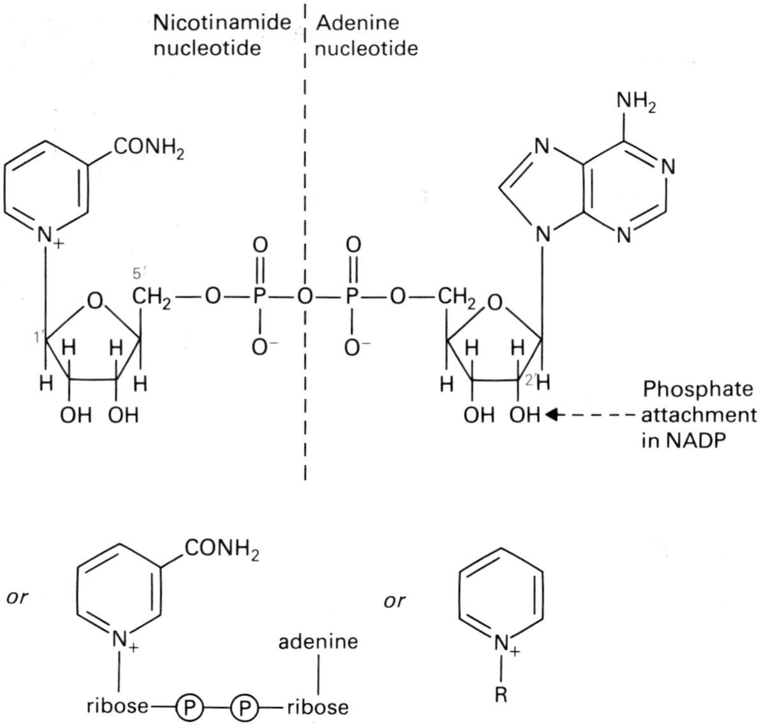

Fig. 7.0.2

NADP$^+$ is the phosphorylated analogue of NAD$^+$. The 2'hydroxyl of the ribose attached to adenine bears the phosphate group (see Fig. 7.0.2). The two coenzymes are quite distinct, that is they participate in different reactions.

Flavin adenine dinucleotide (FAD) and flavin mononucleotide (FMN) are synthesized from the B-vitamin riboflavin. The moiety that undergoes reversible reduction and oxidation is the isoalloxazine ring (Fig. 7.0.4); simpler representations for FAD and for FMN are shown in Fig. 7.0.5. Reversible reduction can take place (Fig. 7.0.6). In some enzymes FAD or FMN is firmly bound by covalent bonds and is better described as a prosthetic group than as a coenzyme. These enzymes are sometimes described as *flavoproteins* and they are generally associated with other proteins in which iron is bound to the sulphydryl groups of cysteine residues. (Iron is thought of as being more naturally bound to porphyrins, so these are called *non-haem iron proteins*.) Proceeding through the series of respiratory coenzymes as they occur in the respiratory chain, there is ubiquinone or coenzyme Q (Q for quinone) which has some analogies to vitamin E but is apparently not in itself a requirement in the diet. It is reversibly reduced as in Fig. 7.0.7.

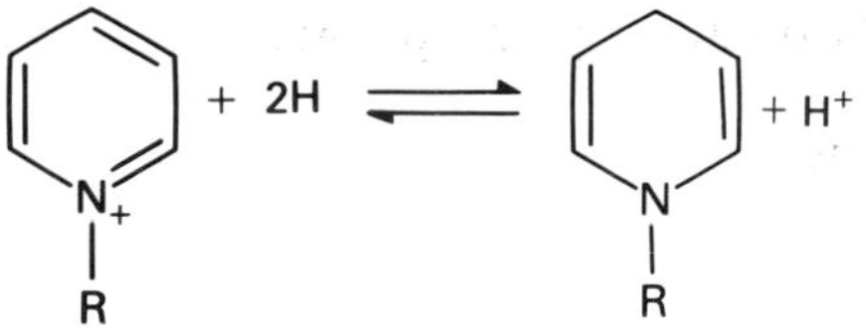

Fig. 7.0.3 Reversible oxidation

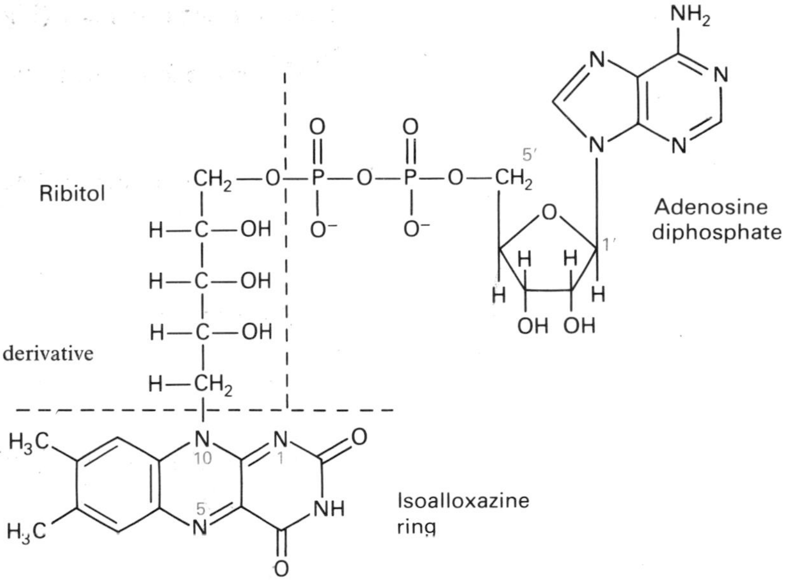

Fig. 7.0.4 FAD contains the alcohol derivative of D-ribose, that is ribitol.

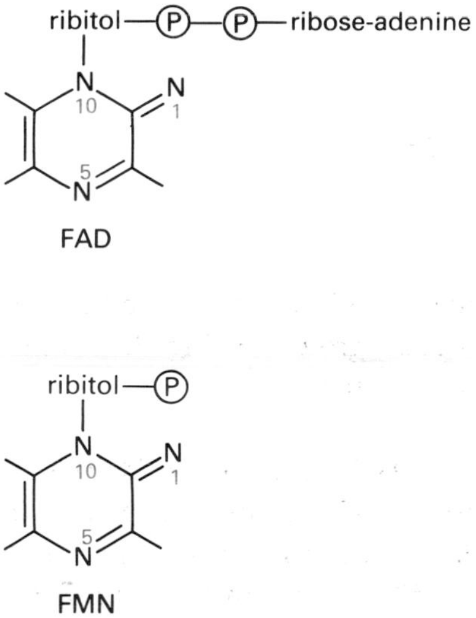

Fig. 7.0.5

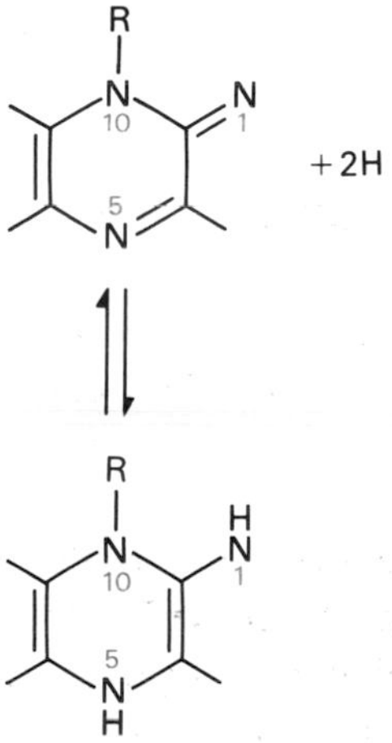

Fig. 7.0.6

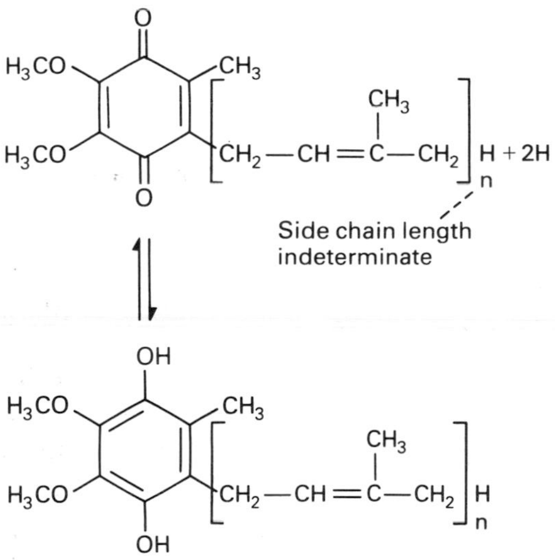

Side chain length indeterminate

Fig. 7.0.7

There is a series of proteins in all cells (bacterial, plant, and animal) which contain porphyrins and iron. These are the cytochromes and, due to their iron content, oxidation and reduction take place in them by single electron transfer. Quite a few distinct cytochromes are now known and they are confusingly named. The *a*, *b*, *c* classification was commenced on the basis of absorption spectra and has no functional significance. The porphyrin of cytochrome *c* is analogous to protoporphyrin IX and so is easy to memorize

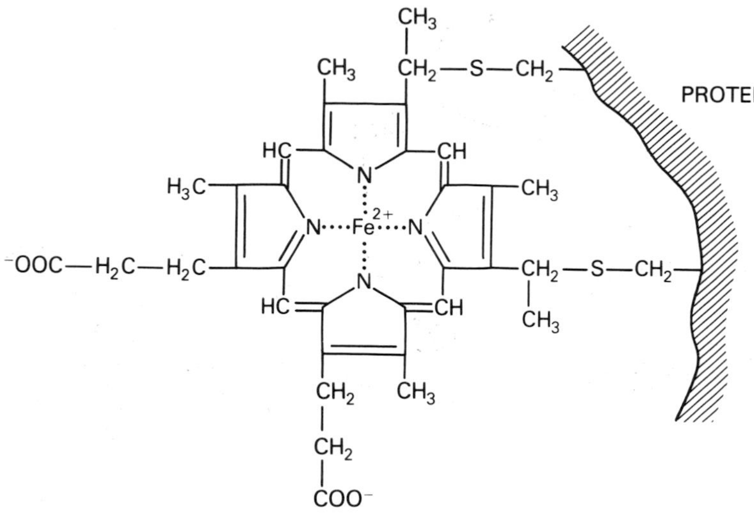

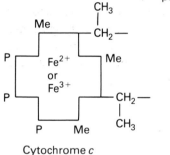

Fig. 7.0.8

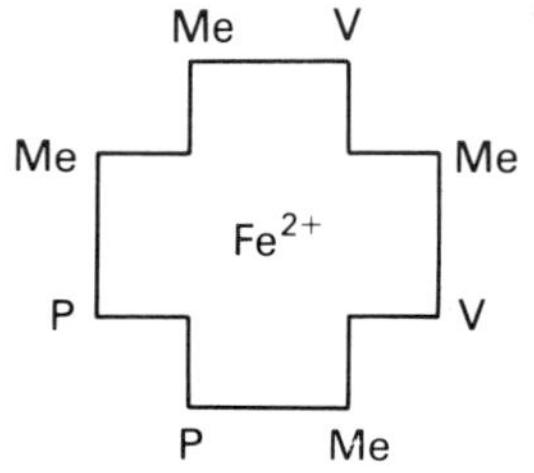

Haem
(protoporphyrin IX + Fe) Me = methyl
V = vinyl
P = propionyl

Cytochrome c

Fig. 7.0.9

(Fig. 7.0.8). Indeed both may be written in shorthand form (Fig. 7.0.9). The reversible oxidation/reduction of cytochrome can be even more simply represented

$$cyt\ Fe^{3+} \rightleftharpoons cyt\ Fe^{2+}$$

The coenzyme sequence is

$$NAD^+ \rightarrow FMN \rightarrow \overset{ubi-}{quinone} \rightarrow cyt\ b \rightarrow cyt\ c \rightarrow cyt\ c_1 \rightarrow cyt\ (a, a_3) \rightarrow O_2$$

This is in order of increasing redox potential, the arrows representing transfer of electrons. There is a two-electron transfer to the ubiquinone stage (when the electrons can be considered as associated with hydrogen ions as 2H). Subsequent to ubiquinone there is one electron transfer. The two hydrogen ions released may be imagined as being removed by the final oxidation involving cytochromes a and a_3 (together known as cytochrome oxidase)

$$2\ electrons\ +\ 2H^+\ +\ \tfrac{1}{2}O_2\ \rightarrow H_2O$$

These are the essentials of the system. Associated with cytochrome oxidase are copper ions which are reversibly reduced and oxidized and non-haem iron proteins intervene between flavin-containing components and subsequent members of the chain, the sequence of which is set out in a little more detail in Fig. 7.1.1.

7.1 Oxidative and substrate-level phosphorylation

Aerobic organisms derive about 95% of their energy from the coupling of the electron transport system to ATP synthesis. The electron transport system can still operate in the absence of phosphorylation, when it is said to be *uncoupled*.

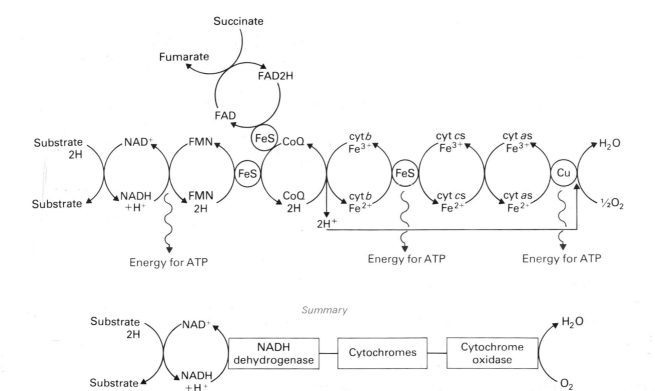

Summary

Fig. 7.1.1 Sequence of the components of the respiratory chain.
Enzymes which contain FAD as hydrogen acceptors (such as succinic dehydrogenase) bypass NAD and as a result only two moles of ATP are synthesized for each mole of substrate oxidized. After coenzyme Q in the sequence, pairs of hydrogen atoms as reductants are replaced by single electrons. FeS represents the non-haem iron proteins.

There are at the moment two theories to explain the phenomenon of phosphorylation linked to the electron transport chain. (You may speculate that with two theories extant, examiners might be uneasy about asking questions about the subject at an elementary level.) The two points of view are set out here at their simplest; the former is obsolescent.

Coupling factor theory

The energy made available when an electron is transferred from one component to another in the chain can only be conserved as chemical energy, that is in bond formation; it cannot jump across a gap, as it were, from

$$cyt_1 \text{ (reduced)} + cyt_2 \text{ (oxidized)} \rightarrow cyt_1 \text{ (oxidized)} + cyt_2 \text{ (reduced)}$$

$$\text{to} \quad ADP + P_i \rightarrow ATP$$

You might suppose that the following might happen, where $\sim P$ represents the attachment of a phosphate; $\sim$ represents the bond conserving the free energy released during electron transfer

$$cyt_1 \text{ (reduced)} + cyt_2 \text{ (oxidized)} + P_i \rightarrow cyt_1 \text{ (oxidized)} \sim P + cyt_2 \text{ (reduced)}$$
$$cyt_1 \text{ (oxidized)} \sim P + ADP \rightarrow cyt_1 \text{ (oxidized)} + ATP$$

There is, however, no evidence for the phosphorylation of the cytochromes or any of the other electron transport chain components. In response to this lack of evidence, hypothetical phosphorylated intermediates are introduced, thus:

$$cyt_1 \text{ (reduced)} + X \rightarrow cyt_1 \text{ (reduced)} - X$$
$$cyt_1 \text{ (reduced)} - X + cyt_2 \text{ (oxidized)} \rightarrow cyt_1 \text{ (oxidized)} \sim X + cyt_2 \text{ (reduced)}$$
$$cyt_1 \text{ (oxidized)} \sim X + P_i \rightarrow cyt_1 \text{ (oxidized)} + P \sim X$$
$$P \sim X + ADP \rightarrow ATP + X$$

Such a sequence, if it exists, takes place at three points, such that three moles ATP are synthesized for every two electrons transferred, and concomitantly one atom of oxygen is used up. In sum:

$$3ADP + 3P_i + 2\epsilon + 2H^+ + \tfrac{1}{2}O_2 \rightarrow 3ATP + H_2O$$

Since the ratio of phosphate esterified to atoms oxygen utilized is three, the P/O ratio is three. The points of phosphorylation can be represented:

$$AH_2 \rightarrow NAD \rightarrow FAD \rightarrow CoQ \rightarrow cyt\ b \rightarrow cyt\ c_1 \rightarrow cyt\ c \rightarrow cyt\ (a, a_3) \rightarrow \tfrac{1}{2}O_2$$

$$\qquad\qquad\downarrow \qquad\qquad\qquad\qquad\quad \downarrow \qquad\qquad\qquad\qquad \downarrow$$
$$\qquad\qquad ATP \qquad\qquad\qquad\qquad\ ATP \qquad\qquad\qquad\quad ATP$$

Chemiosmotic theory

The electron transport chain evidently produces protons (Fig. 7.1.1). They are, moreover, pumped towards the exterior, for it can be demonstrated that the intramembraneous space of the mitochondrion has a lower pH (about 1.4 units) than the matrix. For such a translocation, it is necessary that the components of the electron transport chain be suitably positioned within the inner mitochondrial membrane, that is the chemiosmotic theory implies *vectorial organization* of the system. The protons are pumped at three sites, corresponding to those in Fig. 7.1.2; ATP is not synthesized at these points, but rather by an ATP synthetase as the protons flow back through a specific channel in response to the electrochemical potential difference set up by the pump. This 'force' is now often given the name of *proticity*, or *proton-motive force*.

In the scheme as usually presented (Fig. 7.1.2) the electron transport chain forms three 'loops' each of which translocates $2\ H^+$. One ATP is synthesized for each $2\ H^+$ and the P/O ratio is 3. The apparently excessive number of protons generated (only two are shown in Fig. 7.1.1) suggests that they come from sources other than the substrate, perhaps from the solvent, water, or from proteins undergoing conformational change. (Haemoglobin, for example, can release hydrogen ions when it is oxygenated, p. 194.)

The chemiosmotic hypothesis will probably always be called a hypothesis for historical reasons (recall Avogadro's hypothesis) but is now well accepted in comparison to the coupling factor theory.

Respiratory control and inhibition Naturally all components (respiratory chain coenzymes, oxygen, substrates, phosphate, and ADP) must be present

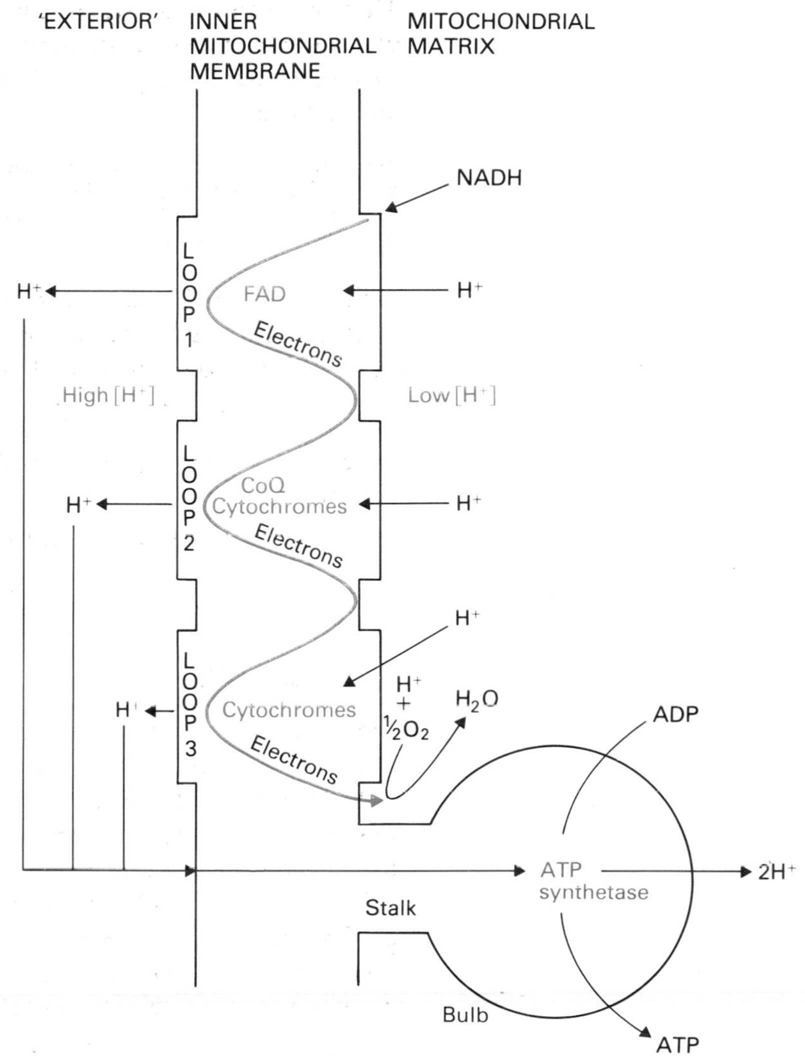

'EXTERIOR' **INNER MITOCHONDRIAL MEMBRANE** **MITOCHONDRIAL MATRIX**

Fig. 7.1.2 The chemiosmotic theory of oxidative phosphorylation.

There are a series of redox loops containing the respiratory chain intermediates in the inner membrane of the mitochondrion. These act as proton pumps, operated by the electron flow, and expel protons taken up from the mitochondrial matrix into the extramembraneous space, thus creating a potential gradient. There is an orientated ATP synthetase on or in the knobs of the cristae which utilizes the free energy difference of the protein gradient to drive ATP synthesis. The possible stoichiometry of H^+ transfer has been omitted here; it is thought however that $2H^+$ are pumped for every one ATP synthesized.

The stalk and knob are supposed to represent the structures identifiable on the inner surface of the inner membrane upon examining in the electron microscope (Fig. 6.0.2).

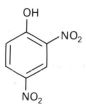

Fig. 7.1.3 2,4-Dinitrophenol

to allow optimal oxidative phosphorylation. It is found that the rate-controlling factor is generally ADP, and the term *respiratory control* is the name given to the control of the rate of oxidative phosphorylation by ADP concentration. Loss of respiratory control leads to increased oxygen consumption and exhaustion of substrates. This situation arises in the presence of *uncoupling agents* which dissociate the phosphorylation mechanism from the electron transport mechanism. The best-known of the uncoupling agents is 2,4-dinitrophenol (DNP), shown in Fig. 7.1.3.

In the presence of an uncoupling agent, the energy released by electron transport, if not conserved as ATP, must appear in some form or other (by the first law of thermodynamics, it cannot merely cease to exist). It is therefore evolved as heat. The uncoupled respiratory chain is thus a means of generating heat and is probably used for this purpose in *non-shivering thermogenesis*. Shivering thermogenesis is heat production by the increased muscular activity plainly visible in a shivering subject; this generally precludes rest and sleep. Newborn mammals and hibernators have a special tissue (brown fat) which conducts electron transport in the absence of ATP accmulation; either there is uncoupling, or ATP is synthesized and hydrolysed by a very active ATPase. These two equally result in heat production. Related, at least in theory, is the maintenance of a constant body weight under conditions of excess energy intake. Put another way, why can some people who remain thin eat large quantities of food? If substrates from the food are oxidized in cells in which the electron transport system is uncoupled then they are not available for conversion to fat. They are literally burned off.

Many inhibitors of oxidative phosphorylation and electron transport are known. They have been useful, in many cases, in working out the details of the component sequence. They are listed here with some of their points of interest.

Cyanide is well known as a poison. It acts by combining with the iron of cytochrome a_3, in the ferric (Fe^{3+}) form, effectively blocking its reduction by electrons from the other cytochromes. Since it rapidly enters all tissues it quickly deprives the nerves and muscles of electron transport and death ensues after convulsions.

Carbon monoxide also combines with the ferrous ion of cytochrome a_3, as does *hydrogen sulphide*.

Rotenone is a plant product, discovered through its use as a fish poison. It blocks the transfer of electrons from NADH. Thus in a system reduced by NADH and inhibited by rotenone, all components except the NADH remain oxidized.

Amytal is a barbiturate with an effect similar to that of rotenone.

Antimycin A is an antibiotic (not used in clinical medicine) that blocks the reduction of cytochrome c_1 by cytochrome *b*.

Substrate-level phosphorylation is distinguished from oxidative phosphorylation in that it occurs when a compound with a large *group transfer potential* participates in an enzyme-catalysed reaction associated with ATP synthesis. (The group transfer potential, a largely self-explanatory term, refers to the free energy decrease, upon hydrolysis, of a compound producing a group normally taken up by another nucleophile.) The group transfer potential for phosphoenolpyruvate is 62 kJ/mol at pH7. It is relevant to the reaction catalysed by pyruvate kinase.

Phosphoenolpyruvate + ADP $\rightarrow$ enolpyruvate + ATP

The significant examples of substrate-level phosphorylation occur in glycolysis (8.2).

7.2   The tricarboxylic acid cycle

The tricarboxylic acid cycle (synonyms: the Krebs cycle, the citric acid cycle) is a device for the total oxidation of two-carbon units; these may be derived

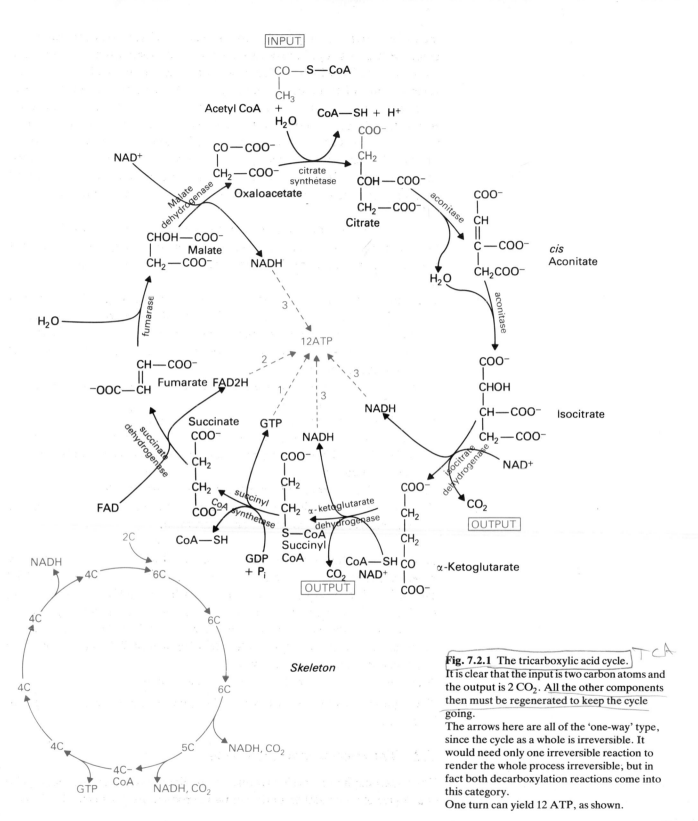

INPUT

CO—S—CoA
|
CH₃

Acetyl CoA
+
H₂O

CoA—SH + H⁺

NAD⁺

Malate dehydrogenase

CO—COO⁻
|
CH₂—COO⁻
Oxaloacetate

citrate synthetase

COO⁻
|
CH₂
|
COH—COO⁻
|
CH₂—COO⁻
Citrate

aconitase

COO⁻
|
CH
||
C—COO⁻
|
CH₂COO⁻

cis
Aconitate

CHOH—COO⁻
| Malate
CH₂—COO⁻

NADH

H₂O

aconitase

3

H₂O

fumarase

12ATP

2

COO⁻
|
CHOH
|
CH—COO⁻
|
CH₂—COO⁻
Isocitrate

CH—COO⁻
||
⁻OOC—CH   Fumarate FAD2H

1   3

3

NADH

succinate dehydrogenase

Succinate
COO⁻
|
CH₂
|
CH₂
|
COO⁻

GTP

NADH

isocitrate dehydrogenase

NAD⁺

FAD

succinyl CoA synthetase

CoA—SH

GDP
+ Pᵢ

COO⁻
|
CH₂
|
CH₂
|
S—CoA
Succinyl
CoA

α-ketoglutarate dehydrogenase

COO⁻
|
CH₂
|
CH₂
|
CO
|
COO⁻

CO₂

OUTPUT

CoA—SH

CO₂   NAD⁺

OUTPUT

α-Ketoglutarate

2C

NADH

4C   6C

4C

6C

4C

6C

4C

6C

Skeleton

5C   NADH, CO₂

4C

4C–
CoA

NADH, CO₂

GTP

Fig. 7.2.1 The tricarboxylic acid cycle.
It is clear that the input is two carbon atoms and
the output is 2 CO_2. All the other components
then must be regenerated to keep the cycle
going.
The arrows here are all of the 'one-way' type,
since the cycle as a whole is irreversible. It
would need only one irreversible reaction to
render the whole process irreversible, but in
fact both decarboxylation reactions come into
this category.
One turn can yield 12 ATP, as shown.

59

from carbohydrates, proteins, or fats, that is any of the principal energy-producing foodstuffs. It operates in the mitochondria, like the electron transport chain, for which it generates electrons.

It is helpful to remember a skeletal form (Fig. 7.2.1) as a frame of reference into which details can later be inserted. The skeleton form shows the following points.

1 There is total oxidation of two-carbon units. For each entry of acetyl CoA, two CO_2 are evolved. The two CO_2 coming off in one turn do not both derive from the particular two-carbon units entering the cycle to initiate that turn. This does not, however, affect the main principle: two-carbon units are completely oxidized by the cycle. By the time oxaloacetate is regenerated, two carbons are removed as CO_2 for every two entering.

2 Since the intermediates of the cycle are regenerated at each turn of the cycle, they need be present only in catalytic quantities; they are not consumed by the cycle, but are renewed as a consequence of oxaloacetate combining with the incoming 2-C units. Conversely, if they are depleted by some means or other, two-carbon units cannot regenerate them, and the cycle 'runs down'.

3 It is 'one-way', that is you cannot add malate to reduced coenzymes, bubble through CO_2 and achieve acetyl CoA plus oxaloacetate. This is due to reactions which are thermodynamically not feasible in the 'backwards' direction at reasonable concentrations of reactants.

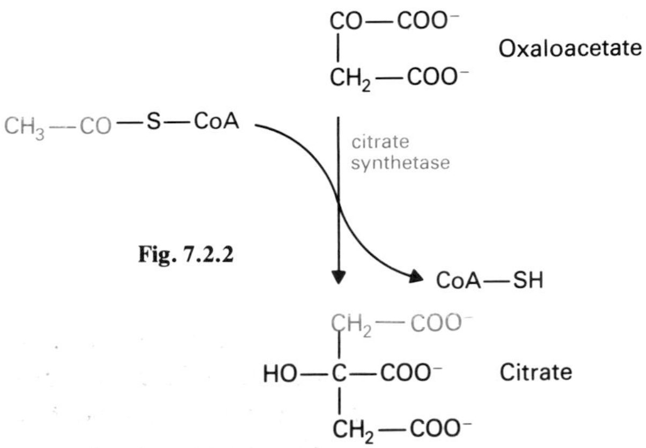

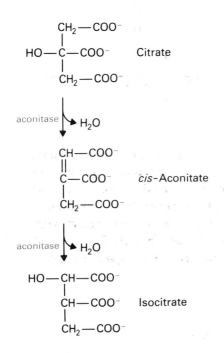

Fig. 7.2.2

Fig. 7.2.3

The intermediates are a series of di- and tricarboxylic acids. Citrate is derived from oxaloacetate and a two-carbon unit (Fig. 7.2.2). There is isomerization of this structure to isocitrate by way of *cis*-aconitate (Fig. 7.2.3). Isocitrate undergoes *oxidative decarboxylation* to α-ketoglutarate (synonym: 2-oxoglutarate). The α-keto acids such as α-ketoglutarate and pyruvate (in glycolysis, 8.2) undergo this reaction in the presence of two more coenzymes derived from members of the vitamin B complex (like NAD^+ and FAD). Thiamine pyrophosphate (TPP) is derived from vitamin

B_1 or thiamine and has the thiazole ring, which mammalian cells are not able to synthesize (Fig. 7.2.4). The reaction with α-ketoglutarate occurs at a reactive carbon in the thiazole ring (Fig. 7.2.5).

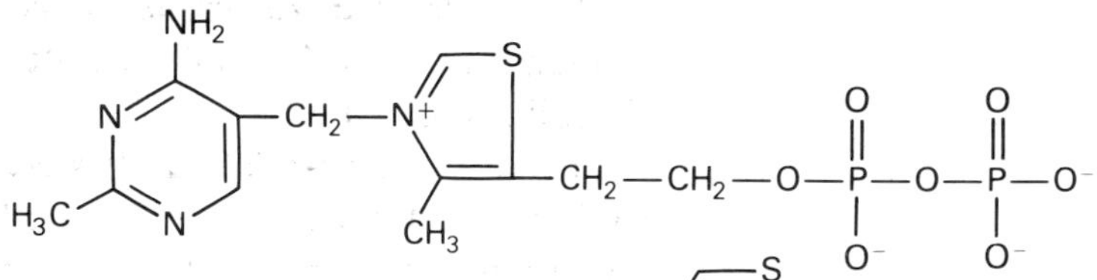

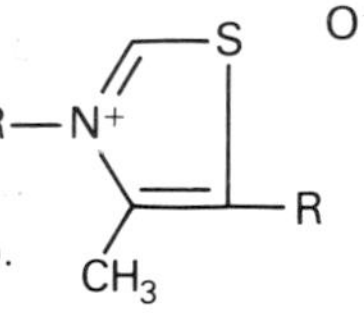

Fig. 7.2.4 Thiamine pyrophosphate (TPP).

Fig. 7.2.5

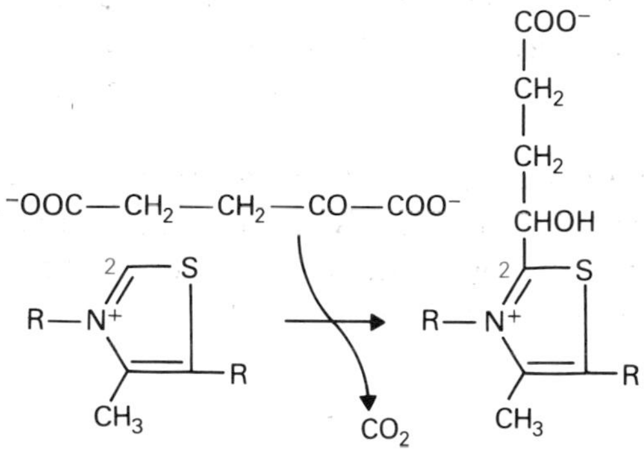

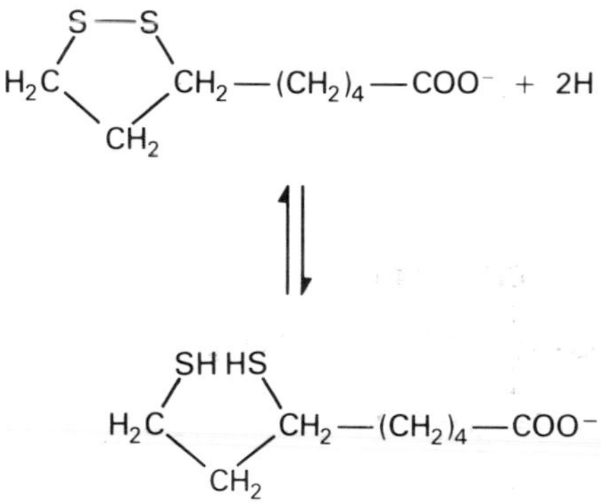

Fig. 7.2.6

The other coenzyme, lipoate (Fig. 7.2.6), is a long-chain dithiol, existing in oxidized and reduced forms, as do NAD^+ and FAD. It can combine with carboxylic acids to yield thioethers of general formula R—S—COR, should R—SH represent lipoate, for example, the thioether derived from reaction with the substituted TTP is shown in Fig. 7.2.7. This thioether reacts with CoA—SH to yield succinyl CoA and the reduced lipoate. The series of reactions is shown in Fig. 7.2.8. The complexity of the sequence suggests

Fig. 7.2.7

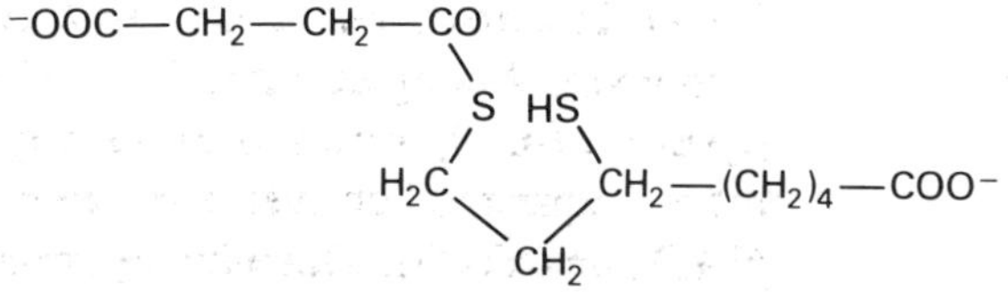

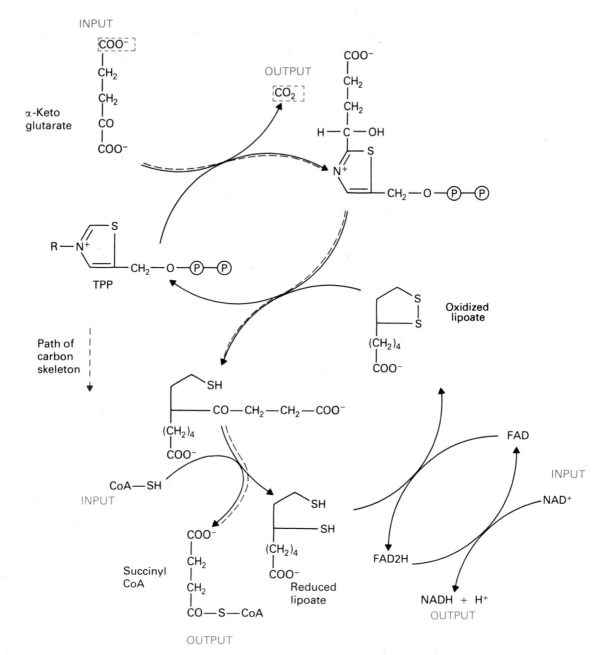

Fig. 7.2.8 The α-ketoglutarate dehydrogenase reaction.
The coenzymes of the multienzyme complex are set out showing input and output.

that there is more than one enzyme; this is the case, but it is thought that the enzymes are associated in a *multienzyme complex*.

The intermediate which is the end-point of the multienzyme mechanism is succinyl CoA, and this is of sufficiently high group transfer potential (43 kJ/mol at pH 7) to allow synthesis of GTP from GDP in a substrate-level phosphorylation. GTP is equivalent to ATP in so far as it releases a large amount of free energy on hydrolysis. In the next reaction succinate dehydrogenase, to which FAD is covalently linked, catalyses oxidation to fumarate

$$\begin{array}{ccc}
\begin{array}{c}
\text{COO}^- \\
| \\
\text{CH}_2 \\
| \\
\text{CH}_2 \\
| \\
\text{COO}^-
\end{array} + \text{FAD}
\xrightarrow[\text{dehydrogenase}]{\text{succinate}}
& \begin{array}{c}
\text{COO}^- \\
| \\
\text{CH} \\
|| \\
\text{CH} \\
| \\
\text{COO}^-
\end{array} + \text{FAD2H}
\end{array}$$

Succinate   Fumarate

(Fig. 7.2.9). The redox potential of FAD2H/FAD is not sufficiently negative to reduce NAD^+, so non-haem iron, in association with the enzyme, is reduced which in turn reduces ubiquinone (cocnzyme Q, Fig. 7.0.7). Bypassing NAD^+ means that there is only the possibility of gaining two ATPs.

Thereafter there are two straightforward dehydrogenase reactions, between which a hydratase reaction is interposed.

Control of the tricarboxylic acid cycle

The rate of oxidation of two-carbon units in the cycle is thought to be regulated by a number of factors. The cycle generates ATP, and accumulating ATP inhibits the enzyme citrate synthetase. (This is one of the examples of feedback control of metabolism, and the inhibition is of the allosteric type.) The converse situation, a shortage of ATP, would imply that the cycle should be stimulated; indeed, ADP allosterically activates the enzyme isocitrate dehydrogenase.

Significance of the tricarboxylic acid cycle

The cycle is of enormous significance in biology, and probably operates in all aerobic cells. The following points are equally important; they include some reiteration of those made at the beginning of the section (they cannot be iterated too often).

1   The cycle is a device for the total oxidation of two-carbon units.

2   The intermediates of the cycle do not accumulate; they are present in catalytic quantities.

3   The cycle is 'one-way'.

4   The cycle operates as a 'common terminal pathway', that is it accepts the breakdown products of carbohydrates, fats, ketone bodies, and amino acids.

5   Certain segments of the cycle are 'pre-empted' for other mechanisms, such as:

(a) the synthesis of malate from aspartate as a means of transporting electrons across the mitochondrial membrane (the malate shunt, 8.2); (b) the synthesis of citrate in the citrate synthetase reaction as a means of transporting two-carbon units from the mitochondria for palmitate synthesis in the cytosol (11.4); (c) the sequence fumarate → malate → oxaloacetate is related to

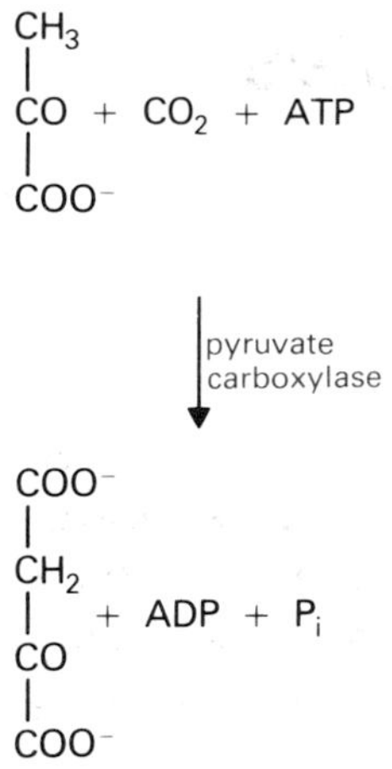

$$\begin{array}{l} CH_3 \\ | \\ CO \\ | \\ COO^- \end{array} + CO_2 + ATP$$

$$\downarrow \text{pyruvate carboxylase}$$

$$\begin{array}{l} COO^- \\ | \\ CH_2 \\ | \\ CO \\ | \\ COO^- \end{array} + ADP + P_i$$

Fig. 7.2.10

the provision of nitrogen for the urea cycle (10.2).

6 The carbon of cycle intermediates may be utilized for the synthesis of various substances such as amino acids (10.8) and porphyrins (10.9). If this happens then obviously there must be replacement of the carbon atoms removed, so that the cycle does not 'run down'. The possibility is avoided by *anaplerotic reactions*, such as that catalysed by pyruvate carboxylate (Fig. 7.2.10), so that oxaloacetate, a cycle intermediate, is made available for reaction with acetyl CoA.

7 The cycle is very productive of ATP. There are four reactions resulting in the synthesis of ATP, and one substrate-level phosphorylation producing GTP. The succinic dehydrogenase reaction has a P/O ratio of 2, rather than 3, since NAD^+ is bypassed as a respiratory chain intermediate. They can be listed:

1	isocitrate dehydrogenase step	– 3 ATP
2	α-ketoglutarate dehydrogenase step	– 3 ATP
3	succinyl CoA synthetase step	– 1 GTP
4	succinic dehydrogenase step	– 2 ATP
5	malate dehydrogenase step	– 3 ATP

Thus the total number of phosphorylations from a single two-carbon unit is 12.

Catabolisim of Carbohydrates

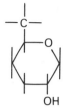

Fig. 8.0.1 Monosaccharide ring structure for active transport.

Fig. 8.0.2 The active transport of glucose. If a gradient can be maintained for sodium such that it tends to diffuse back into cells, then glucose can be carried with it.

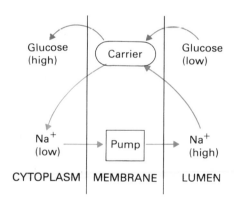

8.0 Digestion of carbohydrates

The principal carbohydrates of the diet are starch, sucrose, and lactose. Some glycogen is ingested in meats and fish but muscle has only about 0.7% of glycogen on a wet weight basis. Free fructose is taken in fruit and honey. Many other carbohydrates are present in the diet in small amounts, including 'fibre', that is non-digestible carbohydrate.

Starch is hydrolysed to maltose, initially by salivary amylase, subsequently be pancreatic amylase. The duodenum synthesizes disaccharide-hydrolysing enzymes (sucrase, maltase, lactase) associated with the brush border of the mucosal cells rather than the space of the lumen. The monosaccharides released (glucose, glactose, fructose) are absorbed by the jejunum by one of two mechanisms. These are *active transport* and *simple diffusion*. Any substance diffuses across a membrane (if the membrane is indeed permeable to it) at a rate proportional to the difference in concentration across the membrane, otherwise known as the *concentration gradient*. Glucose could not be absorbed by passive diffusion, since its concentration in the cells is generally higher than the concentration in the *chyme*. Therefore energy and active transport are required. Glucose and certain of its analogues, as well as galactose, are absorbed by the active process; fructose, ribose, mannose, and others are absorbed by passive diffusion. The structural requirements (Fig. 8.0.1) for binding the carrier molecule involved in active transport are:

1 at least six carbons;
2 a hydroxyl function at C_2, correctly orientated; and
3 a pyranose ring structure.

It is thought that the membrane of the brush border contains a carrier substance binding glucose or galactose at one site, and sodium at a separate site. An ATP-consuming pump maintains a concentration gradient for sodium, continuously ejecting it from the cells such that it tends to diffuse in again in response to the gradient created. If glucose transport is coupled to this by way of the carrier, it can be carried against its concentration gradient (Fig. 8.0.2).

8.1 Glycogenolysis

Glycogenolysis and glycogenesis are completely different, exhibiting the general rule that the breakdown and synthesis of a compound follow different courses, thereby facilitating differential control. Glycogenolysis in

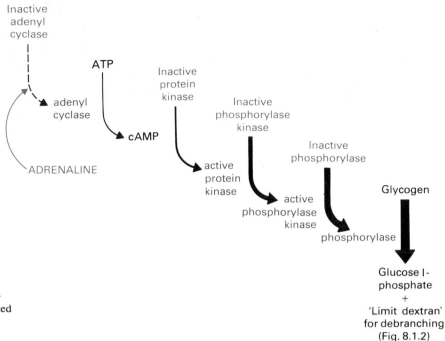

Fig. 8.1.1 The glycogenolysis cascade in muscle.
The progressively thickening arrows are intended to suggest the amplification process. The phosphorylated (active) enzymes are inactivated by phosphatases; cAMP is inactivated by hydrolysis in a reaction catalysed by a phosphodiesterase.

liver supports blood glucose concentration; in muscle it supports contraction. In terms of percentage wet weight, liver has more glycogen than muscle but the mass of muscle is so much greater that muscle glycogen predominates in the whole body. Glycogen occurs in almost all tissues (brain is an exception) but the finer points of its metabolism have been studied only in liver and muscle. There are some differences in glycogenolysis in the two tissues. Here we concentrate on muscle. Below, 'kinase activity' means the catalytic attachment of a phosphate from ATP to a seryl residue of an enzyme. Eight clear steps can be distinguished (Figs 8.1.1 and 8.1.3).

1   Where the initiation of glycogenolysis is hormonal rather than neural (see below) adrenaline attaches to its specific receptors on the muscle cell membrane. (Glucagon secretion by the pancreas following hypoglycaemia results in glucagon attaching to hepatic receptors.)

2   Adenyl cyclase is activated and cyclic AMP (cAMP) is synthesized (Fig. 8.1.2).

3   'Second messenger', cAMP, (the hormone is the 'first messenger') activates a protein kinase

$$\text{Inactive protein kinase} \xrightarrow{\text{cAMP}} \text{active protein kinase}$$

4   The active protein kinase catalyses the attachment of a phosphate from

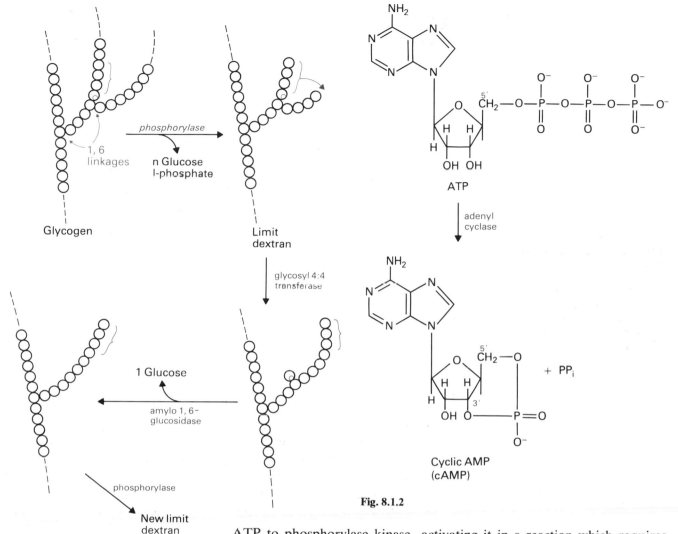

phosphorylase

n Glucose
l-phosphate

1,6
linkages

Glycogen

Limit
dextran

glycosyl 4:4
transferase

1 Glucose

amylo 1,6-
glucosidase

phosphorylase

New limit
dextran

ATP

adenyl
cyclase

Cyclic AMP
(cAMP)

+ PP$_i$

Fig. 8.1.2

Fig. 8.1.3 'Debranching' of glycogen.
The circles each represent a glucose residue.
The residue destined to become the stump susceptible to glucosidase activity is marked with a green spot. The entire process is necessary since phosphorylase is active only to the glycosidic bond four residues away from a 1,6-linkage. After the adjustment it is presented with a new set of hydrolysable 1,4-linkages.

ATP to phosphorylase kinase, activating it in a reaction which requires calcium

$$\text{Inactive phosphorylase kinase} \xrightarrow[\text{ATP} \quad \text{ADP}]{\text{Ca}^{2+}} \text{active phosphorylase kinase}$$

5 Phosphorylase kinase catalyses linkage of a phosphate group to phosphorylase *b* to produce the active phosphorylase *a*

$$\text{Phosphorylase } b \xrightarrow[\text{ATP} \quad \text{ADP}]{} \text{phosphorylase } a$$

Phosphorylase contains the coenzyme pyridoxal phosphate but the significance of this is not known.

6 Phosphorylase catalyses the cleavage of glycogen to glucose 1-phosphate in a reaction utilizing inorganic phosphate, starting at the ends of the branched chains. Phosphorylase is specific for the 1,4-links in glycogen and cannot catalyse the phosphorolysis of 1,6-linkages at 'branch points'. Its activity ceases four residues before the 1,6-linkage.

7 When a branch point is reached, and a 'limit dextran' thus produced, another enzyme, glycosyl-4,4-transferase, catalyses removal of glycosyl residues from one stump to another. This leaves a shorter stump of one residue only.

8 The enzyme amylo-1,6-glucosidase catalyses hydrolysis of the 1,6-linkage, removing the stump as glucose. This leaves a chain exclusively with 1,4-linkages, susceptible to phosphorylase *a*, till the next branch is reached, and the 'debranching' process again occurs.

Significance of the glycogenolysis cascade

The sort of arrangement shown in Fig. 8.1.1 is often termed a 'cascade', on the analogy of a waterfall, or water running down steps.

1 The phosphorolysis of glycogen represents an amplification system. A few molecules of adrenaline arriving at the muscle cell surface trigger off synthesis of large amounts of cAMP, which activate protein kinase, giving many more molecules of active phosphorylase kinase, and so on.

2 The protein kinase in the scheme is the same protein kinase which phosphorylates glycogen synthetase (11.1). However, phosphorylated glycogen synthetase is inactive, not active. The same process (activation of protein kinase), then, both enhances glycogen breakdown and inhibits its synthesis.

3 Active phosphorylase *a* is inhibited by glucose 6-phosphate. Since this is obtained by the isomerization of the product of phosphorylase action, namely glucose 1-phosphate, then the system is effectively inhibited by build-up of its product.

4 The phosphorylase kinase is activated not only by protein kinase, but also by calcium ions. Nervous rather than endocrine (adrenaline) stimulation of glycogenolysis causes a rise of calcium ions, in the muscle sarcoplasm, from about 10^{-7} to 10^{-5} mol/l. Thus the same factor, calcium, triggers off both the muscle activity and the provision of glucose from glycogen to sustain activity.

5 The product of glycogenolysis in liver is glucose, since liver has a glucose 6-phosphatase active in the sequence

Glucose 1-phosphate $\rightarrow$ glucose 6-phosphate $\rightarrow$ glucose + P_i

Muscle does not have this enzyme. Glucose 1-phosphate released in muscle by glycogenolysis is utilized in glycolysis and if the carbon atoms appear in the blood stream, they do so as lactate and pyruvate (8.2).

Glycogenoses

The inherited defects of glycogen metabolism are collectively known as the glycogenoses or glycogen storage diseases, and in four of these it is glycogenolysis which is specifically affected. The conditions were previously named after their discoverers, but the eponyms have been replaced by numbers, and the defective enzymes are as follows:

Type III   — debranching enzyme (all tissues)
Type V   — phosphorylase (muscle)
Type VI   — phosphorylase (liver)
Type VIII — phosphorylase kinase (most tissues except muscle)

The cells of affected tissues become loaded with large vacuoles of glycogen (which in the case of type III is abnormal, having short terminal chains). Despite this common feature, the conditions are of diverse clinical severity.

8.2   Glycolysis and pyruvate metabolism

The splitting of glucose, known as *glycolysis* or the Embden-Meyerhoff pathway, occurs in all cells except for a few bacteria. It is a mechanism for converting six-carbon sugars to three-carbon fragments. There is the generation of a certain amount of ATP. The three-carbon units have much metabolic flexibility. Moreover, glycolysis can be both aerobic and anaerobic. The starting point is glucose 6-phosphate which may be derived from other hexoses or glycogen. The reactions are largely straightforward (Fig. 8.2.1) but the following need some attention and explanation.

The aldolase reaction Fructose 1,6-diphosphate is split so that the systematic name for the enzyme is fructose diphosphate–glyceraldehyde phosphate lyase (Fig. 8.2.2). The two products are interconvertible isomers, but in the glycolysis pathway the glyceraldehyde 3-phosphate is the next substrate, not dihydroxyacetone phosphate, so that in the equilibrium

Glyceraldehyde 3-phosphate $\rightleftharpoons$ dihydroxyacetone phosphate

there is continuous displacement to the left-hand side.

The glyceraldehyde 3-phosphate dehydrogenase reaction This is a combined phosphorylation and dehydrogenation and, as one might expect, the mechanism involves more than one step. For explanatory purposes we can neglect all of the glyceraldehyde 3-phosphate molecule except the aldehyde function and write it as X—CHO. The enzyme has sulphydryl groups at its active site and can be represented as HS—En. The initial step is

$$\text{X—CHO} + \text{HS—En} \rightarrow \text{X}\underset{\underset{\text{OH}}{|}}{\text{—CH}}\text{—S—En}$$

By oxidation a thioester bond with a high group transfer potential is created

$$\text{X}\underset{\underset{\text{OH}}{|}}{\text{—CH}}\text{—S—En} + \text{NAD}^+ \rightarrow \text{X—CO—S—En} + \text{NADH} + \text{H}^+$$

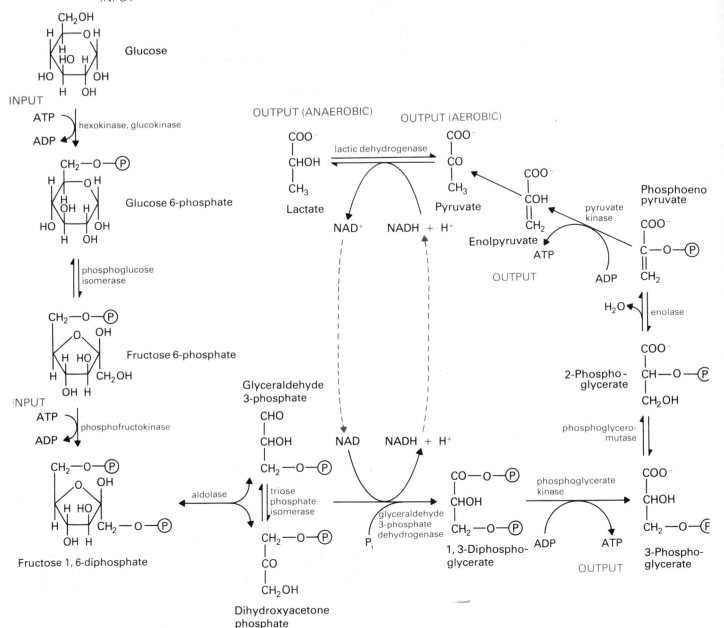

Fig. 8.2.1 Glycolysis.

The initial uptake of glucose by phosphorylation is dependent upon hexokinase in brain and most other tissues, but upon glucokinase in muscle. The K_m of hexokinase for glucose is low, so in the brain, glucose is phosphorylated even at low concentrations; for glucokinase K_m is high, so when glucose is plentiful it is taken up for storage as glycogen.

In this instance, the diagram is carefully constructed to show those reactions involving loss of a large amount of free energy as 'one-way', and those of minimal free energy change as reversible. This is of significance in later consideration of the reversibility of the whole pathway and of gluconeogensis. The dotted green lines indicate that NADH generated in the glyceraldehyde phosphate dehydrogenase reaction can be utilized in the lactic dehydrogenase reaction under anaerobic conditions. Under anaerobic conditions the input, for each six-carbon unit, is two moles of ATP, and the output is four. The net gain, therefore, is two.

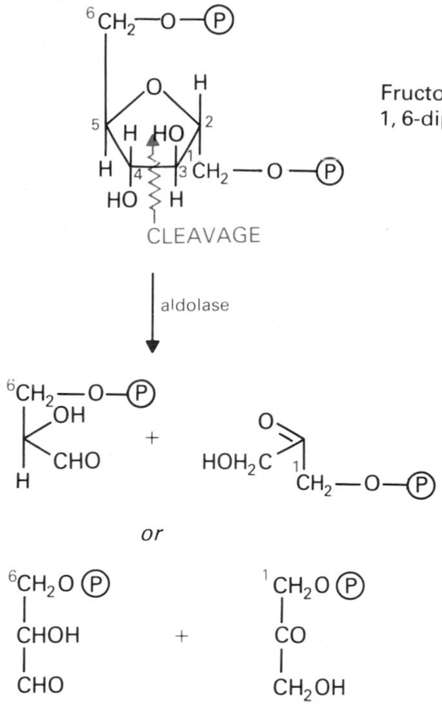

Fructose
1, 6-diphosphate

Fig. 8.2.2

Glyceraldehyde
3-phosphate

Dihydroxyacetone
phosphate

This reacts with phosphate to yield an acyl phosphate plus the regenerated enzyme

$$X—CO—S—En + P_i \rightarrow X—CO—P + HS—En$$

The acyl phosphate hydrolysis in the next step releases enough energy to synthesize ATP. The NADH generated is quantitatively sufficient to reduce to lactate the pyruvate produced by the pathway under anaerobic conditions.

Significance of glycolysis It is an important pathway in all cells. Its noteworthy features are as follows.

1 It can operate under either aerobic or anaerobic conditions. In the former the total reaction is

$$\text{Glucose} + 8ADP + 8P_i + 2O_2 \rightarrow 2\,\text{pyruvate} + 8ATP + 4H_2O$$

The anaerobic stoichiometry is

$$\text{Glucose} + 2ADP + 2P_i \rightarrow 2\,\text{lactate} + 2ATP$$

Thus only 2 ATP are synthesized in anaerobiosis, but as much as 8 in aerobiosis. The latter can be set out as:

(1) glyceraldehyde 3-phosphate dehydrogenase step   6ATP
(3 per 3-C fragment);

(2) phosphoglycerate kinase step   2ATP
(1 per 3-C fragment);

(3) pyruvate kinase step   2ATP
(1 per 3-C fragment).

This comes to ten, so the totals above make allowance for the consumption of ATP in starting off the process (the hexokinase and phosphofructokinase steps).

Rapidly contracting muscle does not obtain a great deal of energy from anaerobic glycolysis; the process is essentially incomplete and pyruvate and lactate are released into the blood stream. But there is no alternative since it cannot be provided with enough oxygen to sustain aerobic glycolysis when it is contracting rapidly. Note that alcoholic fermentation (13.10) is also a form of glycolysis.

2 It occurs in the cytoplasm, in contrast to the citric acid cycle which is intramitochondrial. It is associated with other cytoplasmic pathways, such as gluconeogenesis and fatty acid synthesis. The NADH it generates for oxidation must be specially transported into the mitochondria. The device to achieve this is known as the *malate shuttle* (Fig. 8.2.3). Malate dehydrogenase catalyses the reduction of oxaloacetate to malate

$$\text{Oxaloacetate} + \text{NADH} + \text{H}^+ \rightarrow \text{malate} + \text{NAD}^+$$

Malate diffuses across the mitochondrial membrane by a facilitated mechanism. Mitochondrial malate dehydrogenase catalyses oxaloacetate formation and NADH for oxidation by the electron transport chain in the 'reverse' reaction. Oxaloacetate does not, however, itself diffuse back to the cytosol. Aspartate amino transferase catalyses the transfer of an amino group from glutamate to oxaloacetate and the resulting aspartate diffuses back to regenerate oxaloacetate for another cycle. No energy consumption is indicated although since reducing intermediates are transported against a concentration gradient (the NADH/NAD ratio is generally higher in mitochondria than in cytoplasm) some ATP is probably used up.

3 Glycolysis is not reversible, in common with the majority of pathways. There are three reactions which have equilibria very far to the right hand side if they are written thus:

(1) Glucose + ATP $\rightarrow$ glucose 6-phosphate + ADP
(2) Fructose 6-phosphate + ATP $\rightarrow$ fructose 1, -diphosphate + ADP
(3) Phosphoenol pyruvate + ATP $\rightarrow$ pyruvate + ATP

The alternative 'reverse' pathway, the synthesis of glucose from pyruvate, lactate, and oxaloacetate, is generally termed *gluconeogenesis* (11.3).

4 Glycolysis is subject to a number of control mechanisms, the most studied being at the phosphofructokinase step. This is the 'committed step', that is up to this point there are side reactions possible (from fructose

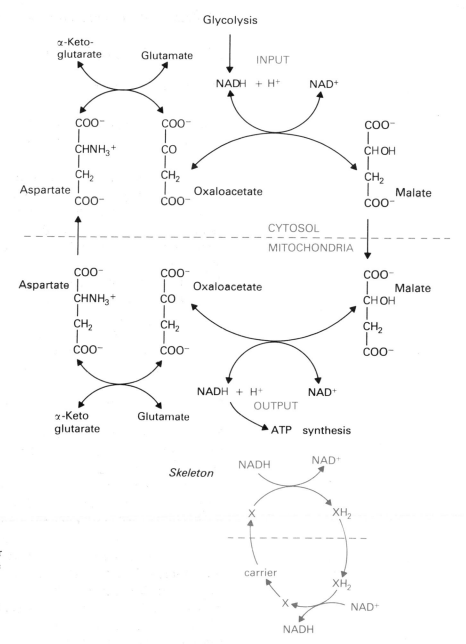

Fig. 8.2.3 The malate shuttle.
It is clear that the process serves only to transfer reducing equivalents across the membrane. The components need be present only in catalytic quantities. All the reactions are readily reversible. The names of the enzymes have been omitted for clarity, but evidently a transaminase and a dehydrogenase are implicated.

6-phosphate, or glucose 6-phosphate) but once fructose 1,6-diphosphate is synthesized, the course of reactions proceeds inevitably to pyruvate. Phosphofructokinase is an allosteric enzyme inhibited by citrate and ATP, and thus glycolysis itself slows down when energy-yielding compounds build up. It also slows down, in adipose tissue and muscle, with any reduction in insulin secretion, since insulin is necessary for its uptake into these tissues.

Fate of pyruvate Pyruvate is diffusible (by means of a carrier) across the mitochondrial membrane. The mitochondria contain the pyruvate dehydrogenase complex, which is analogous to the α-ketoglutarate dehydrogenase complex (Fig. 7.2.8). This being the case, the NADH produced has to be removed, which is impossible in the absence of oxygen. Thus under anaerobic conditions the NADH produced in the 1,3-phosphoglycerate dehydrogenase reaction reduces pyruvate to lactate. Here you can think of oxygen as operating a switch in one direction or the other.

Lactic acid has a pK of about 3.7, and so is almost entirely dissociated at body pH. This is why substantial production of lactate is acidotic ('lactic acidosis'). There is no mechanism to remove lactate other than by reoxidation to pyruvate, so it is one of the 'blind alleys' in metabolism.

Pyruvate can also be transaminated to the amino acid alanine. This is quantitatively significant in muscles when amino acid nitrogen in general is transferred to alanine so that it can be transported to the liver; in the liver the reverse takes place and the pyruvate is utilized for glucose in the gluconeogenic pathways (11.3).

Finally, pyruvate may yield oxaloacetate by carboxylation, catalysed by pyruvate carboxylase. This is an important reaction for two reasons. First of all, some of the intermediates of the tricarboxylic acid cycle may be removed by reactions directed towards the synthesis of special compounds, like porphyrins (10.9) and if the four-carbon oxaloacetate is replenished from pyruvate then all the intermediates are replenished. This topping-up procedure is sometimes called *anaplerosis*. Secondly, in gluconeogenesis there is no direct conversion of pyruvate to phosphoenolpyruvate on the way to glucose; the energetics are unfavourable. The pathway is rather

Pyruvate $\rightarrow$ oxaloacetate $\rightarrow$ phosphoenolpyruvate (Fig. 8.2.4).

8.3 Catabolism of fructose and galactose

Fructose is largely derived from dietary sucrose, and so may become available in quite large quantities to those with a sweet tooth. Sucrase in the brush border of the small intestine hydrolyses sucrose to glucose and fructose. After absorption fructose is rapidly taken up by the liver—this is because fructose uptake, unlike that of glucose, is not dependent upon insulin. The first step in fructose metabolism is phosphorylation by a fructokinase to yield fructose 1-phosphate. (Recall that glucose is phosphorylated by glucokinase to glucose 6-phosphate). An aldolase splits this to dihydroxyacetone phosphate and glyceraldehyde. (Again, in contrast, the pathway for glucose involves aldolase splitting fructose 1,6-diphosphate to dihydroxyacetone phosphate and glyceraldehyde 3-phosphate; it seems odd that fructose too cannot be converted to fructose 1,6-diphosphate, but mammalian intermediary metabolism has just not evolved in that manner.) The triose glyceraldehyde is phosphorylated to glyceraldehyde 3-phosphate, metabolized further as a bona fide glycolysis intermediate (Fig. 8.3.1). Moreover, dihydroxyacetone phosphate is in equilibrium with glyceraldehyde 3-phosphate (as in the glycolysis pathway, Fig. 8.2.1). So, ingestion of fructose

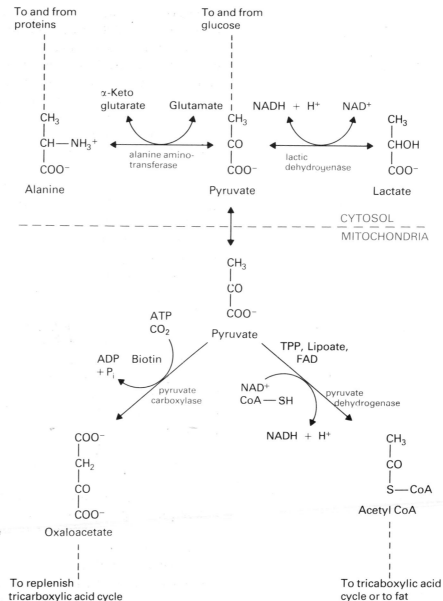

To and from proteins

To and from glucose

α-Keto glutarate — Glutamate

$$CH_3$$
$$|$$
$$CH—NH_3{}^+$$
$$|$$
$$COO^-$$

Alanine

alanine amino-transferase

NADH + H$^+$ — NAD$^+$

$$CH_3$$
$$|$$
$$CO$$
$$|$$
$$COO^-$$

Pyruvate

lactic dehydrogenase

$$CH_3$$
$$|$$
$$CHOH$$
$$|$$
$$COO^-$$

Lactate

CYTOSOL
MITOCHONDRIA

$$CH_3$$
$$|$$
$$CO$$
$$|$$
$$COO^-$$

ATP
CO_2

Pyruvate

ADP + P$_i$ Biotin

pyruvate carboxylase

TPP, Lipoate, FAD

NAD$^+$
CoA — SH

pyruvate dehydrogenase

NADH + H$^+$

$$COO^-$$
$$|$$
$$CH_2$$
$$|$$
$$CO$$
$$|$$
$$COO^-$$

Oxaloacetate

To replenish tricarboxylic acid cycle intermediates

$$CH_3$$
$$|$$
$$CO$$
$$|$$
$$S—CoA$$

Acetyl CoA

To tricarboxylic acid cycle or to fat

Fig. 8.2.4 Fates of pyruvate.
The pyruvate dehydrogenase reaction is exactly analogous to the α-ketoglutarate dehydrogenase reaction (Fig. 7.2.8); CH$_3$—CHOH—TPP replaces $^-$OOC—CH$_2$—CH$_2$—CHOH—TPP as the initial intermediate.
In this diagram reversibility is carefully indicated.

results in a surge of pyruvate synthesis in the liver. This is due partly to the non-insulin dependent nature of fructose assimilation, and partly to the fact that the highly modulated phosphofructokinase step (fructose 6-phosphate to fructose 1,6-diphosphate) is avoided.

Sucrose is not synthesized in mammals, but fructose is found in high concentrations in the seminal plasma. This can be regarded as a survival factor for the spermatazoa; since fructose is not utilized by the cells of the vaginal or cervical mucosa, they are carrying around their own private fuel supply, immune to depredation.

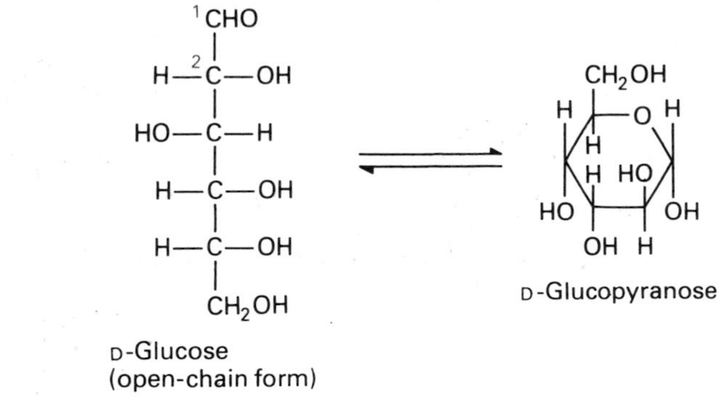

^{1}CHO

H—^{2}C—OH

HO—C—H

H—C—OH

H—C—OH

CH$_2$OH

D-Glucose (open-chain form)

CH$_2$OH

D-Glucopyranose

$H^+ +$ NADPH

reductase

NADP$^+$

^{1}CH$_2$OH

H—^{2}C—OH

HO—C—H

H—C—OH

H—C—OH

CH$_2$OH

Sorbitol

NAD$^+$

dehydrogenase

NADH + H$^+$

^{1}CH$_2$OH

^{2}C=O

HO—C—H

H—C—OH

H—C—OH

CH$_2$OH

D-Fructose (open-chain form)

D-Fructofuranose

Synthesis / Catabolism

ATP

ADP

fructokinase

Fructose 1-phosphate

aldolase

CH$_2$OH

C—OH

H—CHO

Glyceraldehyde

+

Dihydroxyacetone phosphate

HOH$_2$C—C—CH$_2$O$_P$

ATP kinase ADP

isomerase

CHO

CH$_2$OH

CH$_2$O$_P$

Glyceraldehyde 3-phosphate

Glycolysis

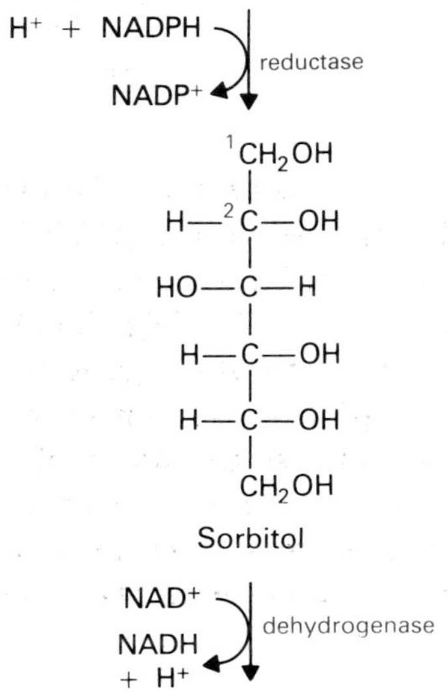

The source of carbon for fructose synthesis is glucose. Since glucose and fructose are isomers, the process is an isomerization, with the polyhydric alcohol D-sorbitol as an intermediate. This is synthesized from glucose in a reaction involving a reductase and NADPH, then oxidized by a dehydrogenase using NAD$^+$ as cofactor. The reacting form of glucose is necessarily the open-chain form, in equilibrium with the pyranose structure (Fig. 8.3.1).

Galactose is a constituent of lactose, or 'milk sugar', and is released at the brush border of the intestinal mucosa by a lactase. It is absorbed by an active transport mechanism (8.0) and, in the liver, phosphorylated by ATP to yield galactose 1-phosphate. You might suspect that this compound is to be converted to a glucose phosphate for entry into glycolysis, especially since glucose and galactose differ only in the orientation about C$_4$, and you would

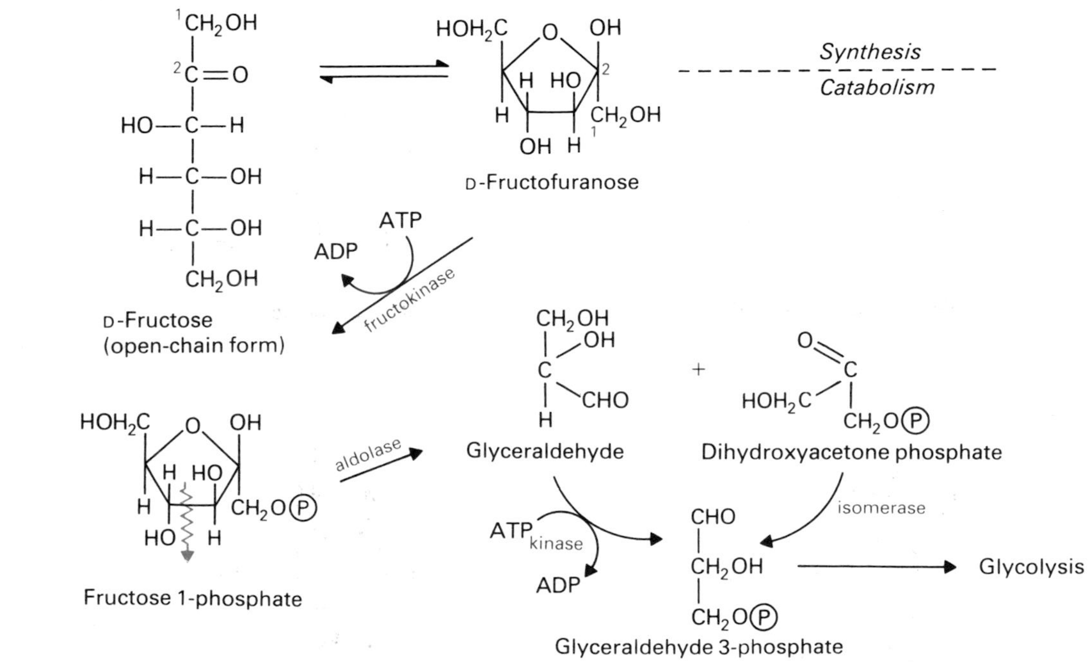

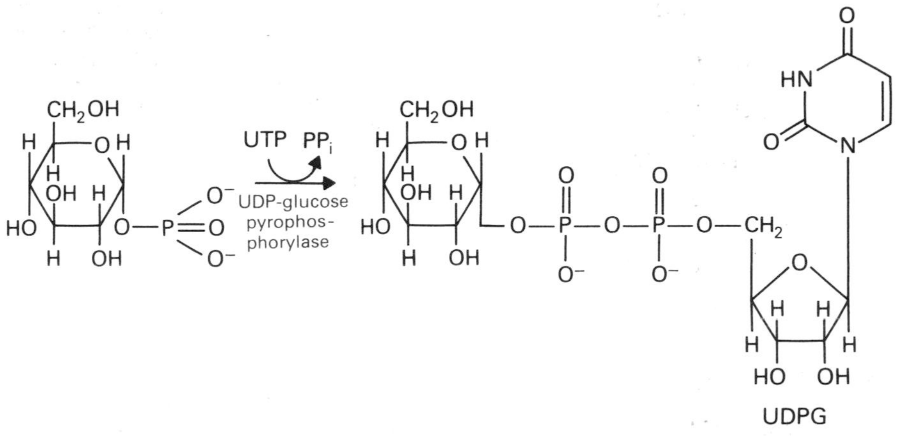

Fig. 8.3.2 Uridine diphosphate glucose (UDPG).

be correct, but the intervention of another coenzyme is necessary for the transformation. This coenzyme is uridine diphosphate glucose (UDPG) which is synthesized in a reaction using as substrate one of the (educationally) basic substances in 2.1, namely uridine triphosphate or UTP (Fig. 8.3.2). In the liver, this reacts with galactose 1-phosphate to form UDPG galactose and glucose 1-phosphate (Fig. 8.3.3). UDP-galactose is epimerized (*epimerization* is the inversion of conformation of a single carbon of a compound) to give UDP glucose (Fig. 8.3.4). UDPG is itself a glycogen precursor, so the pathway from ingested galactose to glucose, through glycogen, is as given in Fig. 8.3.5. UDP-galactose is the precursor of lactose in the mammary gland.

Fig. 8.3.3

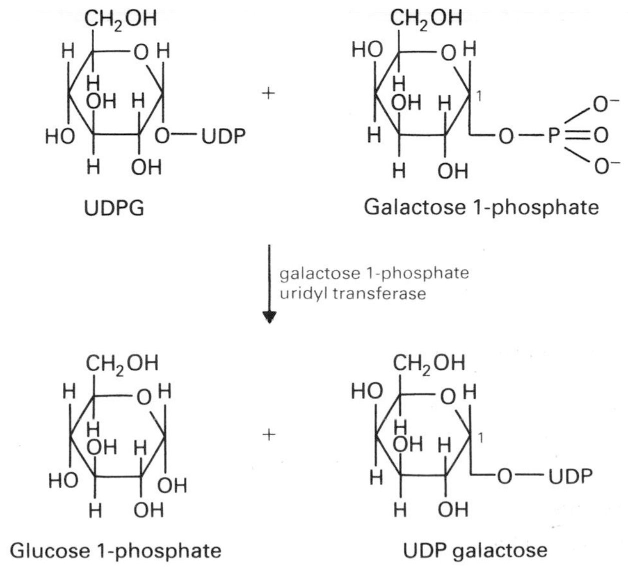

Fig. 8.3.4

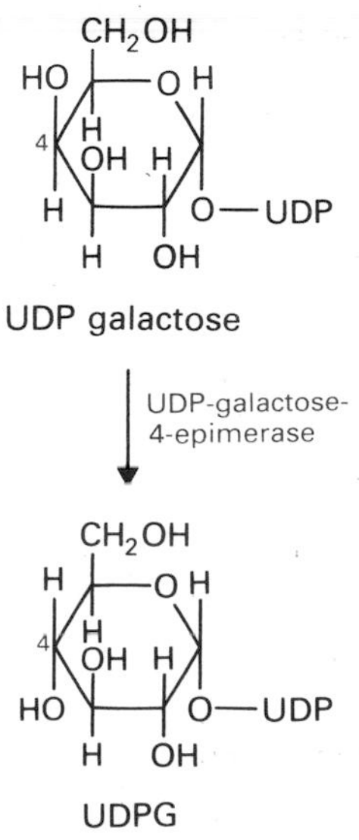

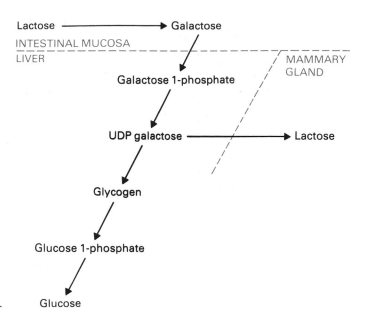

Fig. 8.3.5 Summary of galactose metabolism.

8.4 The pentose phosphate pathway

The pentose phosphate pathway, sometimes called the hexose monophosphate shunt, is an alternative to glycolysis in the catabolism of six-carbon sugars. In mammals it operates principally in the liver, adipose tissue, and erythrocytes. The best way to approach it, in the first instance, is by neglecting the names of the metabolites and enzymes and concentrating instead on carbon flow. At first sight this is puzzling, for it is not a linear sequence; instead there is an interchange of two-carbon and three-carbon units accompanied by CO_2 release (skeleton diagram in Fig. 8.4.1). Since the end products (glyceraldehyde 3-phosphate and fructose 6-phosphate) can be recycled, the release of CO_2 in effect produces a total oxidation pathway. One passage of the pathway or cycle can be represented

$$3 \times (6C) \rightarrow 3C + 2 \times (6C) + 3CO_2$$

With $3CO_2$ released each time, evidently six passages would be required to remove 18 carbons as CO_2, that is quantitatively the equivalent of the three six-carbon units originally taken up. However the pentose phosphate pathway is considered at present to be a system for the generation of pentoses and NADPH for lipid synthesis. The functions of the pentose phosphate pathway can therefore be listed:

1 to generate pentoses for nucleic acid synthesis (when it does not, of course, operate as a total oxidation pathway);

2 in plants, to operate sugar interconversions in the dark phase of photosynthesis (13.15);

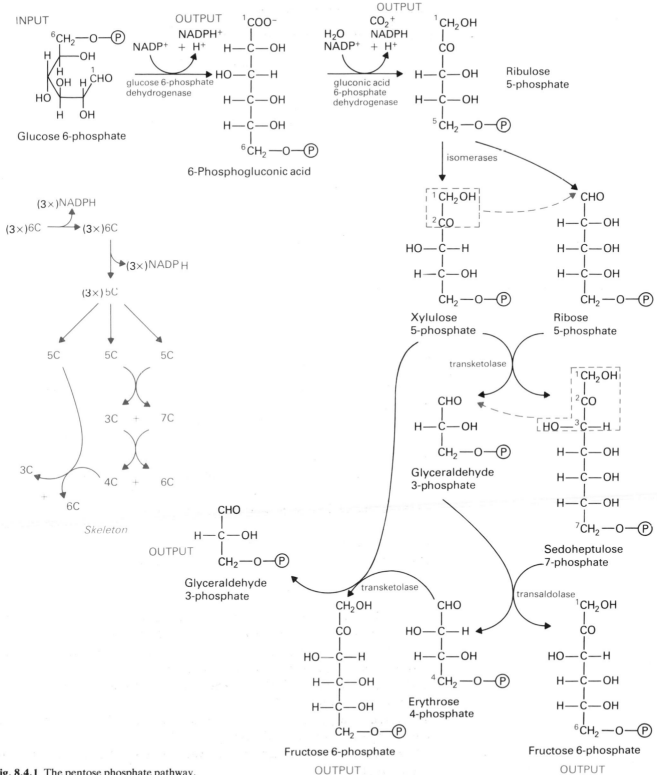

Fig. 8.4.1 The pentose phosphate pathway.

3 to generate NADPH. Its fate is:

(i) utilization for palmitate synthesis; there is a certain coordination here as both palmitate synthesis and the pentose phosphate cycle operate in the cytoplasm;

(ii) maintenance of reduced glutathione (10.0) in the red cell. This contributes to the protection of haemoglobin from oxidation;

(iii) participation in hydroxylation reactions, for example steroid hydroxylations

$$RH + O_2 + NADPH + H^+ \rightarrow ROH + H_2O + NADP^+$$

Evidence has been presented recently that the pathway in Fig. 8.4.1 is significant only in adipose tissue, and it has been designated the 'F-pathway'. An alternative pentose phosphate pathway, introducing rather more intermediates and including two octuloses (eight-carbon sugars), has been proposed for other tissues and designated the 'L-pathway'. The pathway cannot be said to be a minor one. It is estimated to dispose of about 30% of the glucose metabolized by the liver.

9   Catabolism of Lipids

9.0   Digestion of lipids

Lipid is a major energy-producing foodstuff, and under normal circumstances contributes about 35% of the total energy of the diet. Dietary lipid is mostly triacylglycerol, along with some cholesterol and its esters, and phospholipids.

In describing lipid digestion, you have to be clear about the concept of *emulsification*, which is the intimate admixture of two phases, one dispersed in the other as fine droplets or *micelles*. In this context the two phases are water and fat, the latter making up the micelles. Micelles tend to aggregate if they are not stabilized in some way; in the duodenum this role is performed by the bile salts (Fig. 9.0.1). They are cholesterol derivatives with a side chain bearing glycine or taurine in imide linkage. At the prevailing pH of the duodenum the carboxyls of glycine and taurine are ionized and probably in association with sodium ions, hence the term 'bile salts'. There are two sides, as it were, to a bile salt molecule: one is hydrophobic; the other, bearing the hydroxyls and the side chain, is hydrophilic. So one side tends to associate with aqueous phase and the other with lipid phase (such molecules are often said to be *amphipathic* or 'double-feeling') and the micelles are surrounded by a stabilizing shell of water molecules due to the projecting hydroxyls.

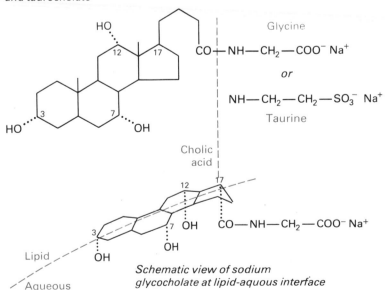

Fig. 9.0.1 Amphipathic nature of the bile salts. In the schematic view the sodium glycocholate molecule is supposed to be almost edge-on to the viewer; it shows that the β-hydroxyls (----OH) and the conjugated glycine are orientated on one face and so can bind a layer of water molecules to stabilize a lipid micelle.

Emulsification by bile salts, aided by mechanical churning, presents an increased surface area for the action of pancreatic lipase and phospholipases. The products of such enzymic reactions are clearly in themselves emulsifying agents, if it is recalled that 'R' represents a long hydrocarbon chain in the formulae shown in Fig. 9.0.2. So the lipases and their detergent-like products result in even smaller micelles, as fine as 1–2 nm. These micelles are taken up by the villi of the intestinal mucosa (Fig. 9.0.3). The triacylglycerols and phospholids need not be completely hydrolysed for absorption; moreover, re-esterification takes place in the mucosa. 2-acyl glycerols are resynthesized to triacylglycerols and cholesterol is also re-esterified to cholesterol esters.

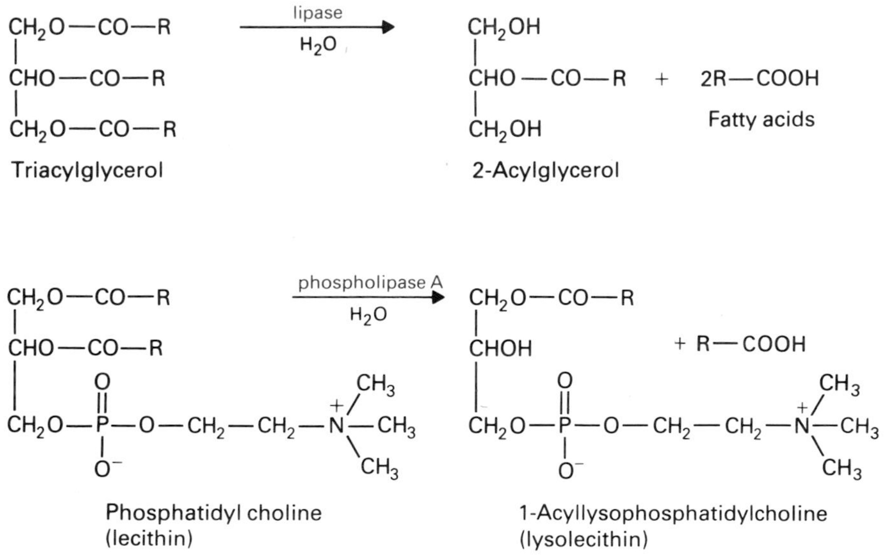

Fig. 9.0.2

The resynthesized lipids are incorporated into another type of particle termed the *chylomicron*. Formation of chylomicrons is an energy-requiring process which also involves protein synthesis; they have about 85% triacylglycerol, and small amounts of cholesterol, cholesterol esters, phospholipids, and protein. The chylomicrons are released not into the portal blood but into the lacteals, the small blind ends of the lymphatic system in the villi. The milky fluid draining away into the lymphatics is the *chyle* (Fig. 9.0.4).

At first sight the resynthesis of what has just been broken down laboriously seems wasteful. However it is probably necessary to allow the introduction of quite large amounts of fatty material into what is essentially an aqueous system. The chylomicron is a stable particle transporting fat in a covenient and harmless form; large quantities of free fatty acids could not be transported easily and would probably be toxic anyway.

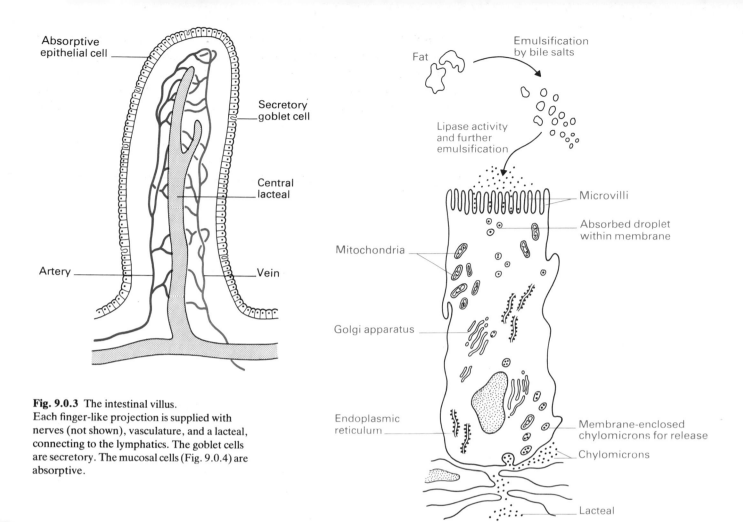

Fig. 9.0.3 The intestinal villus.
Each finger-like projection is supplied with nerves (not shown), vasculature, and a lacteal, connecting to the lymphatics. The goblet cells are secretory. The mucosal cells (Fig. 9.0.4) are absorptive.

Fig. 9.0.4 A mucosal cell.
The fat micelles absorbed are converted to chylomicrons by resynthesis and addition of protein.

Not all the products of lipid digestion appear in the chylomicrons. Short-chain fatty acids do not occur in the diet in large quantities but if they are absorbed, they are released into the portal blood because they are water-soluble. The intestinal mucosa also incorporates a certain amount of newly-synthesized triacylglycerol into the very low density lipoprotein fraction (VLDL), and the VLDL particles, which are similar to the chylomicrons but have more protein, are also released into the portal blood.

9.1   Catabolism of triacylglycerols

Chylomicrons are discharged from the lymph into the venous system; if the meal has been fairly fatty, they are sufficiently numerous to impart a milky sheen to the blood, the *postabsorptive lipaemia*. A creamy layer may even separate out, in some subjects, if the blood is left to stand for an hour or two. The chylomicrons are taken up by the capillary network of the adipose tissue, muscles and liver. Their endothelial cells contain a series of enzymes, the lipoprotein lipases, which catalyse hydrolysis of the chylomicron triacylglycerols. The polysaccharide heparin releases lipoprotein lipase from the endothelia into the circulation, but the physiological significance of this is unknown.

Adipose tissue re-esterifies the released fatty acids for storage; it cannot utilize the glycerol moiety of the hydrolysed triacyglycerols; rather it synthesizes glycerol from blood glucose. The other tissues oxidize released or circulating fatty acids in an energy-yielding pathway termed β-oxidation (Fig. 9.1.1). Essentially β-oxidation is the sequential removal of two-carbon units from the carboxyl end of the fatty acid. It occurs in the mitochondria; since the mitochondrial membrane is not permeable to the thioester derivatives of the fatty acids synthesized in the cytosol, there is a special transport system, based on the synthesis of acyl carnitines, to allow the acyl groups to enter (Fig. 9.1.3).

Figure 9.1.1 shows the system for the oxidation of saturated fatty acids. But the most abundant of the fatty acids stored in human adipose tissue is not saturated. This is oleic acid, which has a double bond between C-9 and C-10. The mechanism for its oxidation is slightly different. Three β-oxidation pathways yield an unsaturated acyl CoA in which the *cis* double bond is between C-3 and C-4 (Fig. 9.1.2). The acyl CoA dehydrogenase step is redundant because the double bond is pre-existing. However, to permit enoyl hydrase activity it should be *trans* and between C-1 and C-3. An isomerase catalyses migration of the double bond to C-2 and at the same time converts it to the *trans* form. Now the unsaturated acyl CoA can be hydrated by enoyl hydrase as though in an ordinary β-oxidation sequence.

Significance of the fatty acid oxidation pathway

1 It represents a means of obtaining energy from fatty acids mobilized to tissues from the adipose tissue.

2 The product of β-oxidation, acetyl CoA, is in itself a substrate for citrate synthetase, the first enzyme of the tricarboxylic acid cycle. Thus β-oxidation plus the cycle represent a means for total oxidation to CO_2 and H_2O.

3 Scission of each two-carbon fragment is accompanied by production of one FAD2H and one NADH. The electrons from FAD2H are introduced into the electron transport chain at the stage of coenzyme Q, thus missing a phosphorylation point (Fig. 7.1.1). The P/O ratio for its reoxidation is 2. For NADH reoxidation, the P/O ratio is 3. Thus 5ATP are formed for each β-oxidation, and there is one β-oxidation for each two carbons in the fatty acid, except for the final two. Since unsaturated fatty acids do not allow so many oxidation steps, they yield less ATP (Table 9.1.1).

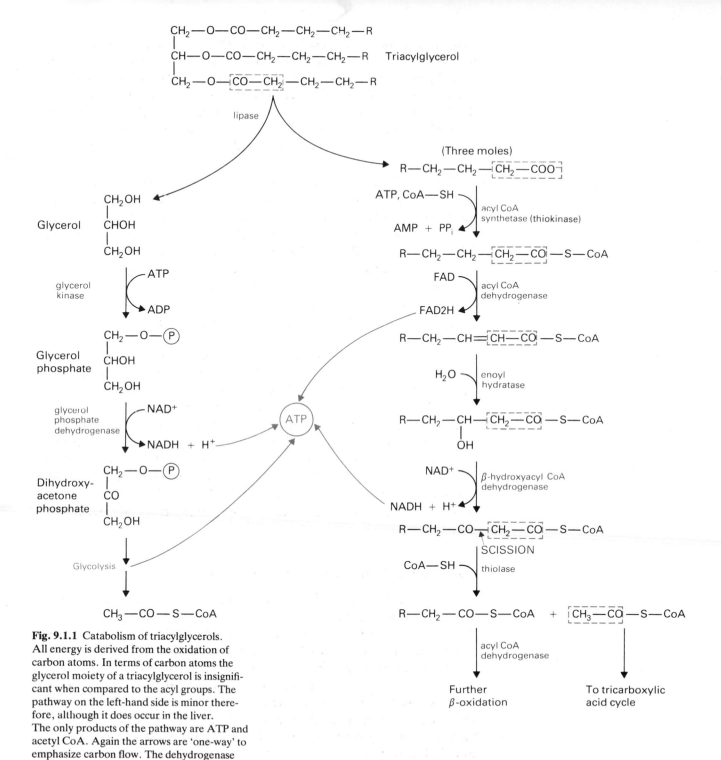

Fig. 9.1.1 Catabolism of triacylglycerols. All energy is derived from the oxidation of carbon atoms. In terms of carbon atoms the glycerol moiety of a triacylglycerol is insignificant when compared to the acyl groups. The pathway on the left-hand side is minor therefore, although it does occur in the liver. The only products of the pathway are ATP and acetyl CoA. Again the arrows are 'one-way' to emphasize carbon flow. The dehydrogenase and hydratase reactions are thermodynamically reversible but this is not so for the synthetase and thiolase reactions, and the system as a whole is unidirectional.

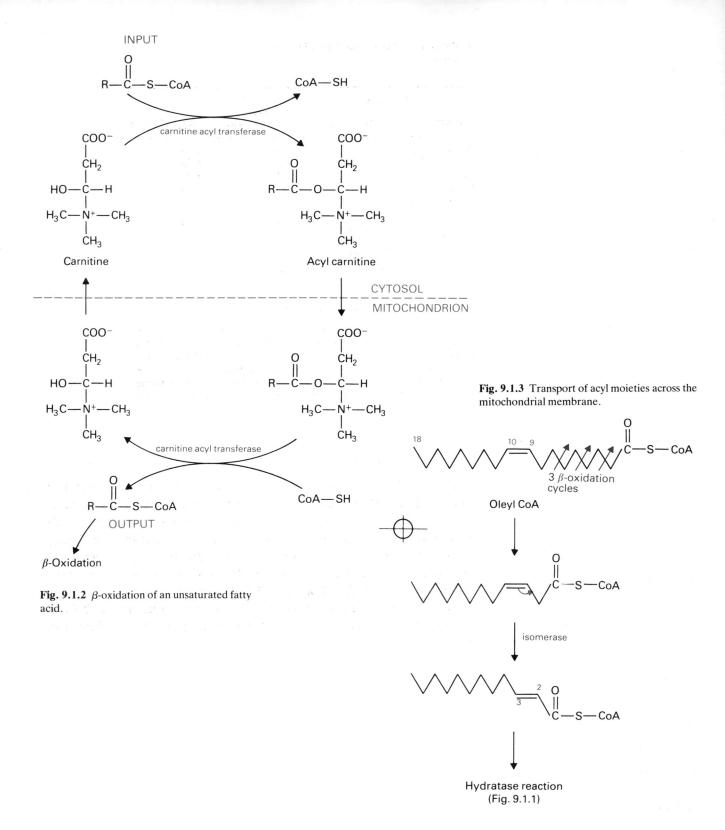

INPUT

carnitine acyl transferase

Carnitine

Acyl carnitine

CYTOSOL

MITOCHONDRION

Fig. 9.1.3 Transport of acyl moieties across the mitochondrial membrane.

carnitine acyl transferase

OUTPUT

β-Oxidation

Fig. 9.1.2 β-oxidation of an unsaturated fatty acid.

3 β-oxidation cycles

Oleyl CoA

isomerase

Hydratase reaction (Fig. 9.1.1)

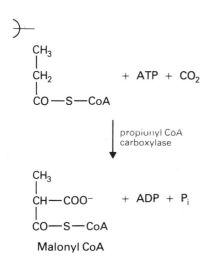

$$CH_3 - CH_2 - CO-S-CoA + ATP + CO_2$$

propionyl CoA carboxylase

$$CH_3 - CH(COO^-) - CO-S-CoA + ADP + P_i$$

Malonyl CoA

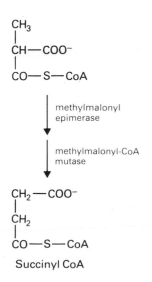

$$CH_3 - CH(COO^-) - CO-S-CoA$$

methylmalonyl epimerase

methylmalonyl-CoA mutase

$$CH_2(COO^-) - CH_2 - CO-S-CoA$$

Succinyl CoA

Fig. 9.1.4

Table 9.1.1 ATP yield from fatty acid oxidation

Fatty acids			Total ATP (mol/mol acid)	mol ATP/C atom
Saturated		Unsaturated		
Acetic	C_2		10	5.00
Butyric	C_4		27	6.75
Myristic	C_{14}		112	8.00
Palmitic	C_{16}		129	8.06
Stearic	C_{18}		146	8.11
		Oleic C_{18} (1 double bond)	144	8.00
		Linoleic C_{18} (2 double bonds)	142	7.88
		Linolenic C_{18} (3 double bonds)	140	7.77
Arachidic C_{20}			163	8.15
		Arachidonic C_{20} (4 double bonds)	155	7.15

For saturated fatty acids each two-carbon unit removed yields 5 ATP, and oxidation of each of these units in the citric acid cycle gives 12 ATP. For unsaturated fatty acids, each double bond means a loss of 2 ATP since there is no fatty acyl CoA dehydrogenase step. Then, two high energy phosphates must be subtracted for the initial activation step, yielding fatty acyl CoA in each case.

Odd-chain fatty acids

Although odd-chain fatty acids are rare in the human diet, there is a mechanism for their metabolism which results in partial conversion to succinate. The β-oxidation sequence successively removes two-carbon units but cannot cope with the three-carbon fragment at the end due to the group specificity of the enzymes involved. So the final fragment is propionyl CoA. This is carboxylated and the resulting methylmalonyl CoA is isomerized (by two enzymes) to succinyl CoA (Fig. 9.1.4). Since propionyl CoA is also an end-product of isoleucine, valine, methionine, and threonine metabolism, these reactions are more important than the paucity of odd-chain fatty acids in the diet would suggest.

10 Catabolism of Nitrogenous Compounds

10.0 Digestion of proteins

The acidic contents of the stomach denature the bulk of the protein systems in the diet, if cooking has not already accomplished this; also in the stomach, pepsin initiates digestion by hydrolysing the interior peptide bonds in the protein chains—it is thus an *endopeptidase*, in contrast to the *exopeptidases*, which hydrolyse the peptide bonds attaching the terminal amino acids. Pepsin is secreted as a *zymogen* or inactive precursor, called pepsinogen. The secretion of a zymogen may be viewed as a precaution against digestion of the digestive organs themselves. The pepsinogens, of which there are about seven, are secreted by the chief cells of the gastric mucosa. The gastric hydrochloric acid hydrolyses specific peptide bonds in each of them to yield the pepsin series. Subsequently pepsin itself catalyses the activation of pepsinogens; in other words, the process becomes *autocatalytic*.

The churned content of the stomach, the *chyme*, is voided into the duodenum, where the alkaline secretions neutralize the acid. A series of endopeptidases are generated from pancreatic zymogens and further break up the protein chains. These are the trypsins and the chymotrypsins. Trypsinogen is activated by the enzyme enterokinase and chymotrypsinogen by trypsin. Trypsin is one of the more specific proteinases, hydrolysing the peptide bonds involving the carboxyl group of lysine and arginine. The pancreas also secretes aminopeptidase, an exopeptidase hydrolysing the N-terminal peptide bond, and carboxypeptidase, hydrolysing the C-terminal. The latter is synthesized as procarboxypeptidase, activated by trypsin. Finally, dipeptidases hydrolyse the smallest protein fragments (Table 10.0.1).

Table 10.1.1

Zymogen	Activating agent	Enzyme	Specificity
Pepsinogen	Gastric HCl then pepsin	Pepsin	Many types of peptide bond hydrolysed
Trypsinogen	Enterokinase* then trypsin	Trypsin	Lysyl and arginyl peptide bonds
Chymotrypsinogen	Trypsin	Chymotrypsin	Peptide bonds involving hydrophobic side-chains
Procarboxypeptidase	Trypsin	Carboxypeptidase	Peptide bond at C-terminal of proteins
—		Leucine aminopeptidase	Peptide bond at N-terminal of proteins
—		Dipeptidase	Dipeptides

*Enterokinase, released from the duodenal mucosa by bile salts, is thus the initiating factor for protein digestion in the small intestine

There is still some doubt as to whether proteins are completely hydrolysed prior to absorption; it is possible that small peptides may enter the mucosal cells and are then hydrolysed. Moreover, the immunologic sensitivity of some individuals to specific dietary proteins (gluten from wheat is a common example) implies that it is possible for some proteins, or large fragments of them, to be absorbed.

The free amino acids are absorbed into the portal blood by an energy-consuming process. The present concept of absorption supposes that the substance glutathione (GSH) or γ-glutamylcysteinylglycine (Fig. 10.0.1) participates in a transpeptidation reaction catalysed by γ-glutamyl transferase. A γ-glutamyl peptide bond is formed with an incoming amino acid, which can then enter the cell. Subsequently GSH must be regenerated in a cycle involving 5-oxoproline as intermediate. The process (known as the *Meister cycle*) is shown in somewhat simplified form in Fig. 10.0.2.

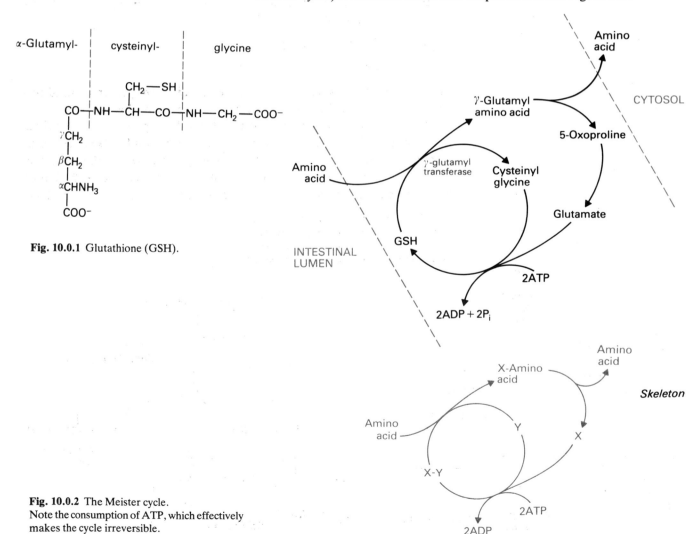

Fig. 10.0.1 Glutathione (GSH).

Fig. 10.0.2 The Meister cycle.
Note the consumption of ATP, which effectively makes the cycle irreversible.

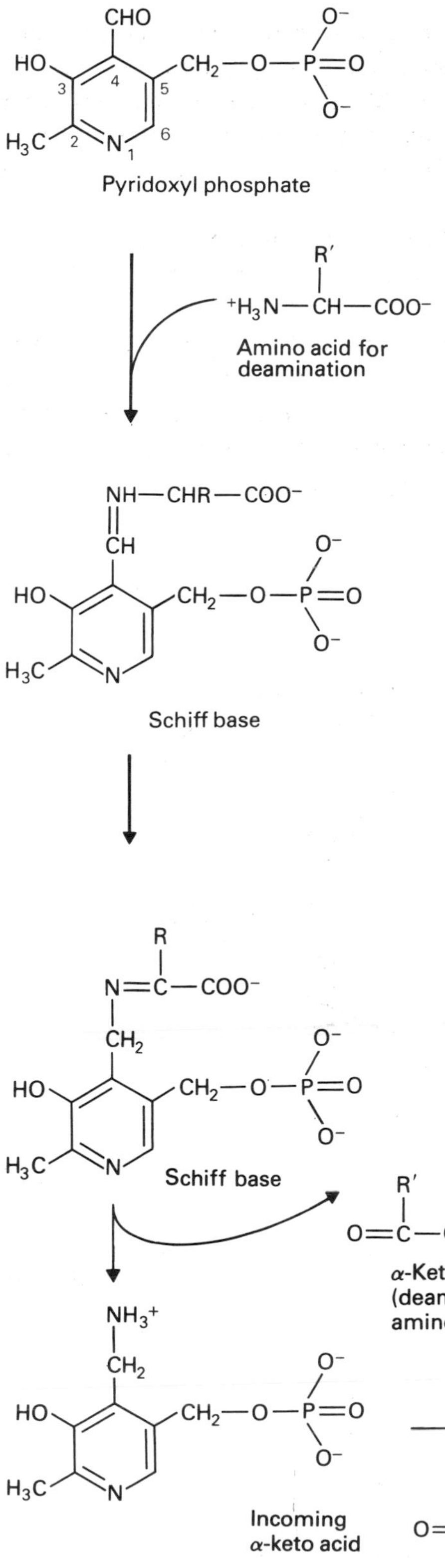

Pyridoxyl phosphate

R′
|
$^+H_3N—CH—COO^-$
Amino acid for
deamination

NH—CHR—COO$^-$
|
CH

Schiff base

R
|
N=C—COO$^-$
|
CH$_2$

Schiff base

NH$_3^+$
|
CH$_2$

Incoming
α-keto acid

R′
|
O=C—COO$^-$
α-Keto acid
(deaminated
amino acid)

R′
|
O=C—COO$^-$

10.1 Deamination and transamination

The individual amino acids in the diet-derived mixture entering the liver from the portal blood are subject to incorporation into proteins, conversion to glucose or ketone bodies, or to oxidation. All but the first are preceded by deamination or transamination reactions. The coenzyme typical of such reactions (although not for all deaminations) is pyridoxal phosphate. This is derived from a variety of closely related compounds in the diet, the vitamins B_6. Indeed 'B_6' is a universal abbreviation for the coenzyme itself. All are derivatives of 3-hydroxy-2-methylpyridine. Pyridoxyl phosphate itself has the formula shown in Fig. 10.1.1 and it has the capacity to form Schiff bases (or imines), as shown. This can tautomerize (tautomerism is the relocation of a hydrogen atom) and release an α-keto acid as well as pyridoxamine phosphate. Evidently this is a deamination reaction. Equally, the pyridoxamine phosphate may react with a second α-keto acid to form a new amino acid and the total reaction is

$$^+H_3N—CHR—COO^- + R′—CO—COO^-$$

$$\rightarrow R—CO—COO^- + H_3N—CHR′—COO^-$$

which is termed a *transamination*. A slightly different presentation of this appears in Fig. 5.3.3 in connection with 'ping-pong' mechanisms.

It may seem that the larger part of the pyridoxal phosphate molecule is redundant, in that only the aldehyde at the 4 position has been shown as reacting; however the phosphate at 5 probably serves to bind the coenzyme to the enzyme surface, and the hydroxyl at 3 may bind a metal ion the better to facilitate the electron shifts needed for Schiff base formation (Fig. 10.1.2).

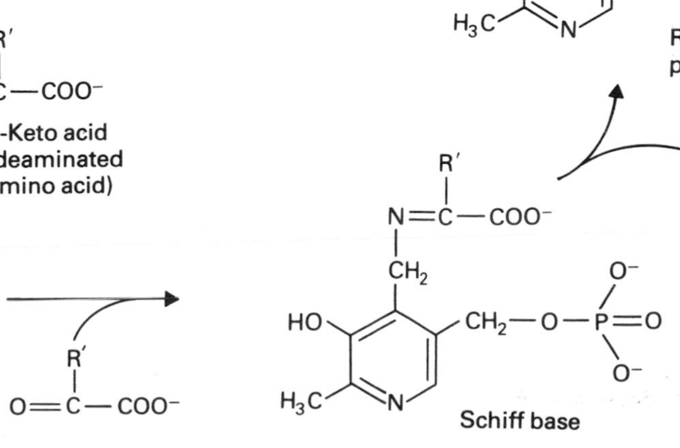

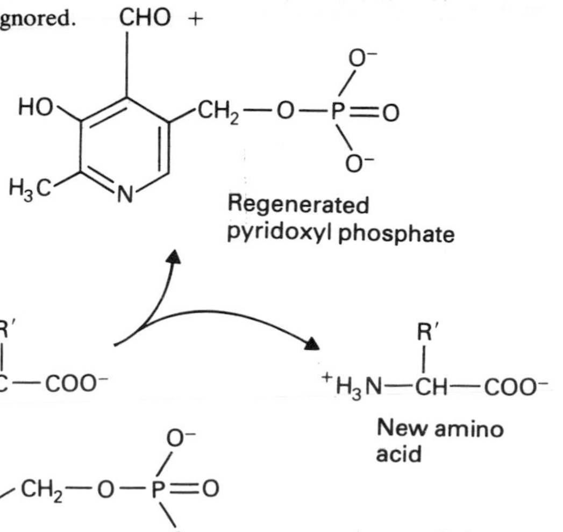

Fig. 10.1.1 Deamination and transamination. Arrows are 'one-way' to show progress of reactions, although each is reversible and balancing hydrogen ions and water are ignored.

Regenerated
pyridoxyl phosphate

R′
|
N=C—COO$^-$
|
CH$_2$

Schiff base

R′
|
$^+H_3N—CH—COO^-$
New amino
acid

Fig. 10.1.2

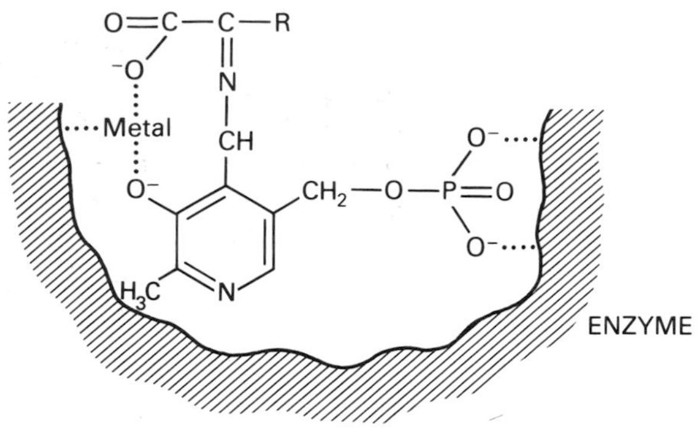

Fig. 10.1.2

The best known of the transaminases are alanine aminotransferase (ALT, formerly named glutamic pyruvic transaminase, GPT) and aspartate aminotransferase (formerly named glutamic oxaloacetic transaminase, GOT).

Liver, skeletal muscle, and cardiac muscle have a very active amino acid metabolism and their cells are rich in the aminotransferases. After severe disturbance of cell function there is some change in plasma membrane permeability, leading to the appearance of enzymes, such as the aminotransferases, in the blood. In general, the enzyme activities in the blood are proportional to the tissue damage. During the first day of a typical myocardial infarction (necrosis of cardiac tissue resulting from cessation of the blood supply to it), the activity of AST in the blood rises precipitately and drops to about normal again after four days. ALT generally shows a negligible rise. In contrast, in liver conditions such as acute viral hepatitis both AST and ALT exhibit prolonged rises. Such observations obviously raise the possibility of differential diagnosis, especially when combined with other tests. (Thus, in myocardial infarction the serum activity of the enzyme creatine phosphokinase also rises, and in liver conditions there is often jaundice.) In diseases of skeletal muscle such as muscular dystrophy the aminotransferase activity in the blood is also often raised, but this condition can be rather easily distinguished from the ones mentioned above on other, clinical grounds.

Deaminases There are several types, not all named 'deaminases' and this may cause some confusion. The following are, then, probably better described as 'enzymes catalysing deamination reactions'. Of the three types listed below, only the first requires pyridoxal phosphate.

1 *The serine and threonine dehydratases* Both of these amino acids bear a 3-hydroxyl function, and are deaminated to form the corresponding α-keto acid. The pyridoxal phosphate participates in the way summarized (for serine) in Fig. 10.1.3. It can be seen that although the total reaction involves no net water loss, water is removed during the course of the reaction.

2 *The amino acid oxidases* They catalyse the general reaction

$$\overset{R}{\underset{|}{^+H_3N-CH-COO^-}} + O_2 + H_2O \rightarrow \overset{R}{\underset{|}{CO-COO^-}} + NH_4^+ + H_2O_2$$

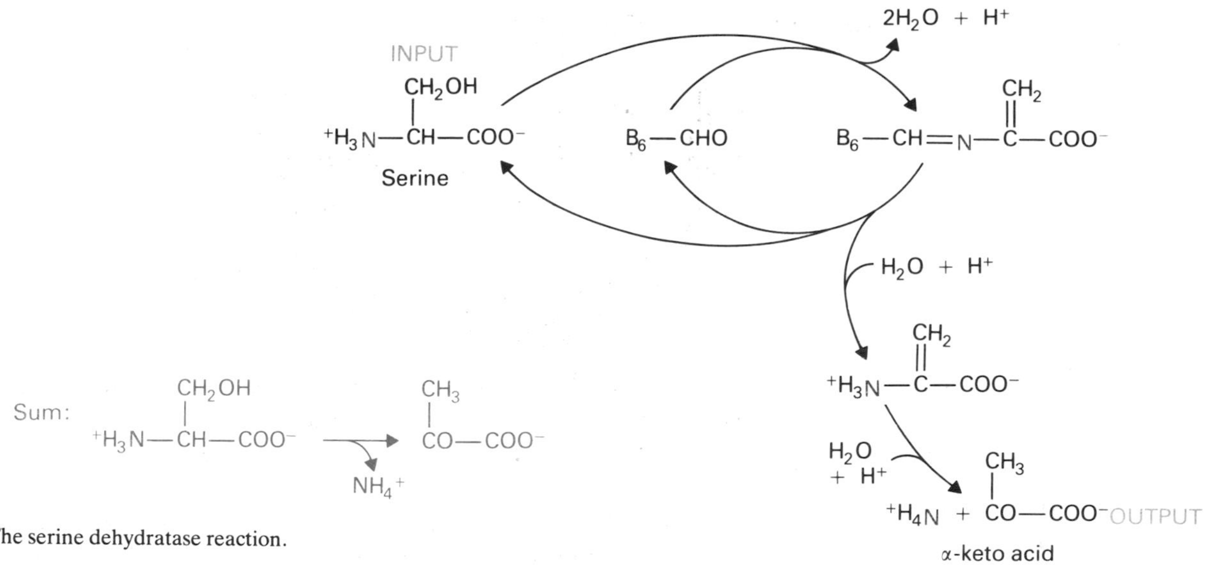

INPUT

$$+H_3N-\overset{\overset{\textstyle CH_2OH}{|}}{CH}-COO^-$$

Serine

B_6-CHO

$$B_6-CH=N-\overset{\overset{\textstyle CH_2}{\|}}{C}-COO^-$$

$2H_2O + H^+$

$H_2O + H^+$

$$+H_3N-\overset{\overset{\textstyle CH_2}{\|}}{C}-COO^-$$

$H_2O + H^+$

$$+H_4N + \overset{\overset{\textstyle CH_3}{|}}{CO}-COO^-\ \text{OUTPUT}$$

α-keto acid

Sum:

$$+H_3N-\overset{\overset{\textstyle CH_2OH}{|}}{CH}-COO^- \longrightarrow \overset{\overset{\textstyle CH_3}{|}}{\underset{\underset{\textstyle NH_4^+}{}}{CO}}-COO^-$$

Fig. 10.1.3 The serine dehydratase reaction.

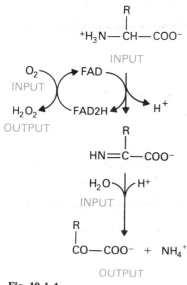

$$+H_3N-\overset{\overset{\textstyle R}{|}}{CH}-COO^-$$
INPUT

O_2 · INPUT
H_2O_2 · OUTPUT
FAD
FAD2H
H^+

$$HN=\overset{\overset{\textstyle R}{|}}{C}-COO^-$$

H_2O · H^+
INPUT

$$\overset{\overset{\textstyle R}{|}}{CO}-COO^- + NH_4^+$$
OUTPUT

Fig. 10.1.4

and contain FAD or FMN. This indicates that there is some sort of oxidation/reduction mechanism in operation and indeed the reaction proceeds by way of the corresponding amino acid as intermediate (Fig. 10.1.4). The D-amino acid oxidases are said to be more active than the L-amino acid oxidases; in other words the non-natural series of amino acids is converted to the corresponding α-keto acids at a much faster rate than is the natural series. (Of course, the D- and L-isomers of a specific amino acid yield the same α-keto acid.) This is viewed as detoxication mechanism, for the D-amino acids absorbed from the gut (products of plant and bacterial metabolism) might otherwise act as antimetabolites. *Antimetabolites* are analogues of natural substrates which compete with the natural substrate for active sites on an enzyme. Many of them may be regarded as competitive inhibitors (5.5).

3 *Glutamic dehydrogenase* This is specific for glutamate but has either NAD$^+$ or NADP$^+$ as coenzyme (Fig. 10.1.5). The reaction is carefully shown with reversible arrows, for the equilibrium allows either formation of ammonia and α-ketoglutarate, when there is an excess of glutamate, or a synthesis of glutamate when ammonia accumulates.

Role of glutamine in nitrogen disposal The amide of glutamate, glutamine, can be formed from ammonia and glutamate (Fig. 10.1.6). Since this reaction requires ATP, we would not expect it to be readily reversible. The reverse reaction is catalysed by an entirely different enzyme, glutaminase (Fig. 10.1.7). Thus ammonia can be taken up not only by α-ketoglutarate to yield glutamate, but also by glutamate to yield glutamine. Glutamine is a means of transporting nitrogen from peripheral tissues to the liver.

Significance of deamination and transamination The whole process can best be visualized as a channelling of nitrogen through glutamate (Fig. 10.1.8). Ultimate disposal is in the form of urea.

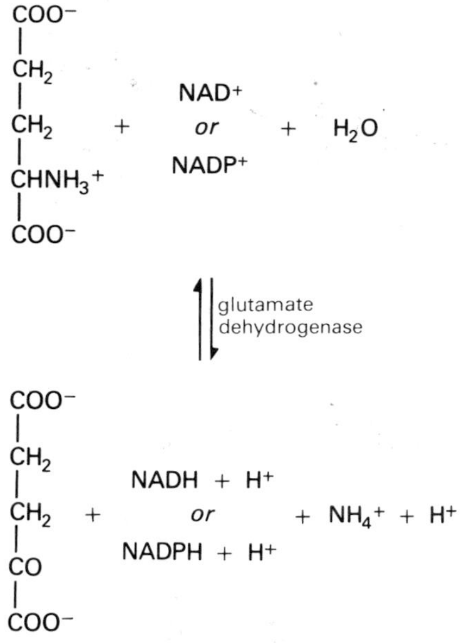

$$\begin{array}{l} COO^- \\ | \\ CH_2 \\ | \\ CH_2 \\ | \\ CHNH_3^+ \\ | \\ COO^- \end{array} \quad + \quad \begin{array}{c} NAD^+ \\ or \\ NADP^+ \end{array} \quad + \quad H_2O$$

glutamate dehydrogenase

$$\begin{array}{l} COO^- \\ | \\ CH_2 \\ | \\ CH_2 \\ | \\ CO \\ | \\ COO^- \end{array} \quad + \quad \begin{array}{c} NADH + H^+ \\ or \\ NADPH + H^+ \end{array} \quad + \quad NH_4^+ + H^+$$

Fig. 10.1.5

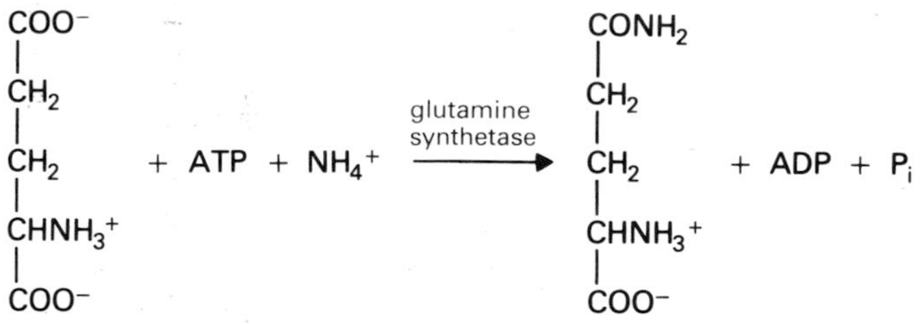

$$\begin{array}{l} COO^- \\ | \\ CH_2 \\ | \\ CH_2 \\ | \\ CHNH_3^+ \\ | \\ COO^- \end{array} \quad + \quad ATP + NH_4^+ \quad \xrightarrow{\text{glutamine synthetase}} \quad \begin{array}{l} CONH_2 \\ | \\ CH_2 \\ | \\ CH_2 \\ | \\ CHNH_3^+ \\ | \\ COO^- \end{array} \quad + \quad ADP + P_i$$

Fig. 10.1.6

$$\begin{array}{l} CONH_2 \\ | \\ CH_2 \\ | \\ CH_2 \\ | \\ CHNH_3^+ \\ | \\ COO^- \end{array} \quad + \quad H_2O \quad \xrightarrow{\text{glutaminase}} \quad \begin{array}{l} COO^- \\ | \\ CH_2 \\ | \\ CH_2 \\ | \\ CHNH_3^+ \\ | \\ COO^- \end{array} \quad + \quad NH_4^+$$

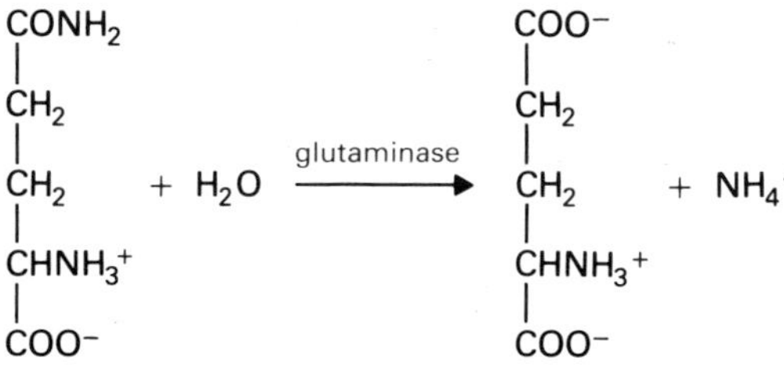

Fig. 10.1.7

Fig. 10.1.8 Flow of nitrogen from amino acid catabolism.
The more quantitatively significant reactions are shown with thickened arrows. 'Reversibility' is carefully indicated, but the flow of nitrogen through glutamate when amino acids are being catabolized is clear. The concentration of aminotransferases and glutamate in the liver is such that it ensures this. All the reactions take place in the liver except the bacterial ammonia production.

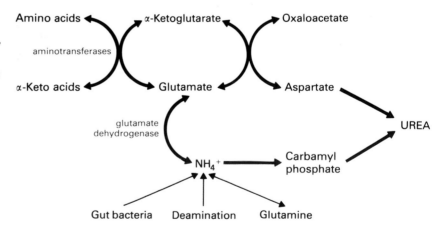

10.2   The urea cycle

The body pool of amino acids is repleted by:

1   the digestion of food proteins;

2   the breakdown of tissue proteins;

3   *de novo* synthesis, in the case of the non-essential amino acids (those which need not be present in the diet).

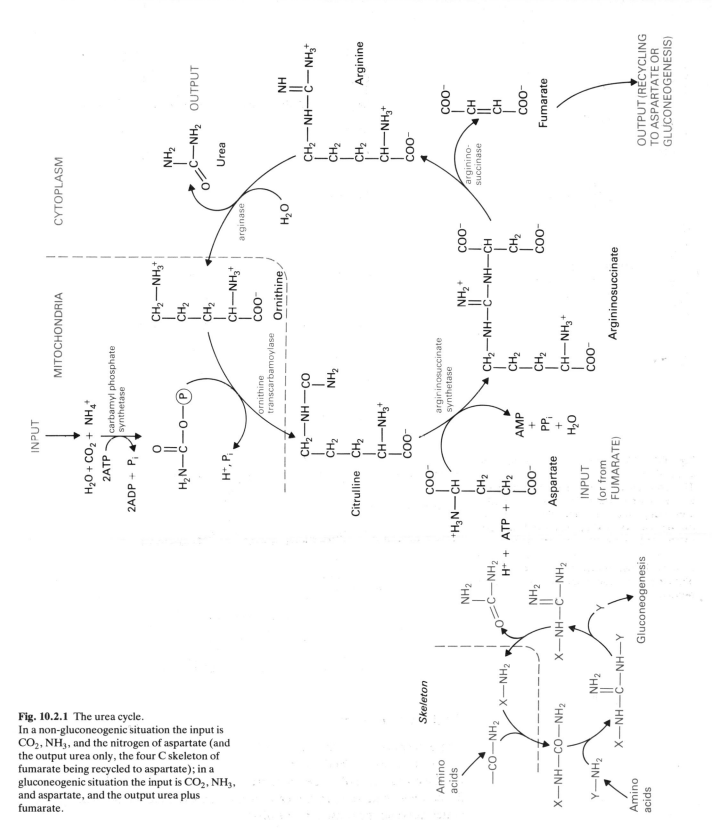

Fig. 10.2.1 The urea cycle.
In a non-gluconeogenic situation the input is CO_2, NH_3, and the nitrogen of aspartate (and the output urea only, the four C skeleton of fumarate being recycled to aspartate); in a gluconeogenic situation the input is CO_2, NH_3, and aspartate, and the output urea plus fumarate.

It is depleted by:

1 the synthesis of the proteins and of amino acid derivatives;

2 the oxidation of the carbon skeletons to yield either energy or glucose.

It is the last of these which releases ammonia, to be converted to urea. Urea, in contrast to ammonia, is relatively non-toxic, but like ammonia it is easily diffusible and therefore excretable if water is available. Organisms with a bountiful supply of water, such as fishes, usually excrete ammonia itself; birds and reptiles conserve water by synthesizing uric acid from ammonia and eliminating it as a semi-solid paste. Lower mammals synthesize allantoin (Fig. 10.5.4).

Sources of nitrogen for urea synthesis Aspartate and carbamyl phosphate serve as the immediate donors of nitrogen for the urea cycle (Fig. 10.2.1). Nitrogen comes to the liver in the amino groups of amino acids like alanine, in the amide of glutamine, and in ammonia produced by gut bacteria and absorbed by the portal system. (Its concentration in the blood, however, remains low.) The incoming amino acids are subject to transamination with α-ketoglutarate, to yield glutamate. Glutamate may also be synthesized from the ammonia arriving in the portal system or from deamination as opposed to transamination. Ammonia not subject to incorporation into glutamate is used for carbamyl phosphate synthesis. Finally, glutamate is transaminated or deaminated. This confusing tissue of statements should be clarified by another look at Fig. 10.1.8, which shows that glutamate is a sort of funnel for the nitrogen flow.

Carbamyl phosphate synthesis Ammonia from glutamate or from the portal blood is incorporated into carbamyl phosphate in a reaction requiring ATP.

$$CO_2 + NH_4^+ + 2ATP \xrightarrow{\text{carbamyl phosphate synthetase}} H_2N-CO-O-\text{ⓟ} + P_i + ADP$$

The reaction (which is not 'balanced' here) requires 2ATP, with the result that the equilibrium is well to the right-hand side. The cofactor required is not biotin, as might be expected for a carbon dioxide-fixing reaction (11.15), but *N*-acetylglutamate, which somehow induces a conformational change in the enzyme.

The urea cycle, like the other cycles, needs only catalytic quantities of intermediates, since these are regenerated at each turn. The fumarate produced by the argininosuccinase reaction can similarly be recycled to aspartate by means of the citric acid cycle enzymes.

Significance of the urea cycle The following points should be made in any essay on the cycle as a whole.

1 It occurs in the liver, and can be viewed as an aspect of the detoxicating function of that organ.

2 It is non-reversible. There are two steps consuming ATP, and this is in effect a biochemical impetus in the direction of urea synthesis, keeping the toxic ammonia at low levels. As urea appears, from deamination or from the portal blood, it is rapidly removed. The normal limit of serum NH_3 is about 0.04 mmol/l, and that of urea about 17.5 mmol/l, so that a 500-fold differential is maintained.

3 The enzymes of the cycle are divided between the mitochondria (carbamyl phosphate synthetase and ornithine transcarbamylase) and the cytosol (all the others). Thus urea synthesis is linked on the one hand to the citric acid cycle in the mitochondria and on the other to the formation of glucose (that is gluconeogenesis) in the cytoplasm. In a gluconeogenic situation, aspartate is stripped of nitrogen by the urea cycle and its carbons reappear as fumarate which, in the cytoplasm, undergoes the sequence:

Fumarate → malate → oxaloacetate → phosphoenolpyruvate → glucose
(for more information on this pathway see 11.3).

Since its carbon skeleton is used for gluconeogenesis, the aspartate cannot be regenerated to maintain the urea cycle at optimal activity; thus in the mitochondria, oxaloacetate from the tricarboxylic acid cycle competes for nitrogen with the carbamyl phosphate synthetase reaction. The sequence

Fumarate → malate → oxaloacetate → aspartate

operating in the mitochondria for the benefit of the urea cycle, obviously pre-empts some of the tricarboxylic acid cycle enzymes. It has been suggested that it is this pre-empting which causes the liver to convert acetyl CoA to ketone bodies for export (11.5). The discussion above may seem tricky at this stage. It relates, in reality, to integration of metabolism and is best read in conjunction with gluconeogenesis (11.3).

4 The urea cycle provides a means of synthesizing arginine from ornithine synthesis. Arginine is therefore sometimes considered a non-essential amino acid.

5 The total reaction can be written

NH_4^+ + CO_2 + 3ATP + $2H_2O$ + aspartate →
urea + 2ADP + 2Pi + AMP + H^+ + PPi + fumarate

Again this is not 'balanced', that is, account is not taken of the charges of the various phosphates. However, the production of one H^+ is significant. It has to be removed from the body in the course of acid-base balance (13.1).

10.3 Catabolism of the amino acids

Early experiments involving animal feeding revealed that when phlorizinized dogs were fed individual amino acids, some of the amino acids caused glucose, while some caused ketone bodies, to appear in the urine (phlorizin inhibits the resorption of glucose by the kidney tubules). These amino acids were called *glucogenic* and *ketogenic* respectively. Some could yield both products, and were referred to as glucogenic *and* ketogenic. It turned out that there was only one wholly ketogenic amino acid, namely leucine. Most of the amino acids, indeed, are glucogenic. This is one of the elementary generalizations one can make about amino acid catabolism. There are about twenty amino acids to consider, and the breakdown of each one is different;

however, there are some correspondences. Many writers have tried to rationalize the treatment of amino acid catabolism, grouping them into families of one sort or another. The 'solution' adopted here is to put them in one massive diagram, showing only key intermediates in their catabolism, and the relationships to the tricarboxylic acid cycle (Fig. 10.3.1). This is not to pretend that such a large format can be assimilated quickly; it is a handy reference while a background of biochemical knowledge is built up. In relation to it, however, the following points can be made.

1 Two amino acids can be directly transaminated to α-keto acids, which are intermediates of the tricarboxylic acid cycle (glutamate to α-keto-glutarate; aspartate to oxaloacetate). A third, alanine, is transaminated to pyruvate.

2 Any amino acid which yields pyruvate or a tricarboxylic acid cycle intermediate is glycogenic, by the pathway

Tricarboxylic acid cycle intermediate/pyruvate $\rightarrow$ oxaloacetate $\rightarrow$ phosphoenolpyruvate $\rightarrow$ glucose

3 All amino acids yielding acetyl CoA or acetoacetyl CoA are ketogenetic. Neither of these intermediates can contribute to the synthesis of glucose (11.3).

4 The pathways for the carbon chain of the branched chain amino acids (leucine, isoleucine, valine) are analogous to β-oxidation for fatty acid oxidation. Indeed, transamination of these amino acids yields α-keto branched-chain fatty acids.

5 Phenylalanine is metabolized by way of tyrosine. Tyrosine can thus be synthesized; phenylalanine, which cannot be synthesized, is said to be an *essential amino acid*. However if there is some tyrosine in the diet, not so much phenylalanine need be hydroxylated to give tyrosine for protein synthesis. Less phenylalanine is then required. In other words, tyrosine *spares* phenylalanine.

6 Similarly, the non-essential cysteine is a product of the catabolism of the essential methionine. Therefore cysteine spares methionine.

Significance of amino acid catabolism Nitrogen is excreted each day, so that amino acid catabolism in the liver is a continuous process. When the nitrogen excreted is greater than the nitrogen intake, the subject is said to be in *negative nitrogen balance*. The reverse situation is *positive nitrogen balance*. Negative nitrogen balance occurs in starvation and in wasting diseases, positive in growth and convalescence.

If, at constant carbohydrate and fat intake, the dietary protein is increased, then more nitrogen is excreted, since the demands on amino acids for tissue replacement can be met and the excess incoming amino acids can only be subjected to oxidation in the tricarboxylic acid cycle after removal of nitrogen. Since energy is made available by this amino acid oxidation, less carbohydrate needs to be oxidized and is converted to fat. To be precise, this only applies to the glucogenic amino acids. The ketogenic amino acids are

Fig. 10.3.1 (opposite) Bird's eye view of amino acid catabolism.
This is merely a summary for reference. NE = non-essential; E = essential; G = glucogenic; K = ketogenic. Note that catabolism is firmly coupled to the tricarboxylic acid cycle and its input (acetyl CoA and pyruvate). Most of the ketogenic amino acids are essential (tyrosine is the exception). The glucogenic amino acids are not necessarily essential or non-essential. Transamination or deamination is the necessary prelude to the catabolism of all of them except probably lysine and glycine. The amides asparagine and glutamine are deaminated before catabolism as aspartate and glutamate respectively. Glycine would not appear to be glucogenic in that it yields CO_2 and NH_3; however, it can be converted to serine.

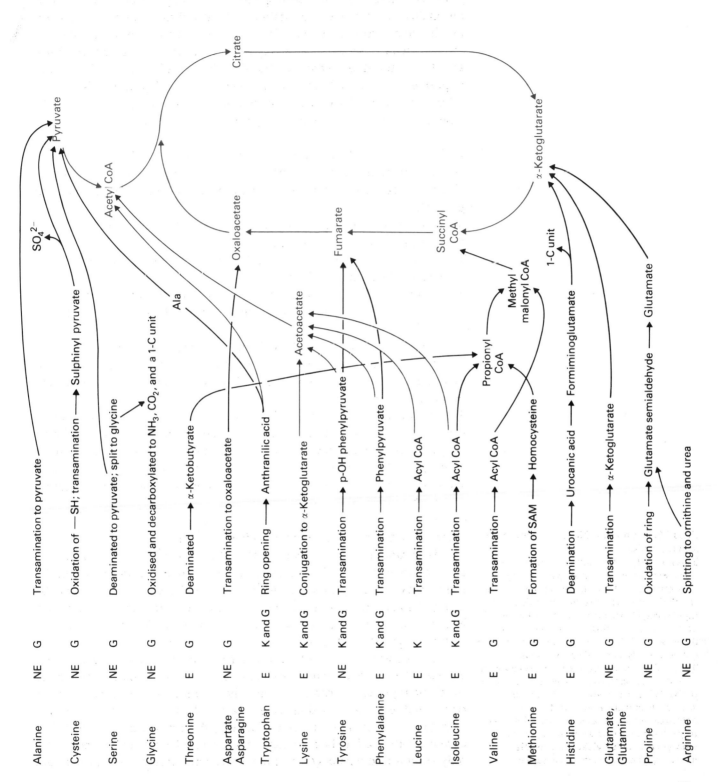

converted to fat through acetyl CoA without any sparing of glucose oxidation.

The corollary of all this is that excess calories in the diet cause lipid deposition whether or not the excess is due to carbohydrate, lipid or protein.

10.4 Bilirubin and its excretion

Accumulation of bilirubin leads to the yellowing of skin and eyes known as jaundice, and the sensations related to the liver conditions causing this have conferred the term 'jaundiced' on the English language. Biochemical studies have been fairly well able to explain the metabolic basis of jaundice, so it is a favourite subject for examinations, especially in the medical course.

The red blood cell has a lifetime of about 120 days, after which it is lysed in the reticuloendothelial system. Thus about 6 g haemoglobin is released every day. Of the three components (as it were) of haemoglobin, two (iron and the amino acids) are conserved while the third (the porphyrin) is excreted after partial catabolism. The first few steps are shown in Fig. 10.4.1. In the endoplasmic reticulum of cells of the reticuloendothelial system, a haem oxygenase catalyses scission of the protoporphyrin ring in a reaction utilizing molecular oxygen and NADPH. The resulting biliverdin is further reduced to bilirubin. The iron passes into the body pools for reutilization. Bilirubin is not a ring structure, though like protoporphyrin IX it is fully conjugated, which explains its red-yellow colour. Although extended in solution, it is convenient to draw it in diagrams in a crypto-cyclic form as an *aide-memoire*—it can then be related to its origins in protoporphyrin IX. Bilirubin is almost insoluble in plasma but binds specifically to albumin. In this form it cannot be removed by the kidney and does not in normal circumstances appear in the urine.

Bilirubin metabolism and the liver Bilirubin arriving in the liver bound to albumin is taken up at the hepatocyte surface and subsequently conjugated with glucuronic acid. This is a typical detoxication-type conjugation (compare 13.21) taking place in the smooth endoplasmic reticulum, and catalysed by a UDP-glucuronyl transferase. Bilirubin is a toxic substance and glucuronic acid conjugation converts it into a compound which is, at any rate, more diffusible and excretable. Initial excretion is into the bile (Fig. 10.4.2). In the intestine there is bacterial deconjugation and bilirubin is reduced to the colourless urobilinogen. Urobilinogen is partially reabsorbed and appears in the urine. Urine left in contact with air becomes darker and this is due to the oxidation of urobilinogen to a series of pigments called urobilins. But some is taken up again by the liver and reappears in the bile. This type of excretion–absorption cycle (also undergone by the bile salts) is called an *enterohepatic circulation*. Urobilinogen passing into the large intestine is known as stercobilinogen ('sterco-': faeces). Its oxidized product in the faeces is stercobilin. The two classes of pigments, 'sterco-' and 'uro-' can be considered identical.

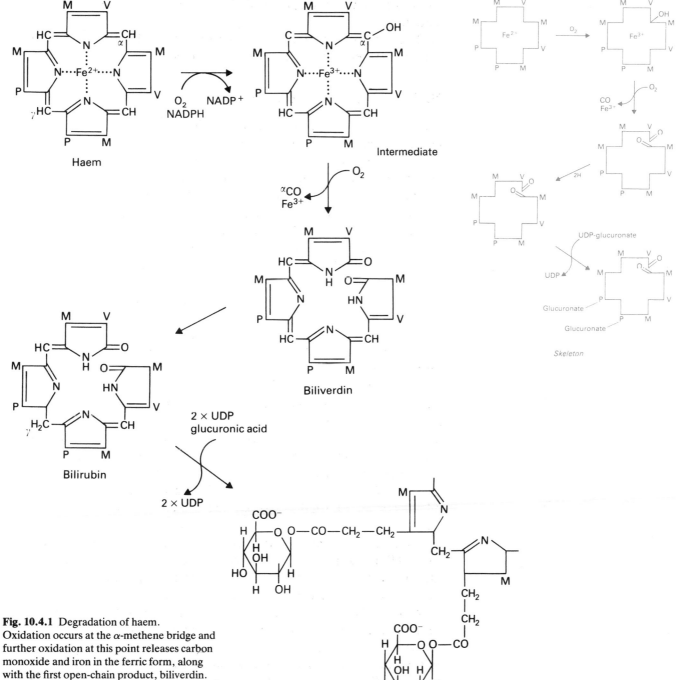

Fig. 10.4.1 Degradation of haem.
Oxidation occurs at the α-methene bridge and
further oxidation at this point releases carbon
monoxide and iron in the ferric form, along
with the first open-chain product, biliverdin.
These reactions occur in the microsomes.
Reduction of the γ-methene yields bilirubin.
This is transported in the blood as a complex
with albumin and is conjugated in the liver to
glucuronic acid. To place the reactions in organ
context see the following figure. V = vinyl; M =
methyl; P = propionyl.

Haem

Intermediate

O_2 NADPH NADP$^+$

^{α}CO Fe^{3+} O_2

Biliverdin

Bilirubin

$2 \times$ UDP glucuronic acid

$2 \times$ UDP

Bilirubin diglucuronide

Fe^{2+} O_2 Fe^{3+}

CO Fe^{3+} O_2

2H

UDP-glucuronate

UDP

Glucuronate

Glucuronate

Skeleton

101

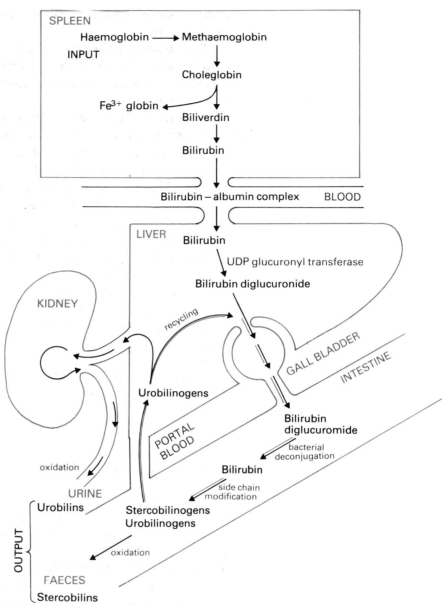

Fig. 10.4.2 The bilirubin excretion system. The system ranges over four organs as well as the blood and the bile. 'Choleglobin' is an ill-defined derivative of methaemoglobin.

Causes of jaundice There are basically three reasons for excess circulating bilirubin.

1 Excessive haemolysis. This releases large amounts of unconjugated bilirubin, and can be due to the haemolytic anaemias (sickle-cell anaemia) or to parasites (malaria).

2 Conjugation failure due to liver damage in, say, cirrhosis (hardening and fibrosing) or to hepatitis (viral infection). Unconjugated bilirubin in the serum is raised.

3   Obstruction of the bile duct, when the conjugated bilirubin excreted into the bile cannot be voided and is taken up in the circulation. (Prolonged obstruction leads to cell damage, however, and a subsequent rise in the unconjugated form.)

In paediatrics an important condition relating to 2 above is a result of the relative immaturity of the neonatal liver. Young and especially premature babies do not have an efficient bilirubin conjugating system and neonatal jaundice can result. The danger of this is that free bilirubin, exceeding the capacity of serum albumin to bind it, can diffuse across the blood-brain barrier and bind to the basal ganglia, causing a toxic encephalopathy (kernicterus). Bilirubin can be decomposed by light of certain wavelengths and so the initial treatment is often phototherapy.

Van der Bergh test It is obviously then of diagnostic value to be able to determine whether jaundice is due to excess free or conjugated bilirubin. In the van der Bergh 'direct' test, conjugated bilirubin reacts with diazotized sulphanilic acid without prior alcohol extraction, because it is water-soluble. In the 'indirect' test, which needs alcohol extraction, it is the unconjugated bilirubin which is detected. Of course the unconjugated form is water-insoluble.

In practice a battery of confirmatory tests is used; notably, transaminases are released from liver cells when they are damaged, and they can also be assayed.

10.5   Purines and pyrimidines

Purines and pyrimidines are both released from nucleic acids but their mode of catabolism is very different. Pyrimidines are catabolized to soluble products, mainly in the liver, whereas purines are converted (in man) to the sparingly soluble urate. In the case of purines the ring structures do not undergo scission. Thus pyrimidines are a group of compounds which are both fully synthesized and fully catabolized by the body; purines are fully synthesized but suffer only very limited catabolism in man. Over the whole range of vertebrates the situation is different, as discussed below.

Figure 10.5.1 indicates pyrimidine catabolism. The methyl group at position 5 of thymine is carried through into methylmalonate which is isomerized to succinate.

Purine catabolism (Fig. 10.5.2) is heavily dependent upon oxidizing enzymes containing FAD, molybdenum, and non-haem iron. The older name, xanthine oxidase, which inaccurately suggests a specificity for substrate, is being replaced by 'iron-sulphur flavoprotein hydroxylase'. In any event the result of the catalytic activity in this context is the oxidation of xanthine and hypoxanthine.

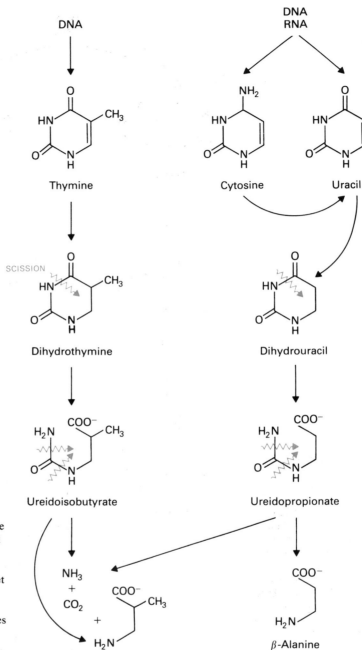

Fig. 10.5.1 Pyrimidine base catabolism. A series of nucleases catalyse hydrolysis of DNA and RNA to mononucleotides. These are hydrolysed to nucleosides by phosphases, and the nucleosides in turn yield the free bases in reactions catalysed by phosphorylases. The ultimate fate of the two end-products is not yet fully documented; by analogy with glycogen phosphorylase, the nucleoside phosphorylase reactions require P_i, and yield sugar phosphates plus free pyrimidines.

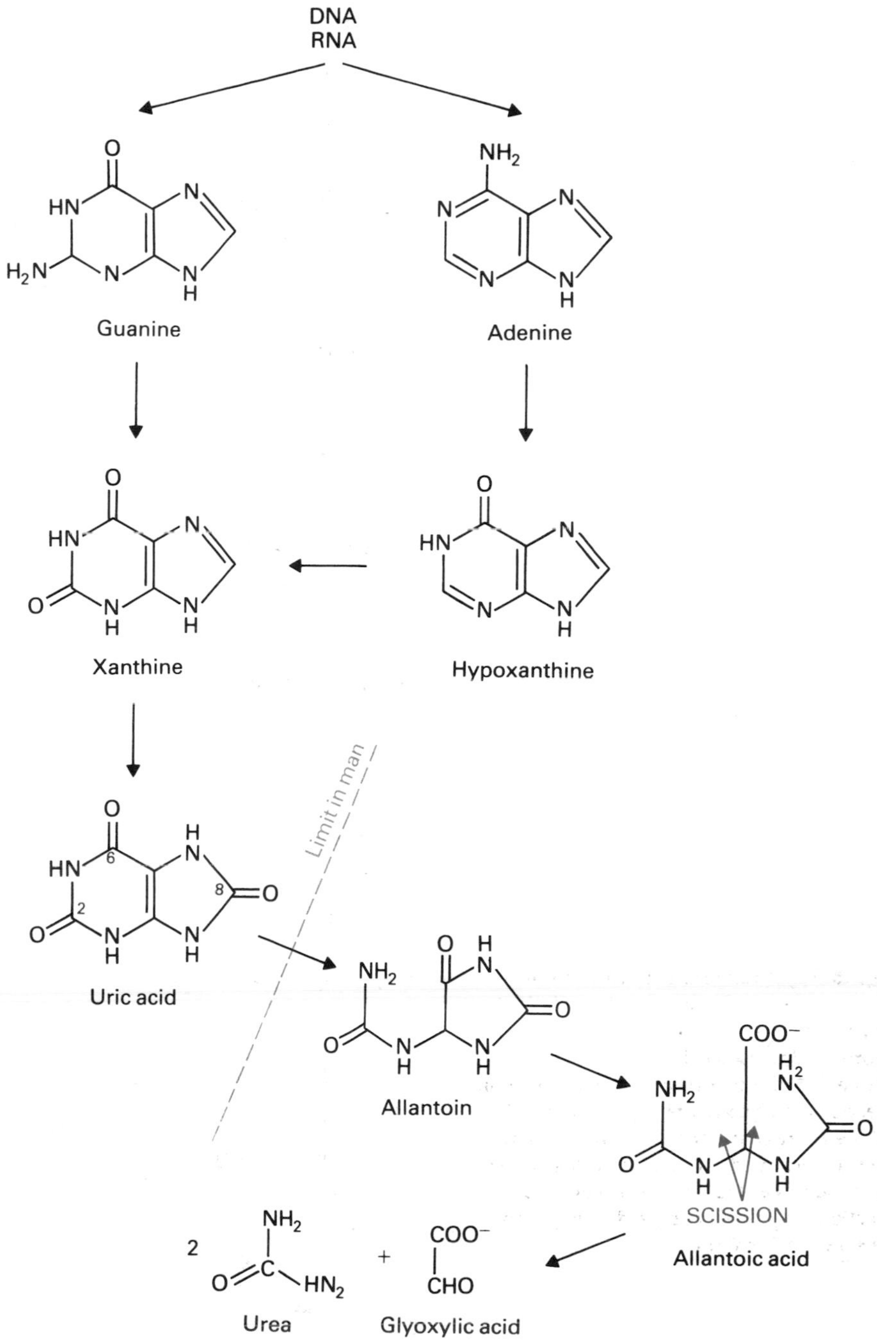

Fig. 10.5.2 Catabolism of the purine bases. It is evident that the process is predominantly deamination and ring oxidation. 'Uric acid' is written here, as usual, in the lactam form, that is ketone functions are written at positions 2, 6, and 8 rather than alcohol functions (the lactim form). In the lactam form, however, the compound does not appear to have a group capable of acid dissociation and therefore does not merit the name of 'acid'. However, it is in equilibrium with the lactim form, the −OH groups of which can ionize. The first ionization pK is between five and six, so uric acid is primarily ionized at the pH of the body fluids. 'Urate' is therefore a more precise term.

Salvage pathways for purines and pyrimidines

The release of pyrimidines and purines from nucleic acids does not commit the body to their wholesale catabolism and excretion. Free circulating purines and pyrimidines can be taken up by cells which do not have the capacity for their *de novo* synthesis and are then incorporated into nucleotides (Fig. 10.5.3). (This occurs in the case of the erythrocytes and polymorphonuclear leucocytes; brain, as a tissue, may also be wholly or partially dependent on circulating purines.) These reutilization pathways are known as 'salvage' pathways and involve the reaction of the bases with phosphoribosyl pyrophosphate. The enzyme hypoxanthine-guanine phosphoribosyl transferase has some controlling role in purine synthesis and utilization as a whole; this is discussed in 11.11.

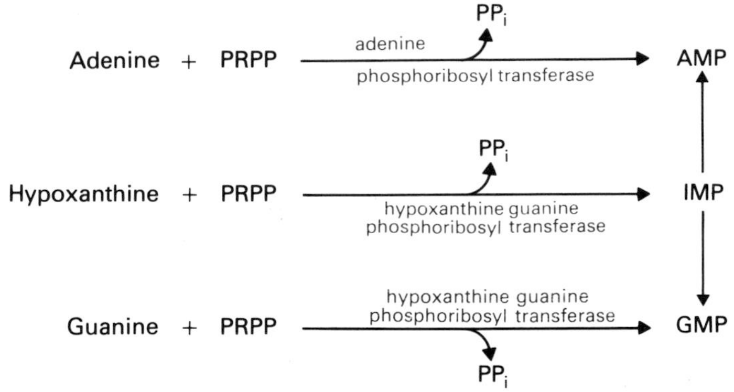

Fig. 10.5.3 Purine salvage pathways. Inosine is the nucleoside monophosphate of hypoxanthine (cf. Table 11.10.1). It can be converted to the principal precursor purine nucleotides of the nucleic acids, ATP and GTP.

PRPP is phosphoribosyl pyrophosphate:

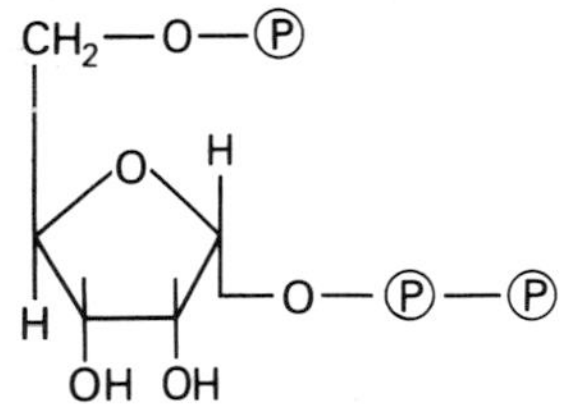

Comparative biochemistry of nitrogen excretion

In many animals urate is further metabolized to allantoin or allantoic acid. The precise end-product seems to be related to the egg-habit, as well as to the availability of water in adult life. The realization of the relationship was an early triumph of comparative biochemistry, and a discussion thereof still occupies a hallowed place in relation to uric acid metabolism.

Urates are insoluble, so they are the favoured mode of purine excretion when water is in short supply; they precipitate in the avian or reptilian egg (which has a strictly finite supply of water) or can be excreted as a relatively anhydrous paste by the adults. If water is very plentiful, as in the case of aquatic eggs and aquatic adults, the urates may be further metabolized to urea or ammonia. Most mammals take urate catabolism to the stage of allantoin, which is much more soluble than urate. Primates, however, retain urates as the excretory product; they do this at some peril, for urate may crystallize in the urinary tract to give kidney or bladder stones. Evidently the mammals operate in a condition where water is more plentiful than for birds and reptiles, but less plentiful than for aquatic forms.

Whereas mammals convert their amino acid nitrogen, released as ammonia, to the less toxic urea, aquatic forms generally have no need to do this. The toxic but very easily diffusible ammonia is removed by the aqueous environment. Some fishes do synthesize urea, but this is to maintain the osmotic pressure of the plasma. Birds and amphibians can take no risks with urea, since they would need moderately large amounts of water for its elimination. Even the ammonia from their amino acids is also converted to urates. There are many peculiarities: the lungfish converts its ammonia to urea when on land, but excretes it unchanged when in the water. Also, water can run very, very short for some mammals: in such straits urea becomes a dangerous substance and there is some evidence that the camel and the hibernating bear can recycle it through amino acids. Some of these points are summarized in Fig. 10.5.4.

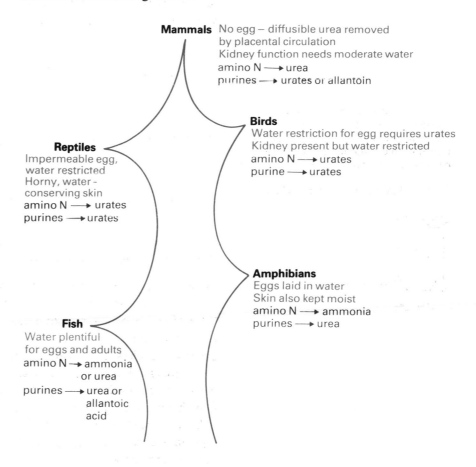

Fig. 10.5.4 Forms of nitrogen excretion. There are many exceptions. Only the general pattern is indicated.

Mammals No egg – diffusible urea removed by placental circulation
Kidney function needs moderate water
amino N ⟶ urea
purines ⟶ urates or allantoin

Birds
Water restriction for egg requires urates
Kidney present but water restricted
amino N ⟶ urates
purine ⟶ urates

Reptiles
Impermeable egg, water restricted
Horny, water-conserving skin
amino N ⟶ urates
purines ⟶ urates

Amphibians
Eggs laid in water
Skin also kept moist
amino N ⟶ ammonia
purines ⟶ urea

Fish
Water plentiful for eggs and adults
amino N ⟶ ammonia or urea
purines ⟶ urea or allantoic acid

11 Synthesis and Resynthesis

11.0 Preamble

It is doubtful if the success of any book depends strictly on its arrangement; few are read from the beginning to the end, and some may even be read backwards. But in order to provide the appearance of unity and integration it is normal to discuss catabolic and anabolic pathways in juxtaposition. With our avowed utilitarian approach here, we can separate anabolism and catabolism on the rationale that they are usually treated separately in examination questions. Thus, an essay question is liable to require a description of 'protein synthesis' rather than 'protein synthesis and breakdown'. But even then separation is no bad thing conceptually for we can make the following general points with regard to biosynthetic pathways.

1　Synthetic processes are very often tissue specific, at least in quantitative terms. Thus the principal tissue of glucose synthesis in man is the liver. Liver, adipose tissue, and the intestine synthesize most of the triglyceride. The brain alone synthesizes a whole range of complex lipids and neurotransmitters. Of course, all cells synthesize protein and nucleic acids.

2　By and large they need energy, and the initial substrates are activated in some way by reaction with compounds of high group transfer potential (the latter point is also true of catabolic pathways).

3　They are different from the corresponding catabolic pathways. Superficially the catabolism and anabolism of a compound may seem to be somewhat the same pathway but in a different direction. This is not so. In many cases there is complete polarity: there is no resemblance whatsoever between protein catabolism and protein synthesis.

4　By and large, they are subject to control by accumulation of their own products, that is by feedback inhibition. This is true, for example, of cholesterol, glycogen, purines, pyrimidines, and many more. (It should be pointed out that some catabolic pathways are also subject to feedback inhibition.)

5　Generally, synthesis of the classes of common cell components starts with rather simple materials, such as:

1-C units, NH_3	— purines, pyrimidines;
2-C units	— fatty acids, steroids;
3-C units	— glucose, amino acids;
4-C units	— porphyrins, amino acids;
5-C units	— glycogen.

There are exceptions in mammals, namely the vitamins and the essential amino acids. To these might be added the essential fatty acids. They are certainly intermediates—they are subject to change in enzyme reactions, but they must be present in the diet in the preformed condition prior to this. Plants and bacteria taken as a whole have more synthetic ability than mammals as a whole. Plants are the original source of most of the vitamins but at least one (vitamin B_{12}) cannot be synthesized even by plants, only by soil bacteria.

6 The compounds which are synthesizable are not necessarily degradable and those which are degradable are not necessarily synthesizable. Thus the steroid ring system is easily synthesized by mammals but although it can be substituted, it is not disrupted prior to excretion. The essential amino acids are not synthesized, but they are oxidized. In this context there is a reference table (14.1).

11.1 Glycogenesis

In quantitative terms, the principal mammalian tissues for glycogen synthesis are liver and muscle. Most cells have recognizable granules of glycogen but the details of synthesis have been intensively investigated only for these two very active tissues. Some tissues, such as brain, contain very little glycogen.

To approach the details of glycogenesis, let us consider muscle in the first instance. The substrate donating the six-carbon unit is UDPG. This is synthesized from UTP and glucose l-phosphate (upper part of Fig. 11.1.1).

Fig. 11.1.1 Glycogen synthesis. The system is conveniently divided into three compartments, as shown. There is the activation mechanism, then the formation of 1,4-glycosyl linkages employing primer, then the branching procedure. Glycogen synthetases 'I' and 'D' are now often referred to as 'a' and 'b' respectively. The protein kinase in the activation mechanism is the same protein kinase involved in phosphorylase activation, although in the case of glycogenesis it operates to inactivate an enzyme except in the presence of large amounts of glucose 6-phosphate.

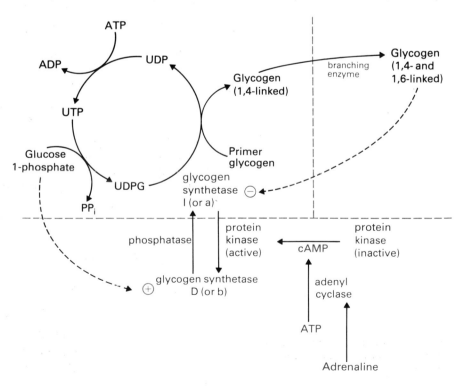

Glycogen synthetase catalyses the formation of a glycosidic link between the primer glycogen and the glucose residue of UDPG. *Primer* is that quantity of preformed product which must be present to initiate a synthetic reaction. The glucose units are attached sequentially to C-4, which is the non-reducing end (Fig. 11.1.2). Of course, glycogen is a branched homopolysaccharide, and there is a branching enzyme to convert some of the 1,4-links to 1,6-links. Glycogen synthesis, like glycogenolysis, is known to be subject to considerable control. In muscle the synthetase exists in two interconvertible forms, called 'I' ('independent') and 'D' ('dependent'). The 'dependency' is on glucose 6-phosphate. The 'D' form requires it for catalytic activity. Glucose 6-phosphate is derived from the phosphorylation of glucose, so there is a tendency towards glycogen synthesis when glucose is plentiful. Glycogen synthesis is also promoted by the conversion of the 'D' form to the 'I' form. The 'D' form is phosphorylated and can be dephosphorylated, and therefore activated, by glycogen synthetase phosphatase, but this enzyme is inhibited in the presence of glycogen. Thus high concentrations of glucose and 6-phosphate promote synthesis, but when glycogen synthesis is well advanced, the activating phosphatase is rendered ineffective.

But there is a more 'urgent' control mechanism. Glycogen synthesis should be inhibited when there is a need for circulating glucose. In such a

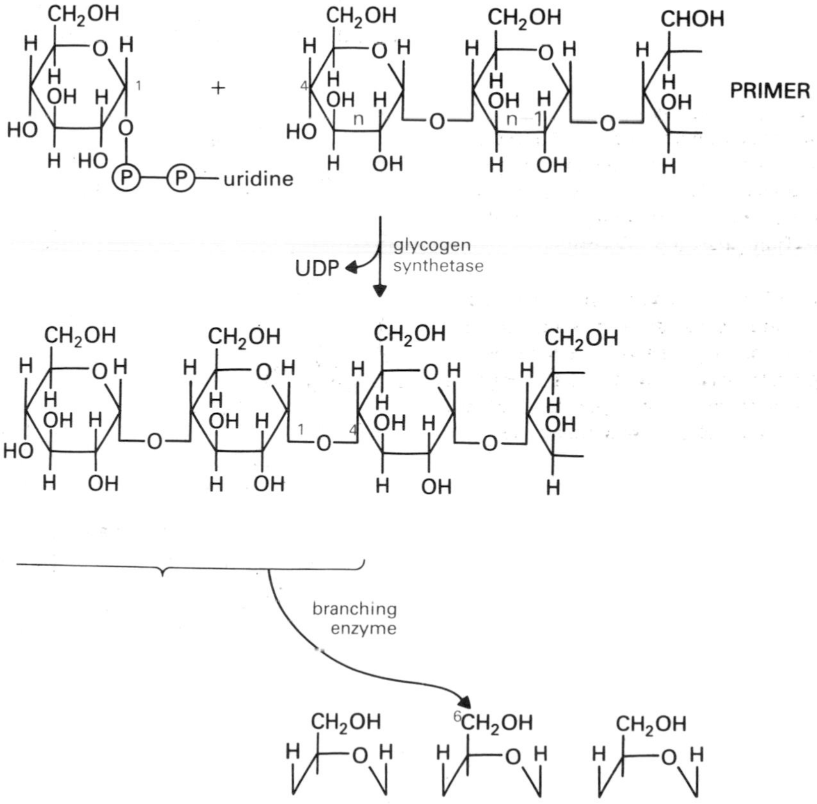

Fig. 11.1.2 Branching enzyme breaks 1,6-linkages and transfers the fragments to 1,6-linkage.

111

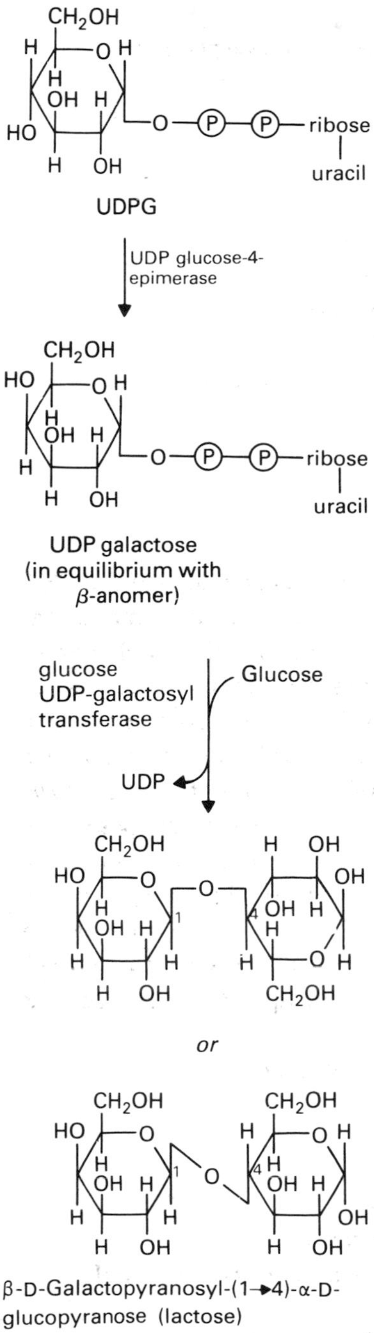

UDPG

UDP glucose-4-
epimerase

UDP galactose
(in equilibrium with
β-anomer)

glucose
UDP-galactosyl
transferase

Glucose

UDP

or

β-D-Galactopyranosyl-(1→4)-α-D-
glucopyranose (lactose)

Fig. 11.2.1

situation adrenaline sets off a cascade mechanism, activating adenyl cyclase at the muscle cell surface. This catalyses the conversion of ATP to cAMP, and the cAMP activates a protein kinase, glycogen synthetase kinase or 'protein kinase', which phosphorylates the glycogen synthetase. Thus synthesis is inhibited because its 'D' form results. The kinase is the same enzyme which phosphorylates glycogen phosphorylase, catalysing glycogenolysis. Thus adrenaline causes both glycogenolysis and inhibition of glycogenesis.

In liver, it is glucagon rather than adrenaline which is the relevant hormone ('first messenger'); the synthetase also exists in an inactive and an active form which are however distinct from the muscle synthetases.

11.2 Disaccharides

Lactose synthesis occurs in the mammary gland. It is not itself utilized by mammalian cells, for example the cells of the mother; after synthesis, its only possible fate is secretion in the milk. This is obviously a means of ensuring a supply of nourishment for the newborn, a sort of metabolic insurance policy taken out by the hormone prolactin. Equally, the baby's tissues do not utilize lactose—it is hydrolysed in the small intestine to glucose and galactose by the enzyme lactase. Lactase persists in the gastro-intestinal tract of Northern Europeans, so that most adults of this origin can digest milk throughout life. In most of the world's populations, however, lactase disappears after about the age of five and the ingestion of large, not moderate, quantities of milk gives rise to flatulence and diarrhoea or, at the least, digestive discomfort (*lactose intolerance*). The flatulence is due to the production, by the bacteria of the large intestine, of methane and hydrogen from the undigested lactose. The persistence of lactase in adults is thought to be due to the development of dairying in ancestral populations about ten thousand years ago; there is still some controversy as to whether the enzyme can be induced in genetically intolerant adults by repeated feeding of small quantities of milk.

The first step in lactose synthesis involves epimerization of UDPG (Fig. 11.2.1). The UDP galactose reacts with glucose. (The galactose component has the β-configuration at position 1 for this reaction.)

Prolactin of the adenohypophysis stimulates milk production in the human mammary gland; it induces synthesis of both the glucose-UDP galactosyl transferase and the milk protein α-lactalbumin. The α-lactalbumin, which is about 2% of the secreted milk proteins, modifies the specificity of the transferase such that it reacts with glucose rather than its alternative substrate, *N*-acetyl-*D*-glucosamine.

Sucrose is a product of green plants; animals do not synthesize it. The pathway involves UDPG and fructose 6-phosphate (Fig. 11.2.2); a phosphatase catalyses the release of free sucrose.

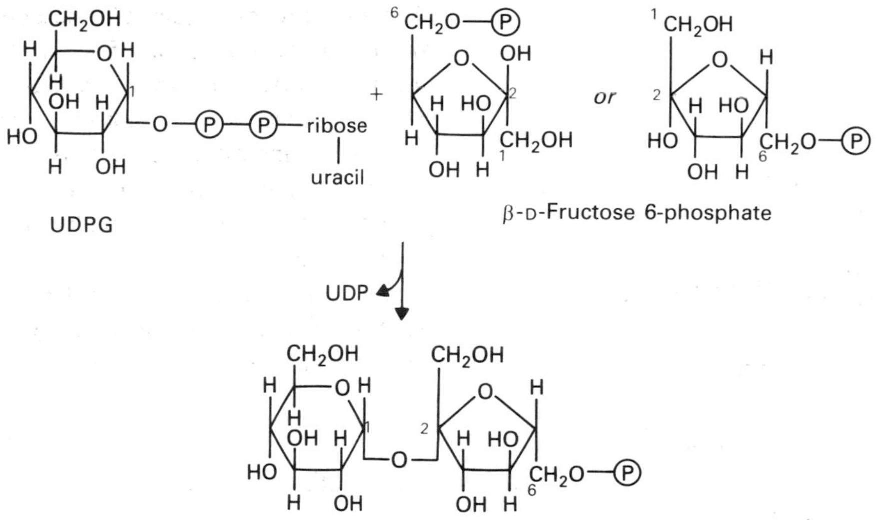

UDPG

β-D-Fructose 6-phosphate

UDP

α-D-Glucopyranosyl (1→2)-β-D-fructofuranose phosphate
(sucrose phosphate)

Fig. 11.2.2

11.3 Gluconeogenesis

Gluconeogenesis might be a good candidate for the nodal topic in metabolism, since it so well encompasses the main pathways, it illustrates the principles of control, and it is redolent with medical and biological interest. It can be defined as the synthesis of glucose from non-carbohydrate sources, namely the amino acids (if gluconeogenic), lactate, propionate, and glycerol. That these particular substances should be gluconeogenic is due to the nature of the glycolysis pathway. Glycolysis is a means of breaking down glucose to three-carbon units, as pyruvate, and it might be supposed that pyruvate could regenerate glucose. This indeed is the case and some gluconeogenic amino acids do yield glucose via pyruvate. However, there are three reactions in glycolysis which are irreversible (the equilibria favour glucose breakdown) and the 'reactions of gluconeogenesis' are those which enable them to be transcended. These and the corresponding gluconeogenic mechanisms are as follows.

1   Glucokinase and hexokinase reactions (irreversible)
    Glucose + ATP → glucose 6-phosphate + ADP
    'Gluconeogenic' reaction:
    Glucose 6-phosphate → glucose + Pi
    The enzyme here is glucose 6-phosphatase.

2   Phosphofructose kinase reaction (irreversible)
    Fructose 6-phosphate + ATP → fructose 1,6-diphosphate + ADP
    'Gluconeogenic' reaction:
    Fructose 1,6-diphosphate → fructose 6-phosphate + Pi
    The enzyme is fructose 1,6-diphosphatase.

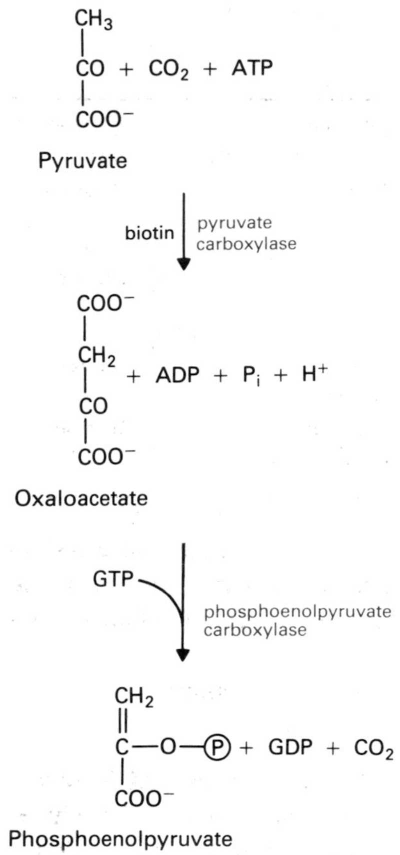

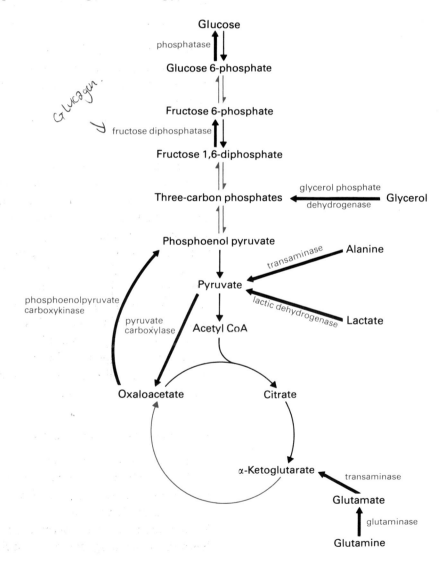

$$CH_3$$
$$|$$
$$CO + CO_2 + ATP$$
$$|$$
$$COO^-$$

Pyruvate

biotin | pyruvate carboxylase

$$COO^-$$
$$|$$
$$CH_2$$
$$|$$
$$CO$$
$$|$$
$$COO^-$$
$$+ ADP + P_i + H^+$$

Oxaloacetate

GTP — phosphoenolpyruvate carboxylase

$$CH_2$$
$$||$$
$$C—O—\text{\textcircled{P}} + GDP + CO_2$$
$$|$$
$$COO^-$$

Phosphoenolpyruvate

Fig. 11.3.1

3 Pyruvate kinase reaction (irreversible)
Phosphoenolpyruvate + ADP → pyruvate + ATP
'Gluconeogenic' reactions:
Pyruvate + CO_2 → oxaloacetate → phosphoenolpyruvate.

The first step of this is catalysed by pyruvate carboxylase in a reaction requiring CO_2, ATP, and biotin. (For biotin, see 11.15.) The second reaction is catalysed by phosphoenolpyruvate carboxykinase, which requires GTP (Fig. 11.3.1).

Thus two moles of nucleotide are consumed in transforming pyruvate to phosphoenolpyruvate but bypassing the pyruvate kinase reaction. The pathways are summarized in Fig. 11.3.2 and Table 11.3.1. Gluconeogenesis, therefore, has an energy requirement. It is of interest to know whether the production of glucose from pyruvate in the liver consumes more ATP/GTP

Fig. 11.3.2 The reactions of gluconeogenesis. Gluconeogenic reactions are shown with thickened arrows. Those which contribute only to glucose breakdown are shown in black, and those common to gluconeogenesis and glucose catabolism are in blue. Quantitatively the most significant sources of carbons for gluconeogenesis in the liver are alanine and glutamine (deriving from skeletal muscle during fasting) and lactate (from muscle glycogen during exercise).

Table 11.3.1 Summary of gluconeogenic pathways

State		Substrate(s)		Tissue(s)		Synthesizing
Recovery from exercise		Lactate Pyruvate		Liver		Glycogen
Starvation	using	Glycerol Amino acids	in	Liver Kidney	yielding	Glucose
Diabetes		Glycerol Amino acids		Liver Kidney		Glucose
Lactation (ruminants)		Propionate Butyrate		Mammary gland		Glucose Galactose

(that is, high energy bonds) than is obtainable on subsequent glycolysis—reversion to pyruvate in some other tissue like the brain. Leaving aside the high energy phosphate bonds for the moment we can represent the stoichiometry:

$$2\,C_3H_3O_3^- + 2NADH + 4H^+ \rightleftharpoons C_6H_{12}O_6 + 2NAD^+$$

Pyruvate   Glucose

The glycolysis 'direction' yields only two ATP (one from each of the three-carbon units undergoing the phosphoglycerate kinase reaction). The gluconeogenesis 'direction' consumes one ATP in the pyruvate carboxylase reaction, one GTP in the phosphoenolpyruvate carboxykinase reaction, and another ATP in the phosphoglycerate kinase reaction, so the total consumption of nucleotide is six for the pair of three-carbon units which is the glucose precursor. In addition, NADH has to be found for the glyceraldehyde 3-phosphate dehydrogenase reaction; true, NADH is produced in glycolysis and the two might be thought to cancel out; however, NADH for gluconeogenesis is produced in the mitochondria in a respiring cell; it has to be transported to the cytosol for the glyceraldehyde 3-phosphate dehydrogenase reaction and this also cannot be done without some energy expenditure. Thus the true energy cost of gluconeogenesis is quite high, but evidently worth paying to ensure a supply of glucose for the brain.

Significance of gluconeogenesis

A good many telling points can be made with respect to gluconeogenesis.
1 It is a device to ensure that circulating glucose does not fall to dangerously low levels, that is sufficiently low to adversely affect brain function. Brain cannot utilize fatty acids.
2 It is largely a hepatic pathway, in humans at least; nonetheless the kidney has some gluconeogenic role, especially in prolonged starvation.

3 It cannot be achieved from fatty acids, acetate, or ketone bodies; in other words, fats cannot be converted to carbohydrate. This is because two-carbon units from these sources can only enter the citric acid cycle, with the evolution of $2CO_2$; they cannot augment the quantity of oxaloacetate or pyruvate.

4 The enzymes of gluconeogenesis are modulated by a large number of factors (Fig. 11.3.3) which can switch the flow of carbons from gluconeogenesis either to the tricarboxylic acid cycle or to fat synthesis. Pyruvate carboxylase has an absolute requirement for acyl or acetyl CoA; when this accumulates, the flow is to gluconeogenesis. If the ADP/ATP ratio rises, so that more ATP is required, then ADP inhibits pyruvate carboxylase and citrate lyase, so citrate is taken up by the tricarboxylic acid cycle rather than the gluconeogenic or lipogenic pathways. If ATP accumulates, pyruvate dehydrogenase is inhibited and the carbon atoms are channelled to oxaloacetate and then glucose.

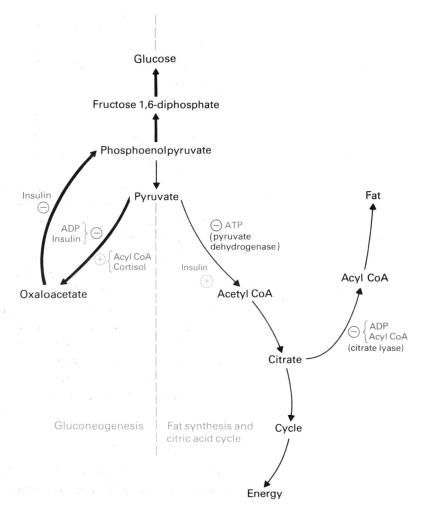

Fig. 11.3.3 Control of gluconeogenesis. Acyl CoA (in effect, fat) and cortisol tend to stimulate the gluconeogenesis part, while insulin tends to stimulate acyl synthesis. High ADP (that is, lack of ATP) inhibits both gluconeogenesis and acyl synthesis and channels carbons to citric acid cycle. The part involving citrate lyase should be fully comprehensible in conjunction with section 11.4.

CH_3
$|$
CH_2
$|$
COO^-

Propionate

$CoA-SH$

CH_3
$|$
CH_2
$|$
$CO-S-CoA$

Propionyl CoA

CO_2

CH_3
$|$
$CH-COO^-$
$|$
$CO-S-CoA$

Methylmalonyl CoA

COO^-
$|$
CH_2
$|$
CH_2
$|$
$CO-S-CoA$

Succinyl CoA

Fig. 11.3.4

Insulin promotes glucose utilization and lipogenesis; it represses synthesis of all the gluconeogenic enzymes. On the other hand, cortisol induces the synthesis of pyruvate carboxylase and thus is a gluconeogenic hormone.

5 Exercising skeletal muscle produces lactate and pyruvate. These diffuse into the bloodstream and are synthesized to glycogen in the liver by the gluconeogenic pathway, which can be represented:

Lactate → pyruvate → oxaloacetate → phosphoenolpyruvate → glucose → glycogen

This is however only a segment of a cycle: the Cori cycle. The glycogen synthesized thereby can be used to replenish muscle glycogen. The full Cori cycle is thus:

Liver glycogen → blood glucose → muscle glycogen → blood lactate → liver glycogen

6 In starvation, there is breakdown of fat and muscle proteins to provide oxidizable substrates as soon as liver glycogen is exhausted. Fatty acids cannot provide glucose but proteins of course do so because a number of the amino acids released from them are glucogenic. It is thus the gluconeogenic pathway from amino acids which provides the glucose for brain function during starvation.

7 Diabetes is superficially similar to starvation in that there is a lack of circulating insulin, which normally represses the synthesis of the gluconeogenic enzymes. In diabetes of course there is no lack of glucose; the defect is the inability of tissues to utilize it. Gluconeogensis in diabetes, therefore, exacerbates the situation.

8 Ruminants, unlike other mammals, cannot convert glucose to fatty acids, so there is not the same depletion of glucose under the stimulating influence of insulin. This spares glucose, which is significant because, in addition, ruminants do not absorb glucose from the intestine. However, since glucose cannot be replenished from the diet, gluconeogenesis is very important, and is achieved from substrates such as propionate and lactate as well as the amino acids. A lactating cow, that is one producing milk, not lactate (though it might do both), can make about 2 kg of glucose a day from non-carbohydrate sources. Propionate is produced by bacteria in the rumen but is also a product of the oxidation of odd-chain fatty acids and of isoleucine. It is gluconeogenic in that it is converted to succinyl CoA (Fig. 11.3.4). Succinyl CoA is a precursor of oxaloacetate and thus of pyruvate.

9 It requires energy, as discussed.

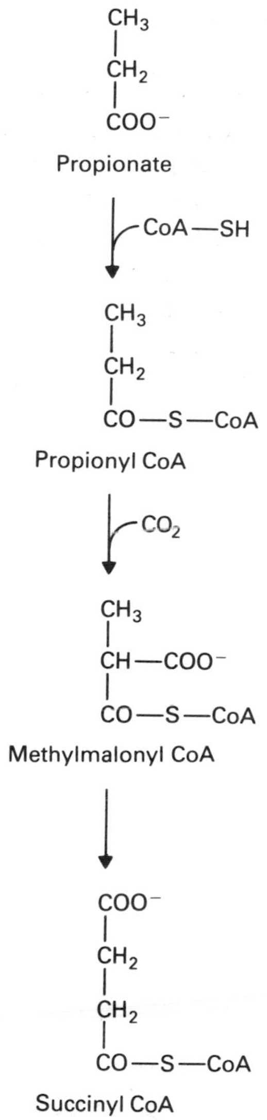

11.4 Fatty acids

Various mechanisms of fatty acid synthesis are known, but for examination purposes at this stage you may concentrate on the palmitate system as it

117

occurs in the mammalian cell cytoplasm. The point to be made straight off is that it is not a reversal of β-oxidation. The enzyme complement, the site, and the control mechanisms differ for breakdown and synthesis. Nonetheless the two are opposite versions of the same process in the sense that palmitate is broken down to two-carbon units (in the mitochondria) and two-carbon units are the basis of palmitate synthesis (in the cytosol). Now the bulk of the two-carbon units are produced in the mitochondria (from the pyruvate dehydrogenase reaction, for example) so that palmitate synthesis, for descriptive purposes, may be divided into:

1 the transport of two-carbon units from the mitochondria to the cytoplasm; and

2 the assembly of two-carbon units into palmitate.

Two-carbon translocation is initiated by the reaction of acetyl CoA with oxaloacetate to yield citrate. Citrate, unlike acetyl CoA, can diffuse through the mitochondrial membrane and this occurs when citrate is not taken up by the enzymes of the tricarboxylic acid cycle. The rundown of the cycle occurs, for example, when excess ATP inhibits isocitrate dehydrogenase.

In the cytoplasm, citrate is split by citrate lyase to yield oxaloacetate and acetyl CoA. Oxaloacetate, which has evidently acted as a carrier, cannot diffuse through the mitochondrial membrane. However, as it must be returned to the mitochondria, it is reduced to malate, which in turn is decarboxylated to pyruvate, a reaction catalysed by malate: NADP dehydrogenase ('malic enzyme') and generating NADPH. The pyruvate diffuses back into the mitochondria where it can give rise to more oxaloacetate by carboxylation. Thus the translocation of more acetyl CoA is possible, and incidentally NADPH has been generated (Fig. 11.4.1).

Reductive polymerization of acetyl CoA is dependent upon carboxylation to malonyl CoA

$$CH_3—CO—S—CoA + HCO_3^- + ATP \xrightarrow{\text{acetyl CoA carboxylase}} \underset{\underset{CH_2—CO—S—CoA}{|}}{\overset{COO^-}{}} + ADP + P_i + H_2O$$

The acetyl CoA carboxylase reaction requires biotin as coenzyme. It is activated by citrate and inhibited by palmitoyl CoA.

The first reaction is between the malonyl moiety and an acetyl group, but before this can take place both exchange —S—CoA for an —SH bearing protein called acyl carrier protein (ACP)

$$CH_3—CO—S—CoA + ACP—SH \rightarrow CH_3—CO—S—ACP + CoA—SH$$

$$^-OOC—CH_2—CO—S—CoA + ACP—SH \rightarrow ^-OOC—CH_2—CO—S—ACP + CoA—SH$$

The mechanism for palmitate synthesis is the repeated reaction of a growing acyl-ACP with malonyl-ACP (Fig. 11.4.2), so the latter can be regarded as

Fig. 11.4.1 Acetyl CoA and NADPH for palmitate synthesis.

Two-carbon units and electrons (as NADPH) are made available in the cytoplasm. Note that malate: NADP dehydrogenase, sometimes called 'malic enzyme', is distinct from the malate: NAD dehydrogenase of the citric acid cycle. Citrate lyase is now often referred to as 'ATP citrate lyase'.

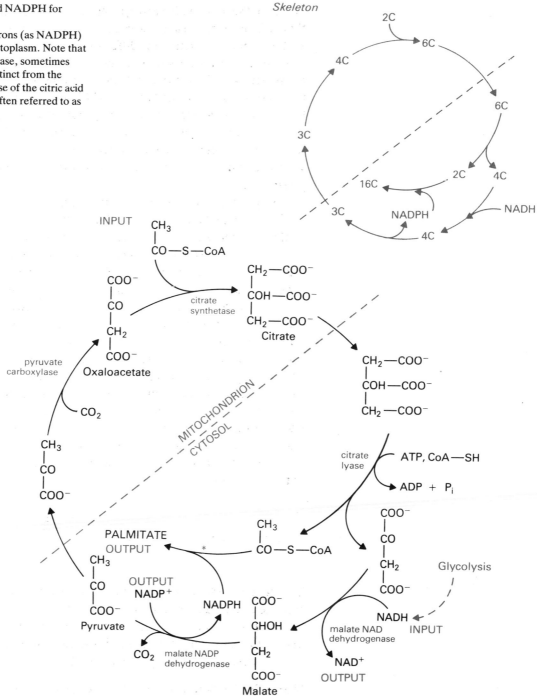

* See Fig. 11.4.2.

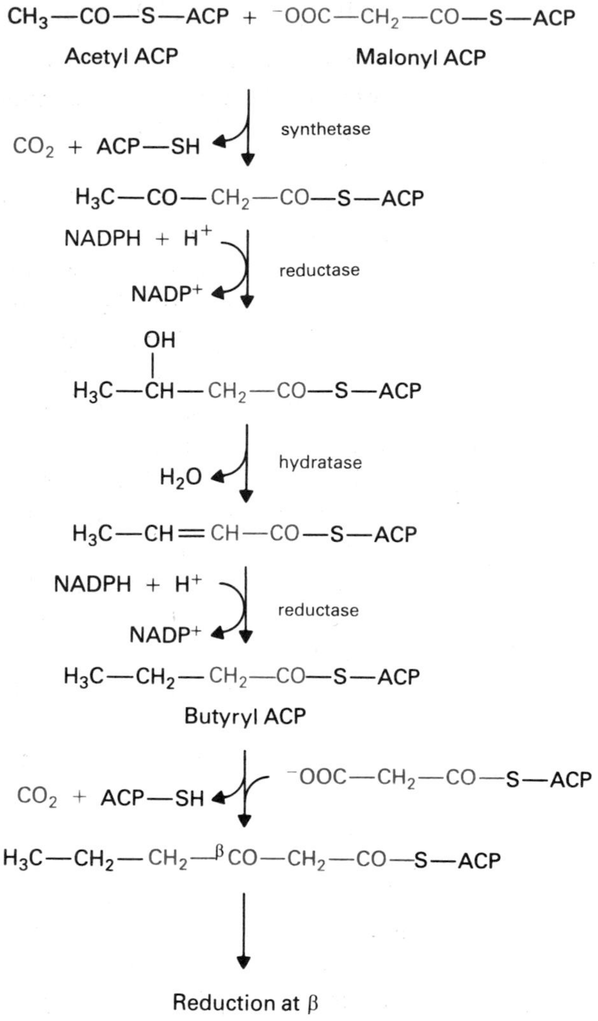
CH3—CO—S—ACP + ⁻OOC—CH2—CO—S—ACP
Acetyl ACP
Malonyl ACP
CO2 + ACP—SH
synthetase
H3C—CO—CH2—CO—S—ACP
NADPH + H+
reductase
NADP+
OH
H3C—CH—CH2—CO—S—ACP
hydratase
H2O
H3C—CH=CH—CO—S—ACP
NADPH + H+
reductase
NADP+
H3C—CH2—CH2—CO—S—ACP
Butyryl ACP
⁻OOC—CH2—CO—S—ACP
CO2 + ACP—SH
H3C—CH2—CH2—βCO—CH2—CO—S—ACP
Reduction at β

Fig. 11.4.2 Palmitate synthesis.
Note that NADPH, not NADH, is the reducing coenzyme in both reducing steps. The arrows are unidirectional here to emphasize carbon flow, although the reduction and hydration reactions are reversible.

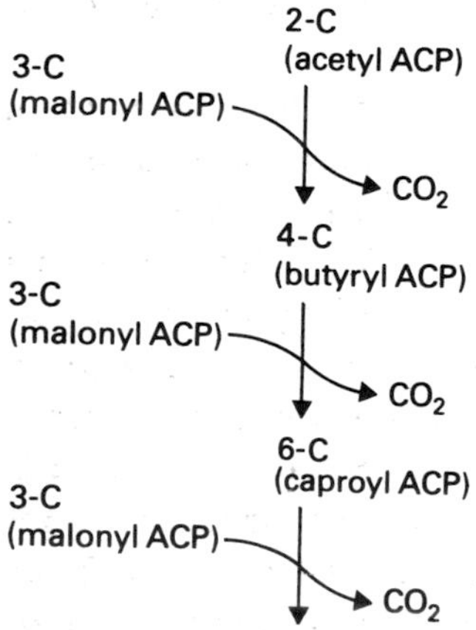
2-C
(acetyl ACP)
3-C
(malonyl ACP)
CO2
4-C
(butyryl ACP)
3-C
(malonyl ACP)
CO2
6-C
(caproyl ACP)
3-C
(malonyl ACP)
CO2

Fig. 11.4.3

the activated form of the two-carbon unit employed in fatty acid synthesis. The reaction of malonyl-ACP with acyl ACP releases CO_2. The first few steps can be summarized as in Fig. 11.4.3. The four-carbon unit is reduced in two stages, with NADPH as coenzyme; then it reacts with more malonyl-ACP. Seven cycles are needed to achieve C-16 (palmitate).

Sources for palmitate synthesis

Two-carbon units are derived from carbohydrate or fat, or ketogenic amino acids, in the mitochondria. However a great deal of NADPH is also required, two for every two carbons incorporated. Some is generated by the 'malic enzyme' reaction as above, and the rest is derived from the pentose phosphate cycle, which is active in the cytosol. Energy is also required: seven cycles require seven ATP for seven acetyl CoA carboxylation reactions.

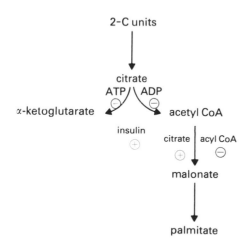

2-C units

↓

citrate
ATP ⟋ ⟍ ADP
 ⊖   ⊖
α-ketoglutarate   acetyl CoA

insulin   citrate | acyl CoA
  ⊕   ⊕   ↓   ⊖

malonate

↓

palmitate

Fig. 11.4.4

Significance of palmitate synthesis

1 It is described as a *de novo* pathway, that is, it is not dependent on preformed fatty acids. In the mitochondria, there is a system for fatty acid elongation. In the microsomes, desaturation of pre-existing fatty acids can occur.

2 The principal sites of palmitate synthesis in mammals are the adipose tissue and the liver.

3 The requirements for synthesis of one mole of palmitate are:

8 acetyl CoA (from the mitochondria)

8 NADH (transhydrogenation to NADPH occurs via malate)

6 NADPH (from hexose monophosphate shunt)

7 ATP (from aerobic glycolysis)

All of these may be derived from carbohydrate; this is the metabolic basis of the fattening effect of bread, rice and cakes.

4 Palmitate synthesis is subject to fine control such that excess carbohydrate is channelled in that direction. Citrate lyase is activated by insulin, acetyl CoA carboxylase by citrate. ADP inhibits the citrate lyase reaction and ATP the isocitrate dehydrogenase reaction. Fatty acids as acyl derivatives (the end-product) inhibit the carboxylase. All influence the flow of carbons in a concerted manner (Fig. 11.4.4). Thus, when ATP and citrate are plentiful, the tricarboxylic acid cycle is inhibited and carbohydrate carbons flow to fat. On the other hand, when ATP is in short supply citrate lyase is inhibited and ATP is generated by the cycle.

5 It illustrates the functioning of the so-called *multienzyme complex*. In Fig. 11.4.2 the reactions are written out as a metabolic sequence as usual. However, the sum of these reactions is catalysed by an aggregate of proteins, not readily dissociated, named fatty acid synthetase. The acyl carrier protein pivots the substrates round this assembly of enzymes catalysing the various transformations. This is an important concept—the multienzyme complex obviously gathers together the enzymes of a sequence into an easily sequestered and controllable assembly—but the proposed mechanism is difficult to draw out on paper and we assume for that reason that you will not be asked to reproduce it. (Crammers can take the short view.) Other important multienzyme complexes are pyruvate dehydrogenase and α-ketoglutarate dehydrogenase.

Essential fatty acids

Although palmitate is the major product of the fatty acid synthesizing system of mammals, the most abundant fatty acid in adipose tissue is generally oleate, which has two more carbons than palmitate, and a double bond between C-9 and C-10. It can be synthesized from palmitate by elongation and desaturation, mechanisms taking place in the endoplasmic reticulum. There are other important unsaturated fatty acids which mammals utilize, but cannot synthesize; although they are necessary in the diet they have failed to be included among the vitamins for historical reasons. They are the *essential fatty acids*, shown in Fig. 11.4.5. They have *cis* double bonds extending distal to the 9 position, and it appears that mammals do not have the capacity for desaturation beyond this point, or at least the rate of synthesis is not sufficient for physiological requirements. In humans, a deficiency of the essential fatty acids is manifested by a scaly skin and (in infants) failure to grow normally.

Fig. 11.4.5

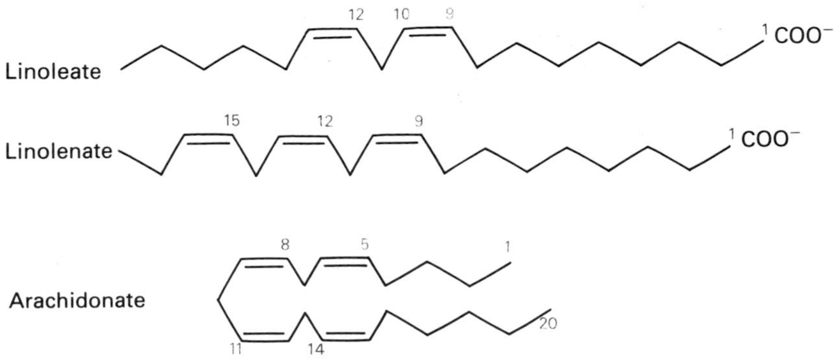

Linoleate

Linolenate

Arachidonate

The essential fatty acids are found in membranes, esterified in phospholipids, and their presence probably contributes to membrane fluidity. But the main importance of arachidonate is as a precursor of the prostaglandins. The synthesis involves cyclization between C-8 and C-13, and there is a variety of substitution on the resulting ring and side-chains to give the series of compounds known as the prostaglandins, the parent compound of which (in the chemical sense) is prostanoate (Fig. 11.4.6).

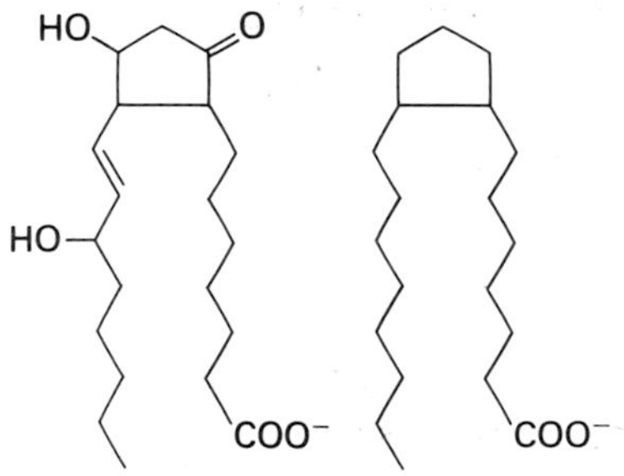

Fig. 11.4.6 Prostaglandin E_1 and its parent compound, prostanoate.

11.5   Triacylglycerol and phospholipid synthesis

Triacylglycerol, or triglyceride, is synthesized in the intestinal mucosa, the liver, and the adipose tissue. In the adipose tissue, synthesis is preliminary to storage. The liver and the intestine operate not only to synthesize triacylglycerols, but to incorporate them in lipoproteins for transport; they are able to modify the composition of the fatty acid mix in the triacylglycerols of the diet. The liver is the only one of these three tissues which can employ glycerol for the synthesis; glycerol is activated via a kinase reaction. The intestinal mucosa and the adipose tissue lack glycerol kinase and the glycerol phosphate for the acyl transferase reaction (Fig. 11.5.1) is derived from glucose. Phospholipid synthesis is thought to occur in most cells, but quantitatively liver is the main producer. The pathway is related to that of triacylglycerol synthesis but requires CTP as cofactor. The following points are customarily stressed in relation to triacylglycerol synthesis and storage.

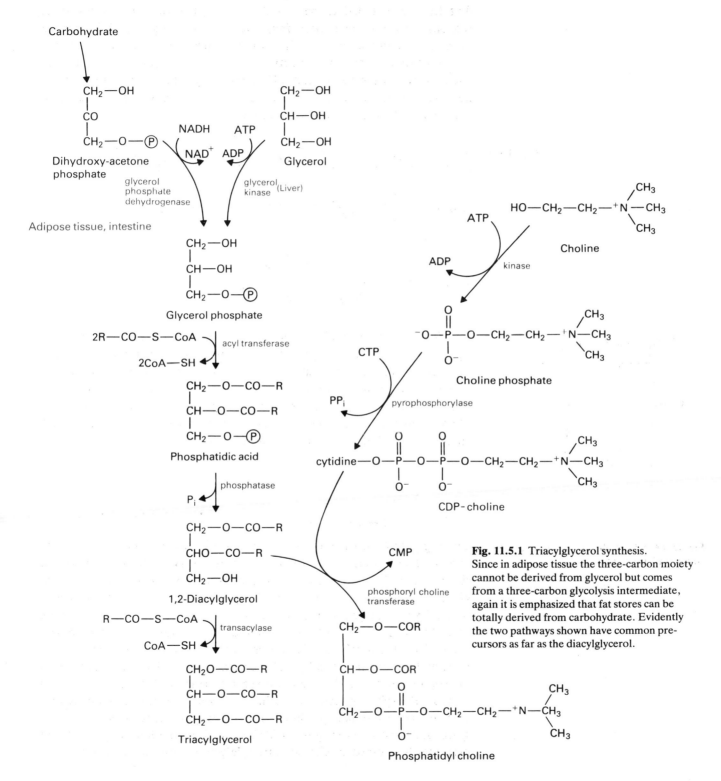

Fig. 11.5.1 Triacylglycerol synthesis. Since in adipose tissue the three-carbon moiety cannot be derived from glycerol but comes from a three-carbon glycolysis intermediate, again it is emphasized that fat stores can be totally derived from carbohydrate. Evidently the two pathways shown have common precursors as far as the diacylglycerol.

1 In some individuals adipose tissue can be progressively laid down with increasing food intake, but most people maintain a stable body weight over long periods whatever they eat. Thus there is some sort of control over lipid synthesis from carbohydrate. That the control is saturable is evident from those who can put on weight rapidly (such as the young ladies in some tribes, who are fattened up for marriage by confinement in huts without exercise while at the same time being force-fed a carbohydrate- and fat-rich diet). There is no doubt that obesity is a powerful survival mechanism (as shown by the equating of plumpness to beauty in many cultures) for when lean wives are all dead of inanation after about thirty days of famine, fat wives will survive to bear more children in better times.

2 The composition of fatty acids within stored triacylglycerol deposits is modifiable over long periods by an altering pattern of dietary fatty acids. Thus the ingestion of unsaturated vegetable oil as the principal dietary lipid causes slow conversion of fatty acids in the carcase to a pattern resembling (not identical with) corn oil. Evidently the intestine and the liver have a limited capacity for the modification of fatty acid mixtures.

3 The more saturated a triacylglycerol mixture the greater is its melting point. Moreover, unsaturated acyl residues yield less ATP on oxidation than do the saturated ones. (They are not as susceptible to the activity of acyl CoA dehydrogenase, yielding FAD2H, since they are already desaturated —9.1.) It appears that triacylglycerols are laid down in a state which is as saturated as possible in respect of the environmental temperature. Eskimos and other denizens of cold regions are said to have subcutaneous fat deposits just sufficiently unsaturated to prevent a very undesirable solidification; thereby they lose some energy-producing potential of more saturated fatty acids.

4 Fat yields more energy on combustion than either carbohydrate or protein. The (rounded-off) figures are:

	kcal/g	kJ/g
Fat	9	36
Carbohydrate	4	16
Protein	4	16

Fat is commonly supposed to be of additional value as a storage fuel because it yields water on combustion, for example a mixed triglyceride might be oxidized according to the following stoichiometry

$$C_{50}H_{100}O_6 + 73O_2 \rightarrow 50CO_2 + 50H_2O$$

This looks like a lot of water, fifty moles for every mole triacylglycerol oxidized, but in reality more water than this is lost by the lungs in respiring the necessary seventy-three moles of oxygen. Thus fat is valuable to desert animals, as it is to other life-forms, because it is a concentrated source of energy, not because it contributes water. It is a concentrated source of energy because it is stored *without* water, which is more to the point. Lipid droplets in cells are essentially anhydrous. Glycogen granules, on the other hand, are very hydrated and if glycogen were laid down under the skin in the same quantities as fat we would be so laden with water that we would be unable to move.

11.6 Ketone bodies

The three compounds collectively known as the ketone bodies (an old-fashioned term which has survived) are synthesized by the liver, which cannot itself oxidize them. The pathway for their synthesis looks rather similar to the pathway for their utilization (Fig. 11.6.1). Again, this is only apparent; the enzymes are different and they occur in different tissues. Thus the ketone bodies are substrates produced in the liver for oxidation in other tissues, notably the brain and the heart muscle. There is nothing untoward about this; glucose, for example, is produced from liver glycogen for peripheral tissue oxidation. If ketone bodies are toxic in excess (see below) then so is glucose (in diabetes). The synthesis of ketone bodies in the liver employs two-carbon units, and utilization in other tissues is by reconversion to two-carbon units, then oxidation by the tricarboxylic acid cycle. The principal ketone body of the blood is β-hydroxybutyrate, which is considered to be a 'blind alley' in metabolism—its only possible fate is reversion to its precursor, acetoacetate. For practical purposes acetone can be regarded as an excretion product, generated by the spontaneous (non-catalysed) decarboxylation of acetoacetate.

One other tissue idiosyncrasy should perhaps be noted: in mammary gland and developing brain it seems that ketone bodies are not oxidized, but are used for the synthesis of lipids (for which acetyl CoA is the 'conventional' precursor).

Significance of ketone body metabolism

The following points should be made in any general discussion of ketone body production.

1   Ketone bodies can be derived from fat, carbohydrate, or protein, indeed any metabolite which is capable of producing acetyl CoA. However carbohydrate is 'antiketogenic'—it readily forms oxaloacetate, which is necessary for ketone body disposal. Quantitatively, the main ketogenic material is fat.

2   *Ketosis* (overproduction), the haematological manifestation of which is *ketaemia*, and the urological *ketonuria*, is caused by mobilization of fatty acids to the liver in such quantities that they cannot be taken up as acetyl CoA by the tricarboxylic acid cycle and other pathways, and they are diverted to ketone body production. There is a threshold over which the ketone bodies cannot be assimilated by the peripheral tissues and ketosis ensues. The most notable causes of excessive mobilization of fatty acids to the liver are as follows.

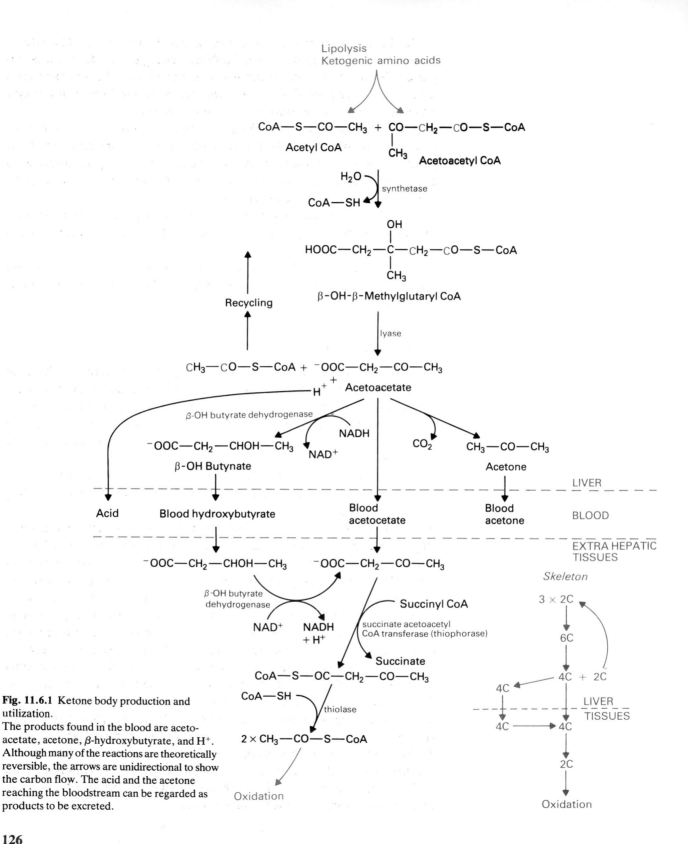

Fig. 11.6.1 Ketone body production and utilization.

The products found in the blood are acetoacetate, acetone, β-hydroxybutyrate, and H^+. Although many of the reactions are theoretically reversible, the arrows are unidirectional to show the carbon flow. The acid and the acetone reaching the bloodstream can be regarded as products to be excreted.

1 Diabetes, wherein there is a lack of circulating insulin (absolute or relative) such that lipogenesis is inhibited, and glucose utilization is impaired. In the absence of insulin lipolysis is increased, the free fatty acids derived from adipose tissue being transported to the liver, and a major proportion of them are converted to ketone bodies.

2 Starvation, wherein the lack of carbohydrate (little is stored and liver glycogen is used up in one day) causes mobilization of extrahepatic lipid and protein. The free fatty acids from the lipids and the ketogenic amino acids from the protein are converted to ketone bodies by the liver, and oxidized by the brain and the myocardium (Fig. 11.6.1).

3 A high fat, low protein-carbohydrate diet, as well as a high protein-fat, low carbohydrate diet, in which equally, carbohydrate is scarce, causes ketosis; such diets are rare.

3 Ketosis has undesirable sequelae since the ketone bodies are acidic and they are excreted with loss of cations such as sodium and potassium. This is dealt with in 13.12.

11.7 Cholesterol and bile salts

Cholesterol is an animal (non-plant) product, but other sterols are found in plants. In mammals cholesterol is a constituent of all cells, but quantitatively the main producers are the liver and the intestinal mucosa. In these tissues the microsomes and the cytosol contain the complement of enzymes for the pathway laid out in Fig. 11.7.1. You should make the following points about cholesterol synthesis.

1 All the carbons are derived from two-carbon units (acetyl CoA).

2 The first few steps are the same as for ketone body synthesis, yielding β-hydroxy-β-methylglutaryl CoA.

3 The step to give the first five-carbon unit, mevalonic acid, is the committed step and is subject to feedback inhibition by cholesterol; however cholesterol does not appear to interact directly with the enzyme (β-hydroxyl-β-methylglutaryl CoA reductase) but possibly alters the fluidity of the microsomal membrane in which the enzyme is embedded. Alteration of the properties of a membrane to which an enzyme is bound would be expected to modulate the activity of the enzyme itself. Starvation of, and feeding cholesterol to, experimental animals decreases the activity, which moreover is subject to a circadian rhythm (rats synthesize much more cholesterol in the early hours of the morning than they do at midday). Since the blood and probably the tissue content of cholesterol is associated with ischaemic heart disease, any factor potentially decreasing the activity of β-hydroxy-β-methyl glutaryl coenzyme A reductase is of interest in the control of the condition.

4 Mevalonic acid yields isoprene units (Fig. 11.7.2), and in plants they are precursors of such substances as rubber, tocopherols (vitamins E), and carotenoids (vitamins A). There is head-to-tail condensation of these five-carbon units to ten- and fifteen-carbon units.

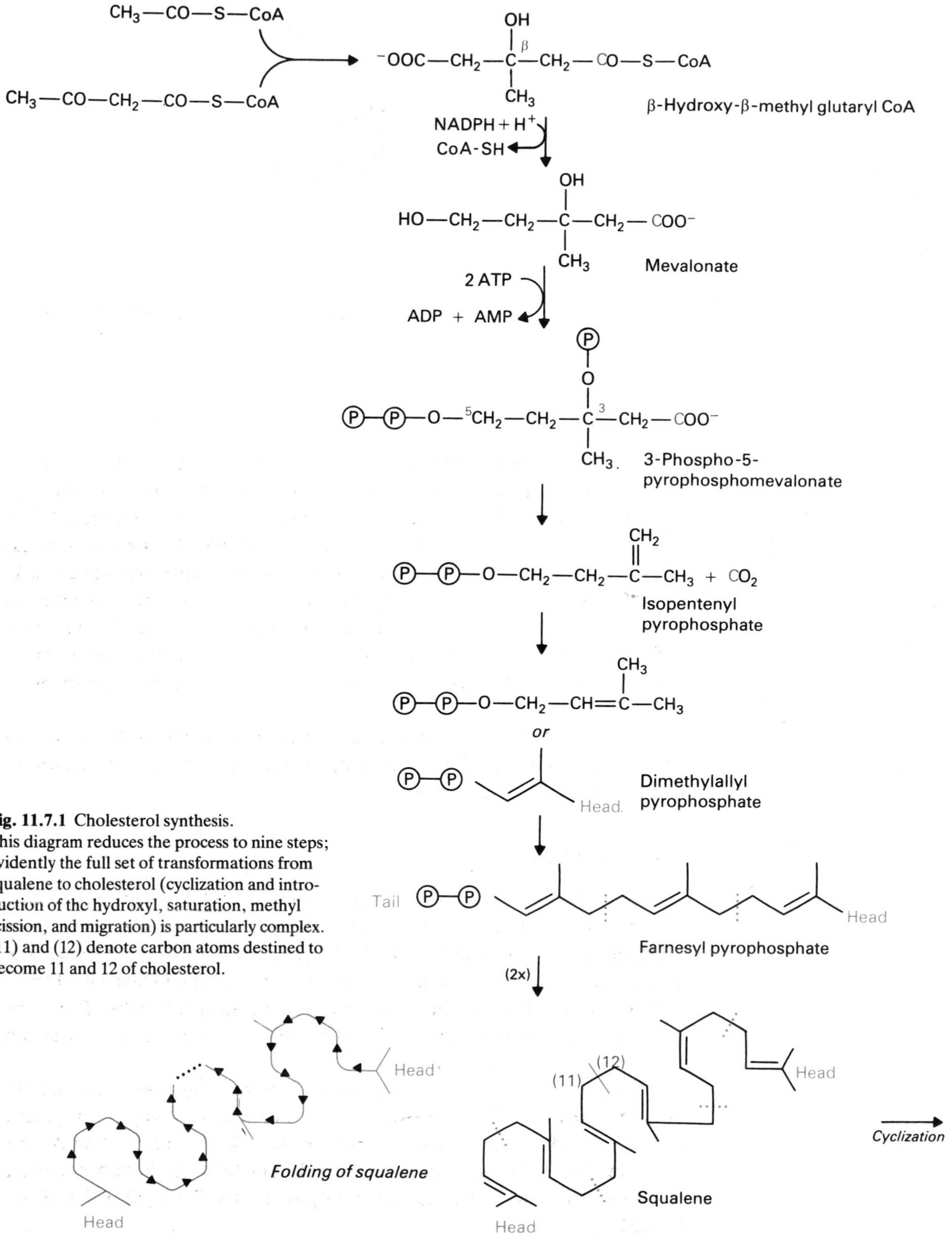

Fig. 11.7.1 Cholesterol synthesis.
This diagram reduces the process to nine steps;
evidently the full set of transformations from
squalene to cholesterol (cyclization and introduction of the hydroxyl, saturation, methyl
scission, and migration) is particularly complex.
(11) and (12) denote carbon atoms destined to
become 11 and 12 of cholesterol.

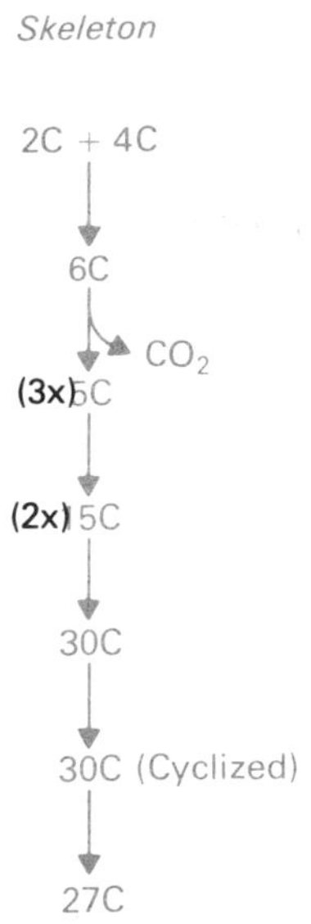

Skeleton

2C + 4C

↓

6C

↓ ⟶ CO_2

(3x)6C

↓

(2x) 5C

↓

30C

↓

30C (Cyclized)

↓

27C

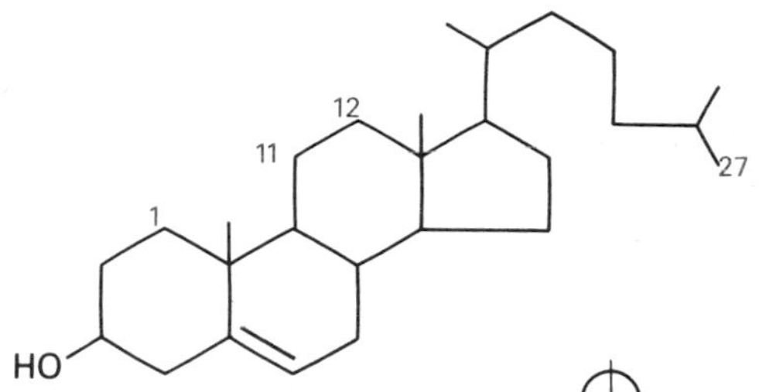

Cholesterol

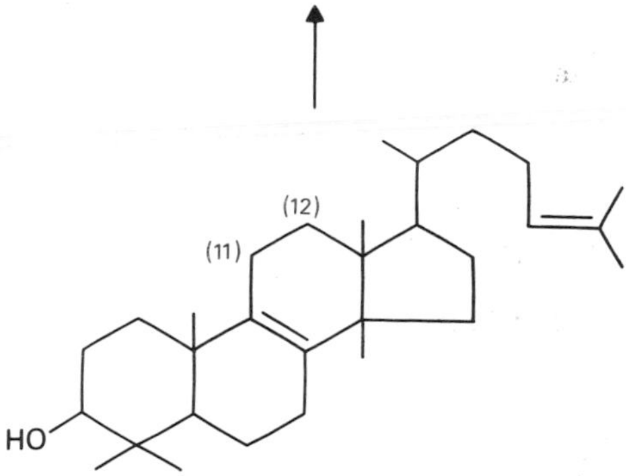

Lanosterol

Fig. 11.7.1 (Cont.)

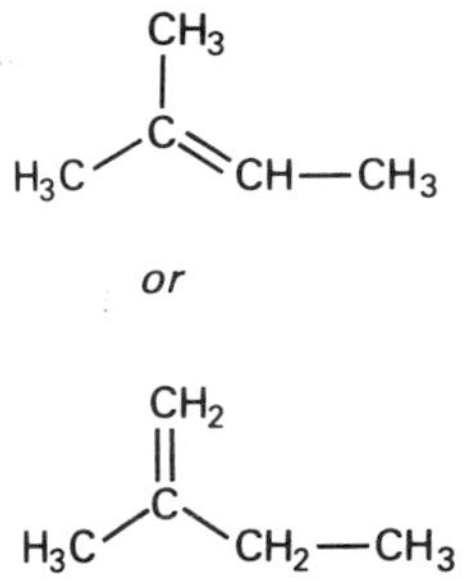

or

Fig. 11.7.2 Isoprene units.

5 Two fifteen-carbon units condense, not head-to-tail but rather tail-to-tail, to yield squalene, which is folded in such a way as to become subject to cyclization. It is perhaps tricky to memorize the mode of squalene folding; there are various ways it can be done to provide the steroid nucleus, although only one is correct. First draw out the cholesterol structure in pencil. Take a pen or a coloured pencil and starting at the bottom left hand corner (position 4) draw in three isoprene units, linked head-to-tail, in a convoluted manner. Then start at the other end and draw three more isoprene units to meet up with the first three between (future) carbons 11 and 12.

6 After squalene cyclization there is desaturation and methyl group re-arrangement removal, then transformation through a series of steroids until cholesterol itself is synthesized.

Bile salt synthesis

The bile salts emulsify lipids in the small intestine, but they also represent an excretory pathway for cholesterol. In animals the steroid nucleus is excreted intact in this form. Bacteria in the large intestine subject the bile salts synthesized by the liver to a series of modifications, without however disrupting the steroid nucleus, and produce the secondary bile salts. These are reabsorbed and re-excreted by the liver into the bile in the enterohepatic circulation for the bile salts.

The principal bile salts, as synthesized by the liver itself, are sodium glycocholate and sodium taurocholate. The synthetic pathway for the parent bile acid, cholic acid, involves scission of the side-chain at carbon 24 and multiple hydroxylations. There is feedback control by cholic acid at the first hydroxylation point, the reaction catalysed by the 7 α-hydroxylase (Fig. 11.7.3).

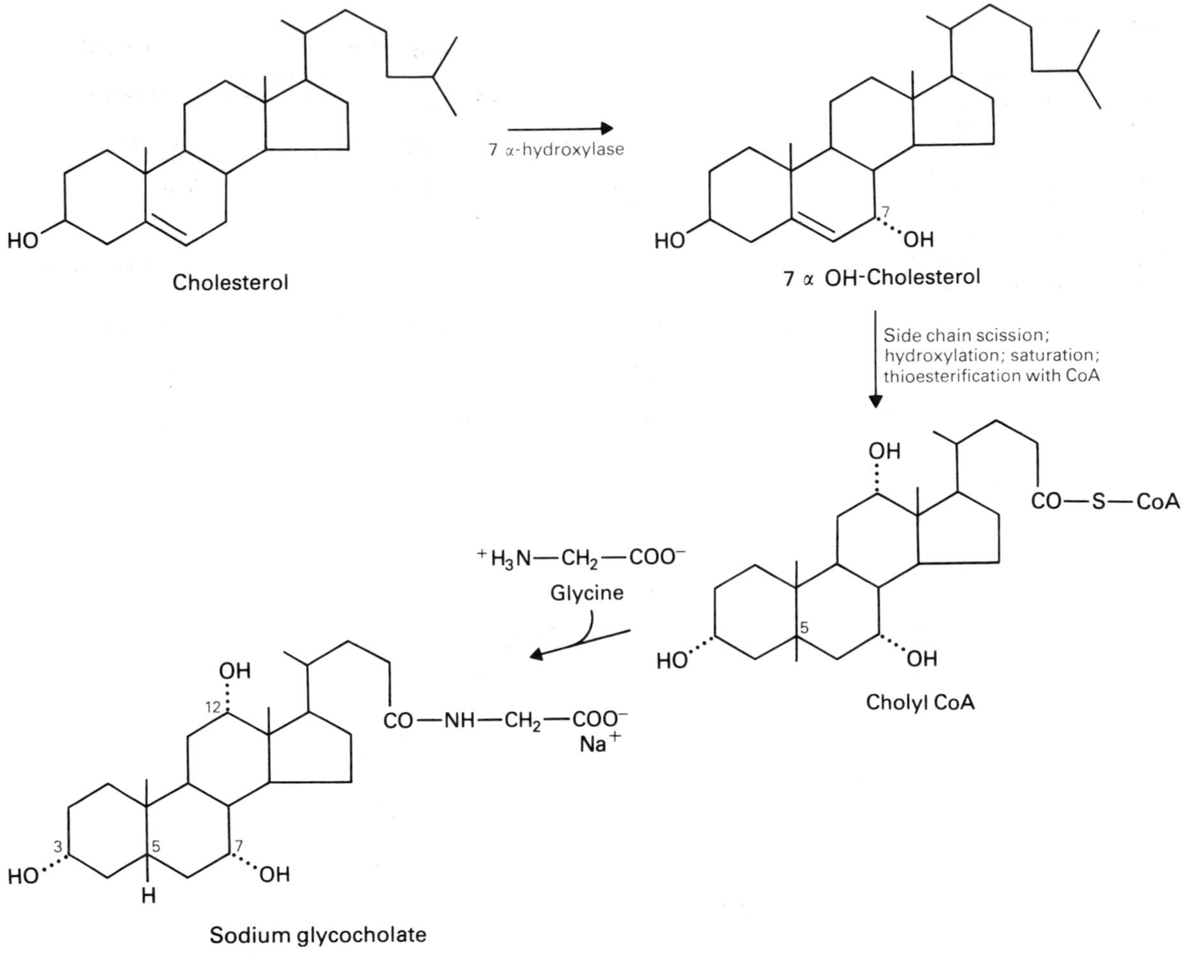

Fig. 11.7.3 Bile salt synthesis.
In this abbreviated version the 7 α-hydroxylase reaction is shown since it is the rate-limiting step, subject to feedback control. During the complex chemical modification of the 7 α-hydroxycholesterol, rings A and B become *cis*; this means that when ring B is saturated the hydrogen at 5 is above the plane of the ring structure as written and is designated as β while being connected with a solid line. The chemistry can be checked against 13.1.

If the reabsorption of bile salts in the enterohepatic circulation could in some way be interrupted, then they would merely be lost in the faeces and this would be a means of clearing cholesterol from the body. The resin cholestyramine has been used for this purpose; it is an anion exchanger, and taken by mouth binds the negatively charged bile salts such that they become unavailable for reabsorption. Any removal of cholesterol is thought to be prophylactic against ischaemic heart disease in those subjects at risk.

11.8 Amino acids and their derivatives

In the protein systems of the various life forms there are about twenty amino acids. Plants and bacteria can generally make all of them but mammals such as the rat and human do not have this synthetic capacity; they are able to synthesize only those amino acids described as *non-essential*. 'Non-essential' refers to the diet: they are decidedly essential to the body. Description of the essential versus non-essential amino acids for man is equivocal. Histidine

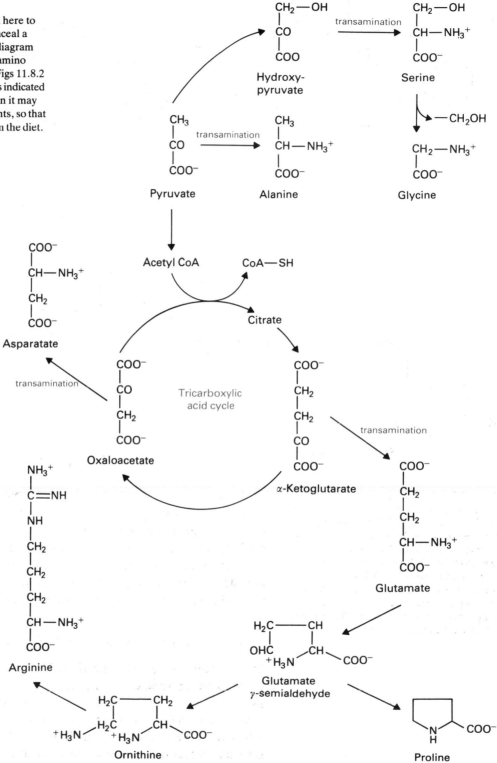

Fig. 11.8.1 Biosynthesis of the non-essential amino acids.
The arrows, which are unidirectional here to represent carbon flow, obviously conceal a plethora of multiple reactions. This diagram covers some six of the non-essential amino acids. For cysteine and tyrosine see Figs 11.8.2 and 11.8.3. The pathway for arginine is indicated here although it is thought that in man it may not be synthesized in adequate amounts, so that some must be derived preformed from the diet.

Table 11.8.1 Essential and non-essential amino acids (for man)

Essential	Non-essential
Histidine	Glycine
Isoleucine	Alanine
Leucine	Serine
Lysine	Glutamate, glutamine
Methionine	Aspartate, asparagine
Phenylalanine	Tyrosine
Threonine	Cysteine, cystine
Tryptophan	Proline
Valine	Hydroxyproline
Arginine	

and arginine are usually included among the essential group because they are probably required in the diet of children, due to the demands of growth, although they are synthesizable. So an 'essential' amino acid is precisely defined as one which cannot be synthesized by an organism at a rate commensurate with its needs.

Table 11.8.1 sets out the two lists. The biosynthesis of the non-essential amino acids is best memorized in conjunction with a skeleton diagram of the tricarboxylic acid cycle (Fig. 11.8.1). The three important keto acids—pyruvate, α-ketoglutarate, and oxaloacetate—yield the corresponding amino acids alanine, glutamate, and aspartate respectively by direct transamination. Further, glutamate is the precursor of histidine, proline, and arginine. Arginine is a constituent of the urea cycle, but there cannot be a net production of it without an input of ornithine (10.2) and this is derived from glutamate. Hydroxyproline is derived from proline. Two others are derived from essential amino acids: tyrosine from phenylalanine (Fig. 11.8.2); and cysteine from methionine via homocysteine (Fig. 11.8.3). Serine and glycine are interchangeable and derived from pyruvate (Fig. 11.8.4).

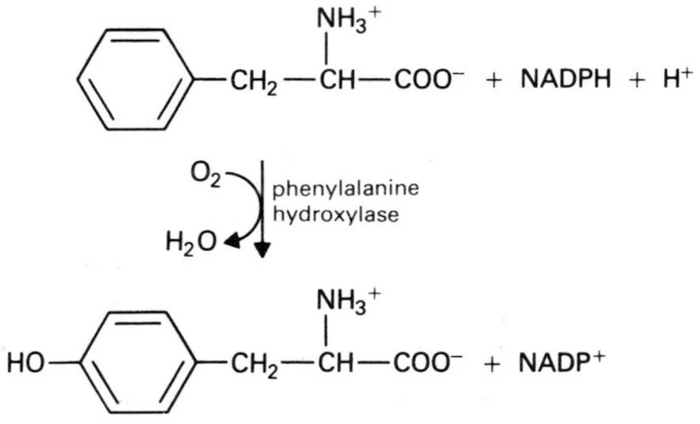

Fig. 11.8.2

The essential amino acids

These are synthesized in a diversity of pathways, worked out for the most part using the bacterium *Escherichia coli*. Consideration of them involves a whole group of novel substances. This is a topic which does not come into many elementary courses, and here we merely note the key substances and make such generalizations as are possible (Fig. 11.8.5). Phenylalanine and tryptophan, the two aromatic amino acids, have a number of intermediates in common, the branchpoint being chorismic acid; they are derived originally from the four-carbon sugar phosphate, erythrose phosphate, and phosphoenolpyruvate. The pathway for tryptophan involves anthranilic acid. Methionine, threonine, and isoleucine are all partially derived from aspartate, with homoserine as a common intermediate. Isoleucine is itself partially synthesized from threonine. Lysine synthesis occurs by a complex pathway

Fig. 11.8.3

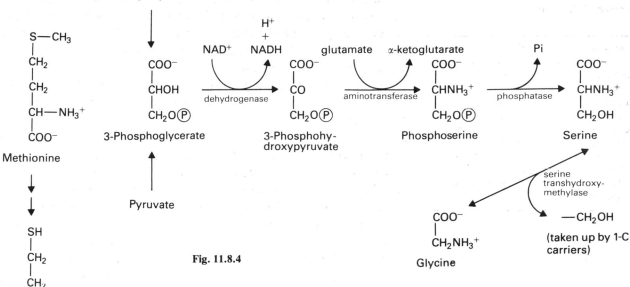

Fig. 11.8.4

which also has its origins in aspartate. Valine and leucine are derived from pyruvate and α-ketoisovalerate, the immediate precursor of valine, is itself a precursor of leucine. Note that most of the amino acids are finally synthesized by transamination of the corresponding α-keto acid.

Syntheses using amino acids

A large number of active substances are synthesized by pathways utilizing various amino acids. They have nothing in common except that they are important in one sense or another and they tend to be favourites for short or 'short notes' questions. One-carbon metabolism, which equally depends on amino acids as a source of raw materials, is dealt with in 11.13 in view of its special importance.

Thyroxine and triiodothyronine

The thyroid gland has the capacity to concentrate iodide for thyroid hormone synthesis; it is a small organ yet has 20–30% of the total body iodine. Within the gland, iodide is oxidized by a peroxidase

$$2I^- + H_2O_2 \rightarrow I_2 + H_2O + \tfrac{1}{2}O_2$$

This reaction is shown as the production of iodine but at some stage prior to this, free radicals of iodine react with tyrosyl residues in thyroglobulin, the protein of the thyroid's 'colloid'. Thyroglobulin is a glycoprotein of relative molecular mass 660 000. Iodination of the tyrosyl residues is initiated at the 3 position (Fig. 11.8.6) and then at position 5. Coupling of two molecules of tyrosyl residues within the thyroglobulin gives triiodothyronine or thyroxine

133

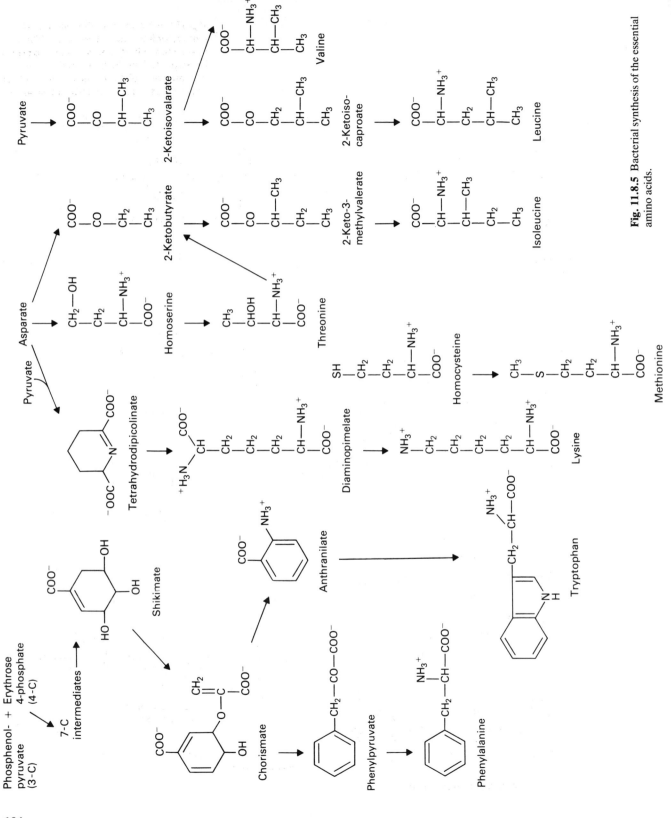

Fig. 11.8.5 Bacterial synthesis of the essential amino acids.

134

Fig. 11.8.6 Synthesis of throxine and tri-iodothyronine in the thyroid gland. This is obviously highly schematic; there is no mechanism suggested for the translocation of an iodinated phenol during the formation of peptide-bound tetraiodo- and triiodothyronines.

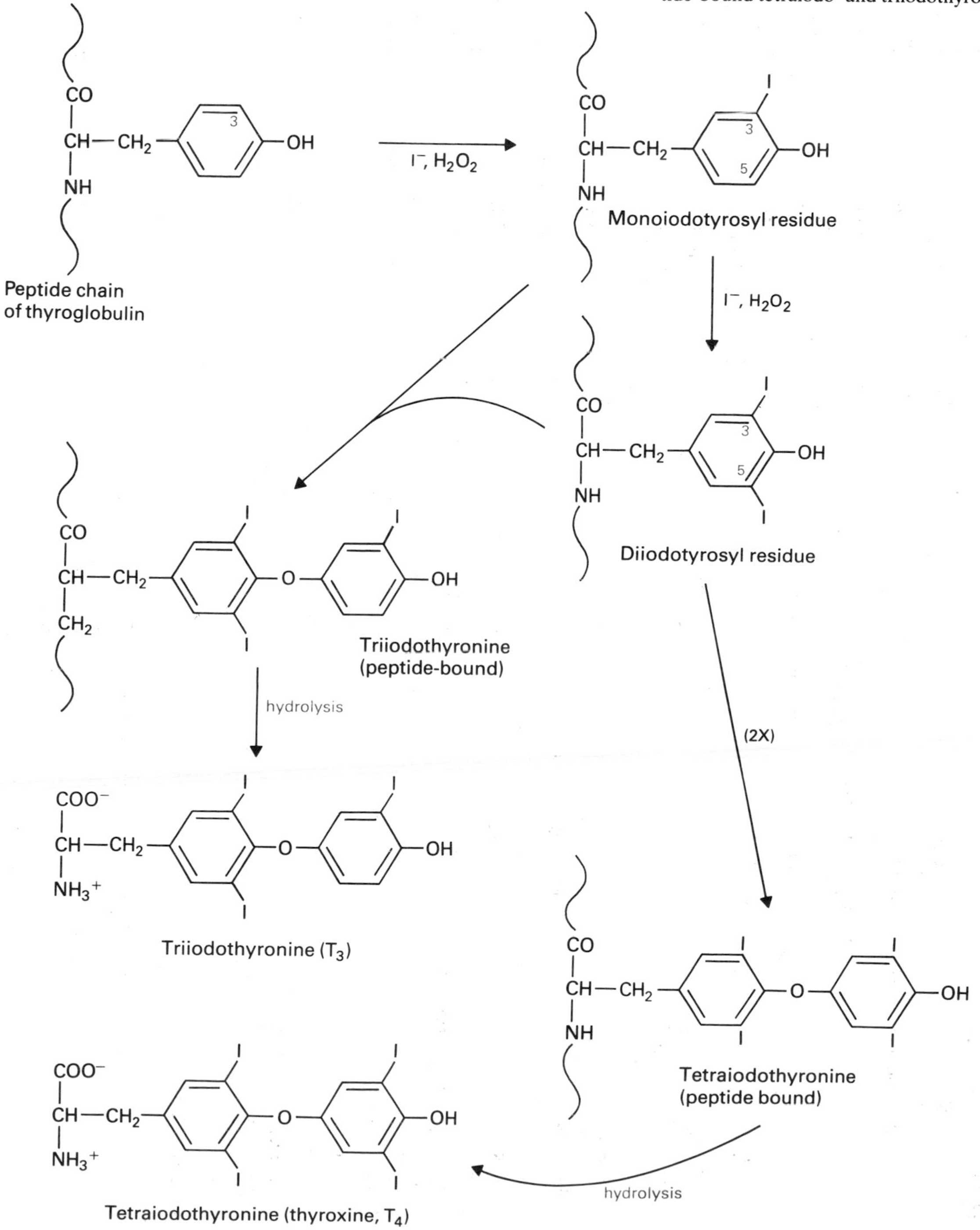

Peptide chain of thyroglobulin

I^-, H_2O_2

Monoiodotyrosyl residue

I^-, H_2O_2

Diiodotyrosyl residue

Triiodothyronine (peptide-bound)

hydrolysis

Triiodothyronine (T_3)

(2X)

Tetraiodothyronine (peptide bound)

hydrolysis

Tetraiodothyronine (thyroxine, T_4)

135

residues. Some free iodinated tyrosine may also be coupled to bound tyrosyl within thyroglobulin. On stimulation by thyrotropin (thyroid-stimulating hormone TSH) or by cold, the iodinated thyroglobulin is hydrolysed by proteases in the lysosomes to yield free thyroxine (T_4) and triiodothyronine (T_3). Thyrotropin also stimulates the uptake of iodide and the synthesis of thyroglobulin.

The catecholamines

Like the thyroid hormones, these are synthesized from tyrosine, and comprise adrenaline, noradrenaline, dopa, and dopamine. They are all otherwise describable as dihydroxyphenylalkylamines. The site of synthesis of adrenaline is principally the chromaffin cells of the adrenal medulla; noradrenaline is synthesized both there and in the neurones of the sympathetic nervous system. In both cases dopa and dopamine are intermediates (Fig. 11.8.7). The initiating enzyme, tyrosine hydroxylase, is subject to allosteric inhibition by the catecholamines themselves; stress, feeding, β-adrenergic stimulation*, pituitary hyperactivity, as well as ACTH and corticosteroids, all in some way stimulate the activity of the enzyme. Adrenaline and noradrenaline are the active principles of the adrenal medulla, with a wide range of effects on metabolism which include muscle glycogenolysis and adipose tissue lipolysis. Noradrenaline and dopamine are both neurotransmitters.

Other active products

Histamine is a ubiquitous substance, released in allergic states, and synthesized from histidine (Fig. 11.8.8).

Serotonin, or 5-hydroxytryptamine, a cerebral neurotransmitter, is a result of the hydroxylation and decarboxylation of tryptophan (Fig. 11.8.9).

Melanin is synthesized from dopa, so that it shares some of the intermediates of adrenaline and noradrenaline synthesis. The sites, of course, are different, for melanin (see 13.18) occurs almost exclusively in the skin and in the eye (iris and retina). In the skin, tyrosine hydroxylation to yield dopa is accomplished by the catalytic action of tyrosinase, a copper-containing enzyme. (Tyrosine hydroxylase, on the other hand, is involved in dopa synthesis in the adrenal medulla.) Tyrosinase also catalyses the second step in melanin formation (Fig. 11.8.10), that is dopaquinone formation. Indole 5,6-quinone, and 5,6-dihydroxyindole spontaneously (non-enzymically) form a series of polymers collectively known as melanin.

Nicotinate is a B-vitamin; therefore, by definition, humans should not be able to synthesize it. Nonetheless a small proportion of the human requirement can be met by a synthetic procedure, using tryptophan, in the liver (Fig. 11.8.11).

*The receptors for the catecholamines are of two types, α- and β- (13.19).

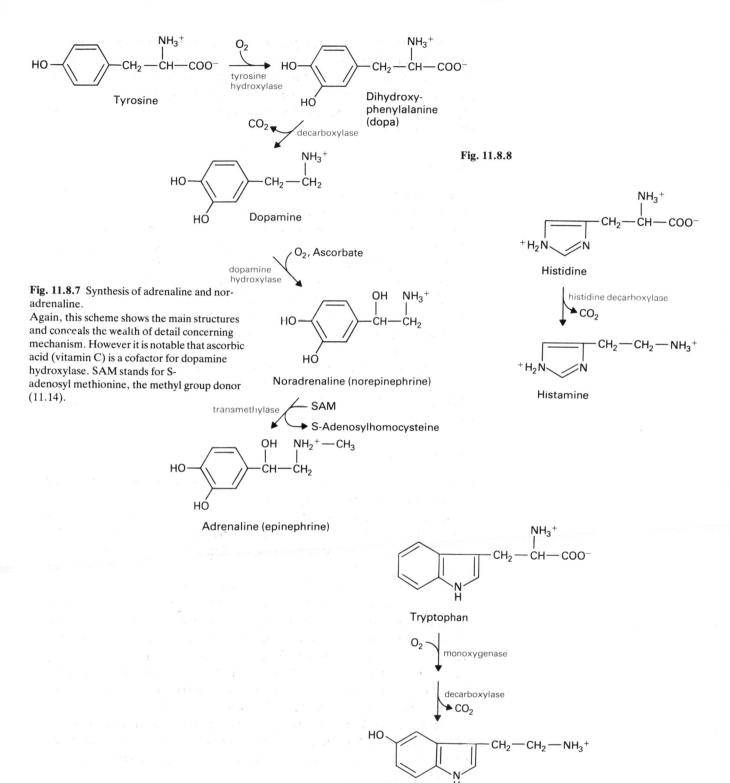

Tyrosine

O_2
tyrosine hydroxylase

Dihydroxy-phenylalanine (dopa)

Fig. 11.8.8

CO_2
decarboxylase

Dopamine

O_2, Ascorbate
dopamine hydroxylase

Fig. 11.8.7 Synthesis of adrenaline and nor-adrenaline.
Again, this scheme shows the main structures and conceals the wealth of detail concerning mechanism. However it is notable that ascorbic acid (vitamin C) is a cofactor for dopamine hydroxylase. SAM stands for S-adenosyl methionine, the methyl group donor (11.14).

Noradrenaline (norepinephrine)

transmethylase — SAM
→ S-Adenosylhomocysteine

Adrenaline (epinephrine)

Histidine

histidine decarboxylase
CO_2

Histamine

Tryptophan

O_2
monoxygenase

decarboxylase
CO_2

Fig. 11.8.9   5-Hydroxytryptamine

137

Fig. 11.8.10 Synthesis of melanin.
The indole structures, which look as though they should be derived from tryptophan, in fact come from tyrosine. It is the lack of tyrosinase which is responsible for the inherited condition of albinism (13.13).

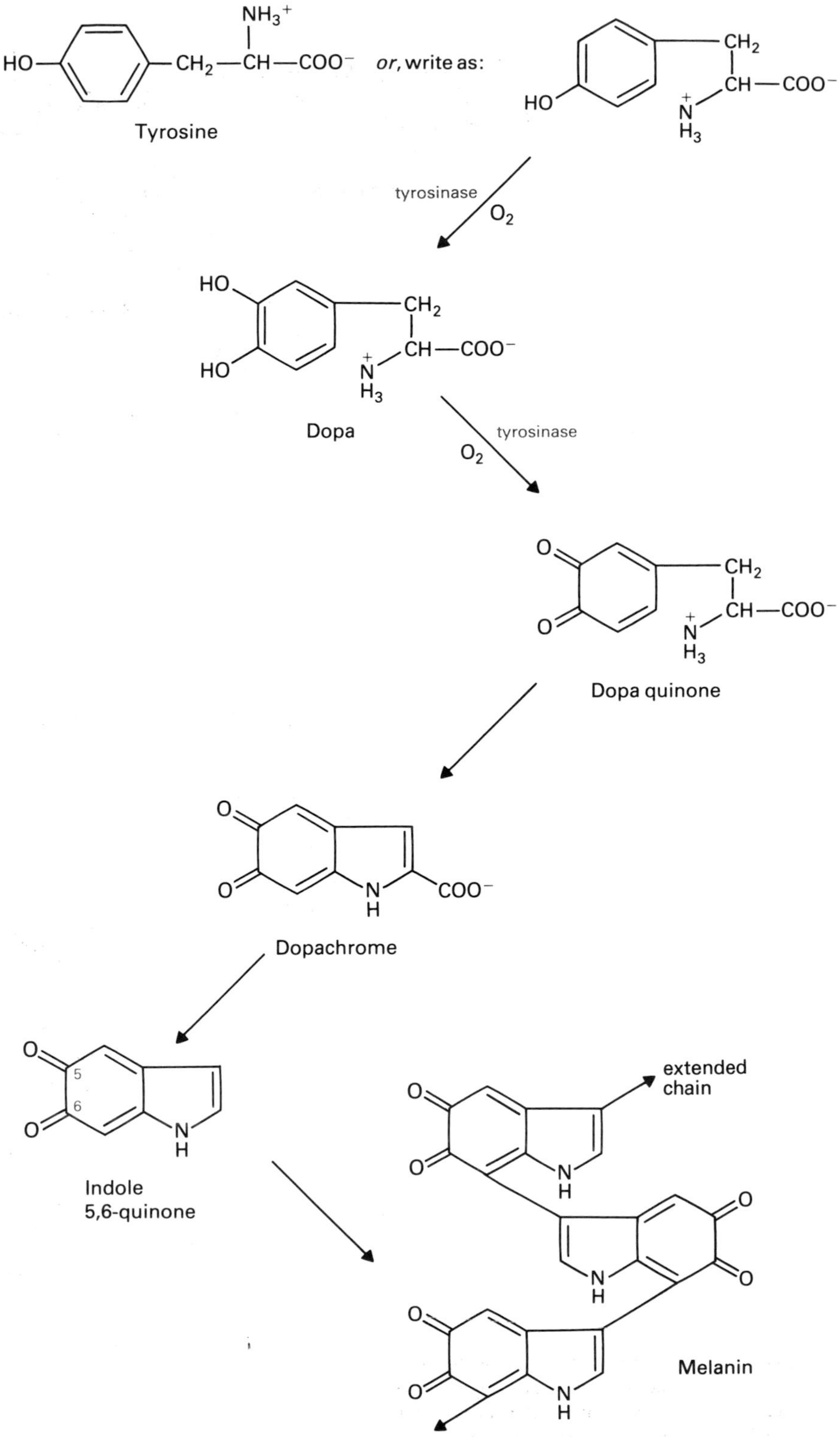

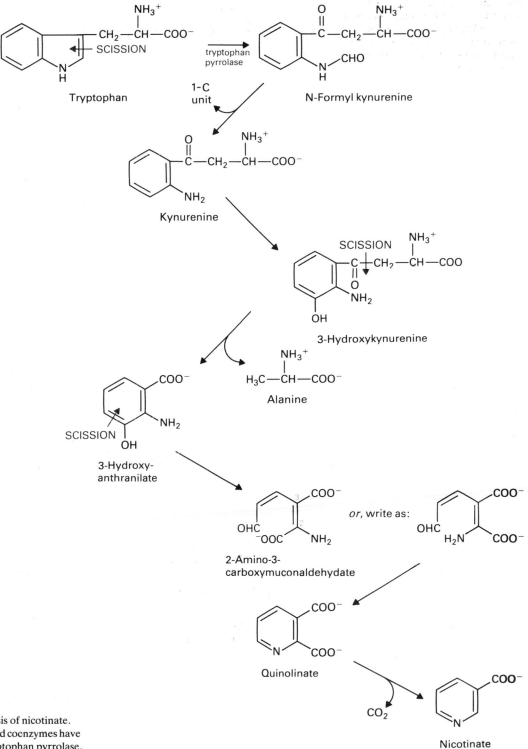

Fig. 11.8.11 Hepatic synthesis of nicotinate. The names of the enzymes and coenzymes have been omitted, except for tryptophan pyrrolase, notable as an inducible enzyme (see 12.1).

Tryptophan

tryptophan pyrrolase

1-C unit

N-Formyl kynurenine

Kynurenine

3-Hydroxykynurenine

SCISSION

Alanine

3-Hydroxy-anthranilate

SCISSION

2-Amino-3-carboxymuconaldydate

or, write as:

Quinolinate

CO_2

Nicotinate (as nucleotide)

11.9 Porphyrins

As components of the cytochromes, porphyrins are found in almost all cells, so that mechanisms for their synthesis are presumably universal, or almost so. One or two bacteria need porphyrins preformed (as 'vitamins') in the medium but all higher forms readily synthesize them. As is common for apparently complicated biological structures, the precursors turn out to be rather simple. Carbon is supplied by succinyl CoA, and both carbon and nitrogen by glycine (Fig. 11.9.1). Since succinyl CoA is synthesized from α-ketoglutarate in the tricarboxylic acid cycle, any intermediate contributing to cycle intermediates, like glucose or aspartate or glutamate, is a potential precursor.

Synthesis is divided between the mitochondria (which provides succinyl CoA from the cycle) and the cytosol. The first step is a condensation of succinyl CoA with glycine to yield δ-aminolaevulinate. Two molecules of δ-aminolaevulinate, in a multistep reaction, are converted to porphobilinogen, the first pyrrole. Evidently four molecules of this must in some way link up to provide the cyclic porphyrin (tetrapyrrole) structure. This process occurs in the cytosol, and the first of these cyclic compounds to be formed is uroporphyrinogen III. The suffix -'ogen' implies that it is in fact a precursor of the porphyrins, and is more reduced than the porphyrins themselves. Because it is more reduced it does not contain alternate single and double bonds, and is colourless. The 'III' refers to its particular arrangement of acetyl and propionyl substituents, from among those possible. Uroporphyrinogen III is converted to protoporphyrinogen III by side-chain modification, then oxidized to protoporphyrin IX. Again, the Roman numeral refers to a particular isomeric arrangement; under an alternative system it is known as protoporphyrin III, but IX is more usual. Finally, in haem (as opposed to cytochrome) synthesis, ferrous iron is introduced in a reaction catalysed by ferrochelatase.

Significance of porphyrin synthesis

The following points may be made about the pathway.

1 In animals it finds its greatest quantitative expression in the bone marrow, synthesizing haem for haemoglobin. On a biosphere basis most of the porphyrin synthesized is incorporated into the chlorophyll of green plants, which contains magnesium rather than iron. In plants the synthetic pathway is rather different; δ-aminolaevulinate is derived from glutamate rather than succinyl CoA.

2 Control of synthesis is exerted at the δ-aminolaevulinate synthetase step, and this is by feedback inhibition by haem itself. Nonetheless the action of haem is not a direct one on the enzyme; it probably combines with some protein to constitute an inhibitor.

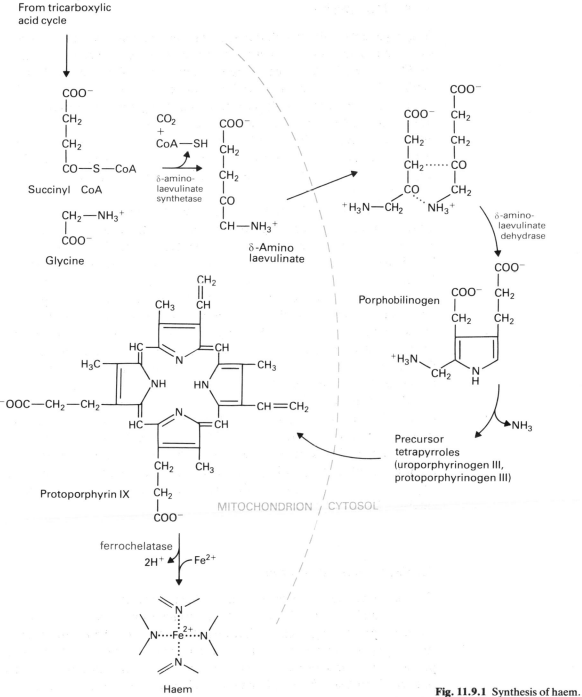

Fig. 11.9.1 Synthesis of haem.
Note that two hydrogen ions are removed
during the ferrochelatase reaction. The haem
could therefore be written with no net charge
on the iron atom. In haem, iron is linked to the
pyrrole nitrogens by coordinate, not ionic,
linkages.

3 A number of defects in porphyrin synthesis, collectively known as the *porphyrias*, are known. Some of these are inherited, some acquired. The most common is porphyria cutanea tarda, which is thought to be an autosomal recessive condition but is in addition associated with liver injury. It may be due to diminished activity of a uroporphyrinogen III decarboxylase, that is, it is the conversion of precursor porphobilinogens to porphyrins which is defective. Patients with the condition are found to have uroporphyrinogen III in the urine; this becomes oxidized to dark porphyrin pigments when the urine is left to stand. Moreover the accumulation of porphyrins in the skin causes an intense photosensitivity; the exposed skin on the back of the hands especially blisters and abrades. Porphobilinogen, rather than porphyrinogens and porphyrins, accumulates in acute intermittent porphyria. There is also photosensitivity in this condition which is usually precipitated by barbiturates or anaesthetics; it appears that the drugs in some way stimulate liver δ-aminolaevulinate synthetase activity.

4 Porphyrin synthesis has a peculiar relationship to lead poisoning: the heavy metal inhibits the enzymes ferrochelatase and δ-aminolaevulinate dehydratase. Haem normally exerts a feedback inhibition on the synthetic pathway through δ-aminolaevulinate synthetase. So in lead poisoning the production of haem is diminished because of inhibition of the first two enzymes mentioned, but the feedback inhibition of the third (the synthetase) is decreased. This is why anaemia is not normally a feature of moderate lead poisoning. Still, there is the potential for assessing its severity by assaying tissue ferrochelatase and dehydratase.

5 In mammals the porphyrin synthetic pathway is not reversible. Haem is catabolized (10.4) but only to the stage of bilirubin; this might well be regarded as an unfortunate biochemical omission, since bilirubin is a toxic substance and the difficulty of disposing of it can give rise to jaundice.

11.10 Naming nucleotides

A full consideration of the various nucleotides cannot be further postponed since they emerge separately in discussions of purine and pyrimidine biosynthesis, as well as of protein biosynthesis. Fortunately nomenclature is subject to rules: the bases adenine, guanine, cytosine, and uracil and thymine may be N-glycosidically linked with *D*-ribose and 2-deoxyribose, as *ribonucleosides* and *deoxyribonucleosides* respectively. Unless stated otherwise, it is assumed that the linkage is at the 1 position of the pentose. Adenine-ribose is termed adenosine, cytosine-ribose is termed cytidine and so on. Adenine-deoxyribose is termed deoxyadenosine, cytidine-deoxyribose is termed deoxycytosine.

Nucleoside phosphates are known as nucleotides. Adenine-ribose-phosphate is termed adenylic acid or adenosine monophosphate, or (most often) as AMP. If it is desired to specify the position of the phosphate this is done

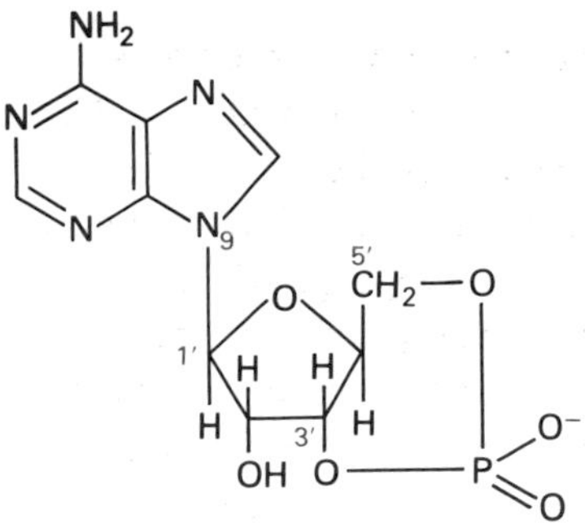

Fig. 11.10.1 Cyclic AMP.

by numbering with a dash. Thus adenosine 5'-phosphate or 5'-adenylic acid is a more precise nomenclature for AMP. It is possible to have a cyclic phosphate arrangement as in 3',5'-cyclic AMP (Fig. 11.10.1) but this is more generally known merely as cyclic AMP or cAMP. If more than one phosphate is esterified, as in AMP-phosphate, this is known as adenosine diphosphate or ADP, again with the site of phosphorylation understood to be 5', and similarly ADP-phosphate is adenosine triphosphate or ATP.

The naming of thymine-containing nucleosides and nucleotides presents a little difficulty because it was thought at one time that thymine was found only in DNA and therefore there was no need to spell out the 'deoxy' in 'deoxythymidine triphosphate'; so thymine-deoxyribose-triphosphate was known as thymidine triphosphate or TTP. This has now been rectified and it is called deoxythymidine triphosphate or dTTP. The prefix 'd' always signifies a deoxyribonucleotide. If one wishes to be very specific one can use the prefix 'ribo-' for ribonucleotides, that is thymine-ribose-triphosphate can be termed ribothymidine triphosphate, but this is seldom done.

There is a small anomaly in that the ribose and deoxyribose nucleosides of hypoxanthine (important as an intermediate in nucleotide synthesis) are termed inosine and deoxyinosine respectively. The nucleoside monophosphates are termed inosinic acid or inosinate (IMP) and deoxyinosinate (dIMP) respectively.

All these synonyms are set out in Table 11.10.1, which can be used as frame of reference against the discussions of DNA replication and protein synthesis below.

Table 11.10.1 Naming nucleosides and nucleotides
(Some of the entries are rarely found in tissues, but illustrate the consistency of the nomenclature)

	Bases	Nucleosides		Ribose nucleotides Monophosphate	Triphosphates	Deoxyribose nucleotides Monophosphates	Triphosphates
Pyrimidines	Cytosine	Cytidine	Deoxycytidine	Cytidylic acid or cytidine monophosphate or CMP	Cytidine triphosphate CTP	Deoxycytidylic acid or deoxycytidine monophosphate or dCMP	Deoxycytidine triphosphate or dCTP
	Uracil	Uridine	Deoxyuridine	Uridylic acid or uridine monophosphate or UMP	Uridine triphosphate UTP	Deoxyuridylic acid or deoxyuridine monophosphate or dUMP	Deoxyuridine triphosphate or dUTP
	Thymine	Thymidine	Deoxythymidine	Thymidylic acid or thymidine monophosphate or TMP	Thymidine triphosphate TTP	Deoxythymidylic acid or deoxythymidine monophosphate or dTMP	Deoxythymidine triphosphate or dTTP
Purines	Adenine	Adenosine	Deoxyadenosine	Adenylic acid or adenosine monophosphate or AMP	Adenosine triphosphate ATP	Deoxyadenylic acid or deoxyadenosine monophosphate or dAMP	Deoxyadenosine triphosphate or dATP
	Guanine	Guanosine	Deoxyguanosine	Guanylic acid or guanosine monophosphate or GMP	Guanosine triphosphate GTP	Deoxyguanylic acid or deoxyguanosine monophosphate or dGMP	Deoxyguanosine triphosphate or dGTP
	Hypoxanthine	Inosine	Deoxyinosine	Inosinic acid or inosine monophosphate or IMP	Inosine triphosphate ITP	Deoxinosinic acid or deoxyinosine monophosphate or dIMP	Deoxyinosine triphosphate or dITP

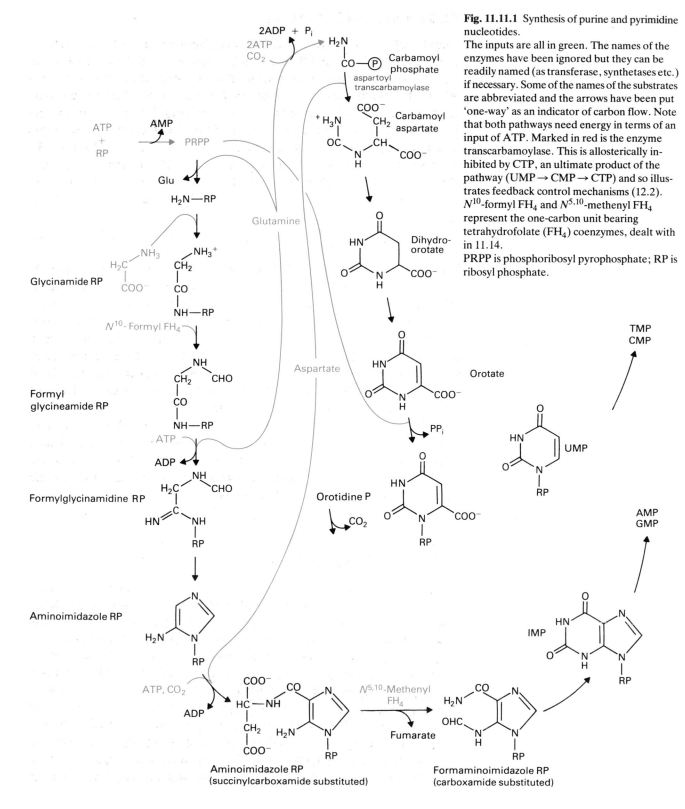

Fig. 11.11.1 Synthesis of purine and pyrimidine nucleotides.

The inputs are all in green. The names of the enzymes have been ignored but they can be readily named (as transferase, synthetases etc.) if necessary. Some of the names of the substrates are abbreviated and the arrows have been put 'one-way' as an indicator of carbon flow. Note that both pathways need energy in terms of an input of ATP. Marked in red is the enzyme transcarbamoylase. This is allosterically inhibited by CTP, an ultimate product of the pathway (UMP → CMP → CTP) and so illustrates feedback control mechanisms (12.2). N^{10}-formyl FH$_4$ and $N^{5,10}$-methenyl FH$_4$ represent the one-carbon unit bearing tetrahydrofolate (FH$_4$) coenzymes, dealt with in 11.14.

PRPP is phosphoribosyl pyrophosphate; RP is ribosyl phosphate.

RNA-directed DNA synthesis

Many viruses, such as the polio virus and foot and mouth disease virus contain RNA rather than DNA. Evidently, where genetic information is contained in RNA sequences, then RNA should direct the synthesis of new mRNA for protein synthesis. This involves the enzyme(s) RNA-directed RNA polymerase. In the case of some oncogenic (tumour-producing) animal viruses, however, this is not the case; RNA in the infecting virus directs the synthesis of double-stranded DNA, which in turn directs the synthesis of progeny RNA. The original viral RNA is destroyed after the DNA copy of it has been synthesized by *reverse* transcriptase or RNA-dependent DNA polymerase.

Fidelity of DNA replication

The replication process is one of high fidelity and it seems that systems have been evolved to ensure this. One of them is based on the ability of one of the DNA polymerases to catalyse both the polymerization of nucleotides and the removal of nucleotides, by hydrolysis, from the 3′ end of a polydeoxy-ribonucleotide chain. (There is nothing mysterious about this; all enzymes equally catalyse the 'forward' and 'back' reactions.) When a new DNA strand is synthesized it should be base-matched to the template strand and then form a double helix. If there is a mismatch, say a cytosine rather than a thymine lined up against an adenine, the double helix formation is interrupted at that point. The mismatched base is then excised by DNA polymerase hydrolysis—it is a catalysis of the back reaction because the product of the forward reaction is not removed by incorporation into a double helix. DNA polymerase certainly cannot excise nucleotides from an intact double helix.

11.13   Protein biosynthesis

The majority of metabolic transformations can be described in terms of a linear pathway or sequence. The synthesis of proteins is not like this, being more of a sort of assembly process. (It must be admitted that this in itself causes students few conceptual difficulties.) As an assembly process, there are six components to be brought together:
(1) the building blocks themselves (amino acids);
(2) the carrier for the amino acids (tRNA);
(3) the site available for the assembly (the ribosome);
(4) the carrier of information for the assembly (mRNA);
(5) the providers of energy for the process (ATP and GTP);
(6) the control factors (signallers for initiation and termination).
The mechanism of protein synthesis was first worked out for prokaryotes but

information for eukaryotes is now becoming more plentiful. Here are some necessary details of the components.

1   The amino acids These are the amino acids found in protein hydrolysates, except that some amino acid residues found in proteins result from modification of other residues after the peptide chain has been put together. This *post-translational modification* (*translation* is the incorporation of amino acids into a peptide, in a sequence defined by mRNA) occurs, for example, in collagen. Collagen is secreted by the fibroblasts as procollagen, bearing numerous prolyl residues. Some of the prolyl residues are then post-translationally hydroxylated to hydroxyprolyl residues, as a stage in the formation of mature collagen.

2   *Transfer ribonucleic acid (tRNA)* is a relatively low molecular weight RNA which serves to bind amino acids and locate them in the correct sequence in the nascent peptide chain. There is at least one tRNA, and there is one enzyme catalysing the synthesis of the tRNA-amino acid complex, for each of the twenty amino acids. Evidently each tRNA must have:

(1) a site for recognition by aminoacyl-tRNA synthetase;

(2) a site for the attachment of the amino acid itself; and

(3) an anticodon, namely a site for the recognition of the codon on mRNA.

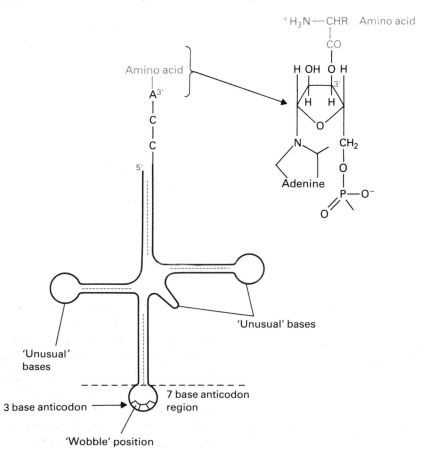

Fig. 11.13.1 General features of the tRNA molecule.

This is evidently a gross distortion to allow representation in the plane of the paper; indeed the molecule is really L-shaped. The 3′ end, bearing the amino acyl residue, is invariably CCA. Some of the 'unusual' bases are in the loops. Although the general shape of all the tRNAs is similar, the base composition and sequence confer high specificity for recognition by the corresponding amino acyl-tRNA synthetase. The 'wobble' position indicates the base of the anticodon which can bind to more than one base of the mRNA codon. Thus the codons for arginine can be AGG or AGA and the 3′ A or G equally will associate with 5′U on the anticodon. This is a manifestation of the degeneracy of the genetic code.

The stringency of these requirements results in the various tRNAs having many features in common (Fig. 11.13.1). The 'unusual bases' such as inosine (abbreviated to 'I') and pseudourine (ψ) occur in the regions of the molecule which are not base-paired, so they may well contribute to the structural identity of each individual tRNA. The anticodon loop consists of seven bases, including of course those three constituting the anticodon complementary to the codon triplet in mRNA. That is, if the tRNA anticodon is UCU, it recognizes AGA (or AGG) on the mRNA. (These are two of the codons for the amino acid arginine.) The activation reaction itself requires ATP, so there is an input of energy at this point. Such an input is of course expected for a synthetic process.

$$\text{Amino acid } + \text{ tRNA } + \text{ ATP } \xrightarrow[\text{synthetase}]{\text{aminoacyl tRNA}} \text{ aminoacyl-tRNA } + \text{ AMP } + \text{ PP}_i$$

There can be more than one tRNA for a given amino acid and similarly there can be more than one aminoacyl-tRNA synthetase, but they must be specific for each other. In other words, all the components of the reaction above, except ATP, specifically recognize each other. The relationships may be summed up as in Fig. 11.13.2.

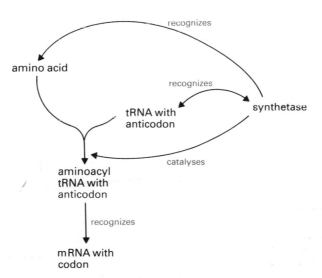

Fig. 11.13.2

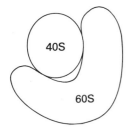

Fig. 11.13.3 Cross-section through a typical ribosome.

3 *The ribosomes* are composed of many discrete proteins as well as another type of RNA, ribosomal or rRNA, which is synthesized in the nucleolus. The functional ribosomes comprise two segments which sediment at different rates in the ultracentrifuge, so they differ in mass. As a reflection of their sedimentation behaviour they are termed the 60S and 40S components (S being the Svedberg constant). Prokaryotic ribosomes are somewhat smaller, the subunits being designated 70S and 30S. In cross-section the typical ribosome may look something like Fig. 11.13.3. The individual ribosomes are arranged in *polysomes*, or *polyribosomes*. These form a sort of train which passes over the length of the mRNA so that each individual successively encounters the codons. tRNA, associating with the ribosome as it encounters the codon, recognizes this codon and so a specific amino acid is made available for a peptide chain (Fig. 11.13.4).

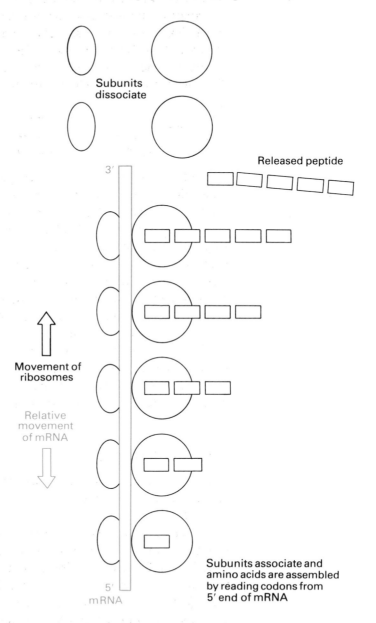

Fig. 11.13.4 Polyribosomes.
In this highly schematic view a peptide of five amino acids is shown being assembled as five ribosomes move down a length of mRNA from the 5′ end.

4 *Messenger RNA* or mRNA is a direct copy of the base sequence of the genomic DNA. The copying is specifically termed *transcription*, so this is the polymerization of ribonucleotides, with uracil lined up against chromosomal adenine, cytosine against guanine, guanine against cytosine, and adenine against thymine (just recall A = U or T, G = C). The DNA double helix

must unwind, but only one of the unwound strands, the *sense* strand, is transcribed (the other is termed the *antisense* strand). Polymerization occurs due to DNA-dependent RNA polymerase and results in large RNA sequences with many three-base *codons*, complementary to the sense strand anticodons. The first products of transcription are heterogeneous nuclear RNAs (hnRNAs), precursors of the functional mRNA, which in mammalian cells has about 1000 nucleotides. There is further modification (post-transcriptional processing) to give a 'cap' and, often, a polyadenylic acid 'tail'. Thus mRNA in eukaryotic cells can be represented

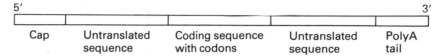

Cap	Untranslated sequence	Coding sequence with codons	Untranslated sequence	PolyA tail

The 'cap' consists of a methylated guanosine residue joined to the next nucleotide by a triphosphate linkage (in RNA the normal linkage is of course monophosphate). It is presently thought that the cap in some way controls or signals the binding of mRNA to the ribosomes. The two sequences which do not code for amino acid synthesis may also be signals of some sort. The polyA terminal is thought to confer stability in the chemical sense. (Prokaryotic mRNAs do not have these appendages.)

5 *ATP and GTP* hydrolysis provide the energy for the synthesis. ATP is involved at the stage of aminoacyl-tRNA formation; GTP is involved both in the initiation process and in peptide elongation.

6 *Signals and controls*, as known, are becoming numerous. It is clear that there should be a signal for the start of a peptide chain (an initiator) and one to signal the end of it (the terminator). Only in this way will a specific peptide with a definite number of amino acids be synthesized.

The stages of protein synthesis

If there are six recognizable participants or classes of participants in protein synthesis, there are three generally recognized stages in the synthetic process itself.

1 *Initiation* In some way the methyl guanosine cap of mRNA facilitates the recognition of mRNA by the ribosome, but it is the recognition of a codon by a methionyl tRNA which initiates protein synthesis. '*A* methionyl-tRNA' is written, rather than just 'methionyl-tRNA', since there is something of a complication here. There is a tRNA combining with methionine to signal initiation and another tRNA combining with methionine which is to be included somewhere along a polypeptide sequence. In bacteria, initiation requires a tRNA bearing a *formylated* methionine. The tRNA for initiation in eukaryotes is generally abbreviated to $tRNA_i^{met}$ (i for initiator) and in prokaryotes to $tRNA_f^{met}$ (although there are several other proposed designations). Additionally, the 'charged' aminoacyl-tRNAs for initiation are usually written, for eukaryotes and prokaryotes, as met $tRNA_i^{met}$ and f met $tRNA_f^{met}$ respectively. (These details are laid out systematically in Table 11.13.1.)

Table 11.13.1 Nomenclature of methionine-bearing tRNAs. Some alternatives found in the literature are given in green

	Prokaryotes	Eukaryotes
Initiator tRNA	$tRNA_f^{met}$ $tRNA_f$	$tRNA_i^{met}$ $tRNA_f^{met}$
tRNA for 'internal' methionine	$tRNA^{met}$ $tRNA_m$	$tRNA^{met}$ $tRNA_m^{met}$
Charged initiator tRNA	f met $tRNA_f^{met}$ f met $tRNA_f$	met $tRNA_i^{met}$ met $tRNA_f^{met}$
Charged tRNA for 'internal' methionine	met $tRNA^{met}$	met $tRNA^{met}$

The ribosomal subunit, met $tRNA_i^{met}$, and GTP form a complex, which is stabilized by one or more *initiation factors*. This complex binds to the untranslated sequence near the 5′ end of the mRNA. The 60S subunit binds to the 40S—met $tRNA_i^{met}$—GTP—mRNA complex so that the two ribosomal subunits form an integral structure. This intact ribosome bears two sites which have been called 'peptidyl' and 'aminoacyl' (sometimes 'P' and 'A'). (Fig. 11.13.5.) Elongation can now commence.

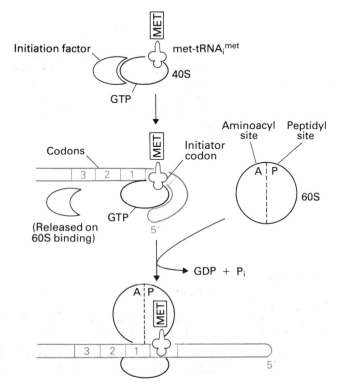

Fig. 11.13.5 Initiation of translation. The tRNA is represented as red clover leaf merely to make it stand out. It binds to the 40S ribosomal subunit, along with initiation factors and GTP. The 5′ end sequence of mRNA binds to the complex and the met $tRNA_i^{met}$ to the initiator codon. Association of the 60S subunit completes the functional ribosome. Thus the interaction of six components has to be described even in this simplified version of initiation

Incidentally methionine is an essential amino acid and a dietary deficiency of it has an exceptionally severe effect on protein synthesis, necessary as it is for the crucial initiation process.

2 *Elongation* The aminoacyl-tRNA specified by the next codon (called 1 in Fig. 11.13.6) binds to the aminoacyl site. A peptidyl transferase catalyses the formation of a peptide bond between the new amino acid and the methionine from the met tRNA$_i^{met}$ which is sitting at the peptidyl site. The free tRNA is released. Translocation takes place with the hydrolysis of GTP to provide energy so that the new dipeptidyl-tRNA occupies the now free peptidyl site, while mRNA moving relative to the ribosome brings up the next codon specifying the next aminoacyl-tRNA to be bound at the aminoacyl site. Thus the nascent peptide chain grows from the amino end.

3 *Termination* Eventually the termination codons of mRNA arrive at the aminoacyl site. In eukaryotic cells, and in many prokaryotes, these are UAA, UAG, or UGA. No tRNA is available to bind to these codons. Instead a protein release factor becomes associated with the ribosome in the presence of GTP and allows hydrolysis of the ester linkage between the peptide chain and the last tRNA. The completed polypeptide is released. If methionine is not to be the amino terminal residue of the completed, physiologically active polypeptide or protein (and usually it is not, in eukaryotes, although many bacterial proteins do have N-terminal methionine) it is removed at some stage during the post-translational modification (Fig. 11.13.7).

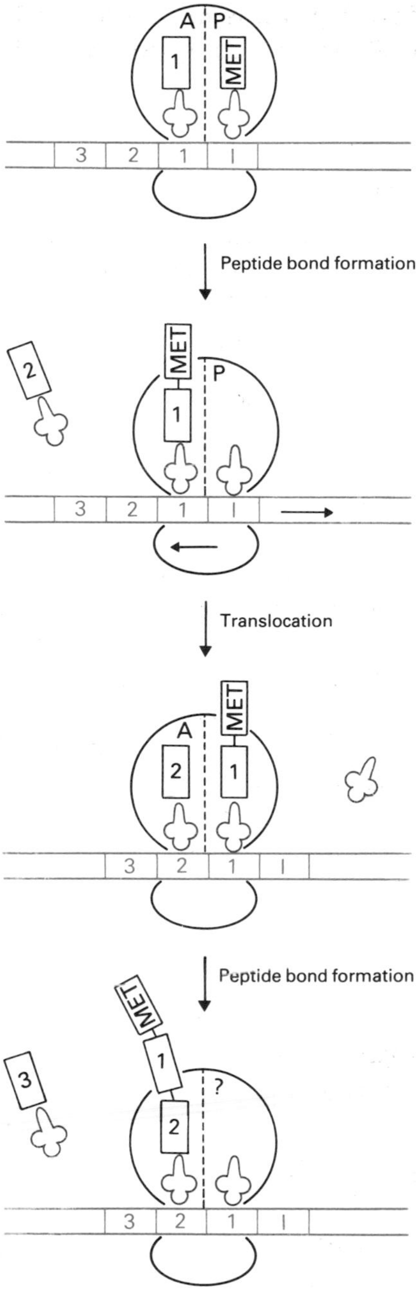

Fig. 11.13.7 The termination of protein synthesis.
It is imagined that a peptide of 200 amino acids is being synthesized. Arrival of the UAA codon signals termination, and with the aid of the termination factors the peptide is released from the last tRNA. The termination factors are not shown.

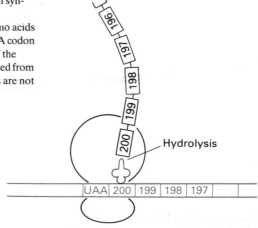

Fig. 11.13.6 Elongation phase of protein synthesis.
This can be represented quite economically in three steps. On the mRNA 'I' represents the initiator codon, and 1,2,3.... the codons for amino acids 1,2,3.... The GTP necessary for translocation is not shown, nor are the elongation factors.

Antibiotics

An antibiotic is not precisely definable but is usually thought of as a product of microbial metabolism which inhibits growth of other, pathogenic organisms, without substantially affecting any host of the latter. Thus they represent a class of metabolites having a structure, or a structural feature, similar

155

to that of a natural metabolite and inhibiting, usually in a competitive manner, enzymes for which the natural metabolite is a substrate. Antibiotics which affect most phases of the cell cycle have been identified, but some are of special interest since studies of their modes of action helped to elucidate the details of protein synthesis. Under headings indicating area of action, some important ones are as follows.

1 *Inhibition of transcription* Actinomycin D is a complex peptide incorporating a ring structure which fits ('intercalates') between G and C base pairs in DNA. Thus it blocks synthesis of all types of RNA. It is of low toxicity to microorganisms and is mainly employed in human medicine as an anti-tumour agent.

2 *Inhibition of initiation* Streptomycin binds to the 30S ribosomal subunit in microorganisms, interfering with the formation of a normal initiation complex in such a way as to disrupt codon-anticodon interactions.

3 *Inhibition of translation* Puromycin is a nucleotide analagous to aminoacyl tRNA and so it competes with this in the transpeptidation process; it combines with the growing peptide but it lacks a free carboxyl group and so translation is aborted. It is too toxic to use in humans, and is primarily a research tool.

Tetracyclines block the binding of aminoacyl-tRNA to the bacterial ribosome; they do not readily penetrate into mammalian cells.

Mutations

Mutations are alterations to chromosomal DNA. They affect subsequent generations if they occur in germ cells, whereas in dividing somatic cells the relevant mature tissues, if they survive, are affected. The biochemical concept of mutations relates entirely to the triplet structure of the genetic code, and the linear manner in which it is read. The two most striking modes of mutation are 'point' and 'frame shift'.

1 *Point mutation* These result from the replacement of one base by another. Thus in the transcribing sequence TAT...ATA...TAT...ATA... which specifies isoleucine...tyrosine...isoleucine...tyrosine... if the first base becomes C instead of T, as CAT...ATA...TAT...ATA... then the peptide synthesized would be methionine...tyrosine...isoleucine...tyrosine... .

A point mutation is often less harmful than might be supposed: (a) a similar amino acid (say leucine for isoleucine) may be the one substituted, little affecting the nature of a large protein molecule; and (b) the degeneracy of the genetic code means that certain substitutions merely result in another codon for the same amino acid. Thus, in the hypothetical sequence above, if the first base becomes A rather than T, isoleucine will still be the first amino acid (DNA codons for isoleucine are TAT, AAT, and GAT).

2 *Frame shift mutations* In this type deletion of a base anywhere along the sequence means that it is read out of register. If in TAT...ATA...TAT...ATA the initial base is deleted rather than substituted, the sequence becomes

ATA...TAT...ATA...TAX where X is whatever base came next in the original sequence, and the resulting peptide would be tyrosine...isoleucine...tyrosine...?... .

Frame-shift mutations may or may not be lethal, depending of course on where they occur. A well-characterized and not too serious result, in humans, of frame-shift mutation is the abnormal haemoglobin HbWAYNE; this is a consequence of deletion of a base in DNA such that the terminator codon is misread and a short additional amino acid sequence is added to the α-chain.

Summary

It will be obvious, from a glance through the major biochemistry textbooks, that they make quite a hoo-ha about protein synthesis. However, if all speculative aspects are avoided, it is quite possible to present the topic in a concise manner; three neat diagrams, about initiation, elongation, and termination, are quite enough to answer any general examination question. Alternatively, if one wanted to describe protein synthesis in a few words it would probably go like this: mRNA and the assembled ribosome come together in such a way that the mRNA, at its 5' end, presents the initiator codon. This is recognized by met $tRNA_i^{met}$ which attaches at the peptidyl site of the ribosome. A tRNA bearing a specific amino acid arrives at the aminoacyl site, recognizing the next codon for that amino acid, and methionine is transferred to the aminoacyl site to yield a dipeptidyl-tRNA. The dipeptidyl-tRNA translocates to the peptidyl site, allowing a new tRNA to recognize the next codon in line as the mRNA also shifts. The dipeptide is transferred to form a tripeptide with the incoming amino acid, and so the protein grows from the amino terminal.

Inevitably these topics have thrown up many neologisms, some of which are collected in the glossary below, which may also be used in conjunction with 13.22.

Glossary of terms used in connection with protein synthesis and biochemical genetics (some are not mentioned in the text)

Aminoacyl site Site on a ribosome, binding incoming aminoacyl-tRNA.

Anticodon A *triplet* of bases complementary to a *codon* triplet.

Base pairing Association, by hydrogen bonding, of adenine with thymine or uracil, and guanine with cytosine.

Cap (catabolite activator protein) A protein which binds to cAMP in bacteria and, in so doing, stimulates *transcription*.

cDNA *Complementary* DNA.

Circular DNA DNA covalently self-attached to give a closed loop.

Cistron Synonymous with *gene*, that is, a set of DNA *codons* defining a functional polypeptide chain.

Clone A group of cells all derived from one cell.

Complementariness Relationship of one nucleic acid strand to another (DNA-DNA, RNA-DNA, RNA-RNA) such that there is *base pairing*.

Codon A set of three adjacent nucleotides in mRNA, functioning as a signal for a specific amino acid in protein synthesis.

Deletion A type of *mutation* via elimination of one or more *base pairs* in a *replicating* nucleic acid.

DNA cloning Production of identical DNA from *cloned* cells.

DNA ligase An enzyme catalysing closure of one DNA-chain to yield *circular* DNA, or of several chains to yield an extended length of DNA.

Duplex A segment of DNA or RNA in which there is *base pairing*.

3′-end In DNA and RNA, the end which has a free 3′-hydroxyl on deoxyribose or ribose.

5′-end In DNA and RNA, the end which has a free 5′-hydroxyl on deoxyribose or ribose.

Exon In distinction to *intron*, a DNA sequence represented in mature RNA.

Elongation factors Proteins necessary for the phase of protein synthesis characterized by growth of the polypeptide chain on the *ribosome*.

Frameshift Addition or deletion in DNA of 1, 2, 4, 5, or 7 bases (that is, other than multiples of 3) so that a new set of *codons* is transcribed to mRNA.

Gene In molecular biology, a sequence of *codons* adequate to define a functional polypeptide chain, enzyme or protein, or a subunit of one of these.

Genome The full set of *genes* in a eukaryotic cell.

hnRNA Heterogeneous RNA, formed in the nuclei and probably a precursor of mRNA.

Initiation factors Proteins necessary to bind initiator tRNA for commencement of protein synthesis.

Insertion Mutation caused by addition of one or more base pairs to DNA.

Intron In contrast to *exon*, a DNA sequence in eukaryotic cells, which is transcribed but then removed during the processing of RNA.

Lagging strand The DNA strand which is synthesized from *Okazaki fragments*.

Leading strand The DNA strand which is synthesized continuously during its replication.

Missense strand The DNA strand against which mRNA is not synthesized during *transcription*.

mRNA Mature RNA which constitutes the template for *translation*.

Mutation In molecular biology, any alteration to the nature or sequence of the paired bases in DNA.

Nick An interruption in one strand of a DNA *duplex*, by the action of *endonuclease*.

Okazaki fragments Newly-synthesized fragments of DNA, joined up to give the *lagging strand*.

Overlapping genes In viruses, the *transcription* of the same nucleotide sequences within different *reading frames*.

Peptidyl site Site on the ribosome to which peptidyl-tRNA is *translocated*.

Plasmid A double-stranded DNA molecule, in both eukaryotic and prokaryotic cells, capable of being inherited by progeny cells without (in the former case) being integrated into the chromosomes.

Polysome (polyribosome) A group of ribosomes connected to an RNA strand, each one separately synthesizing a polypeptide chain.

Post-translational modification A modification to a protein subsequent to the linkage of its amino acids in a chain.

Primary transcripts The RNA molecules newly synthesized by RNA polymerases, to be subjected in eukaryotes to RNA *processing*.

Promoter A site on DNA at which *transcription* is initiated.

Reading frame The length of DNA within which *transcription* occurs.

Recognize React specifically with.

Recombination (of DNA) Formation of a novel DNA molecule by breakage of existing DNA and rejoining.

Replication fork In an unwinding DNA *duplex*, the point where the DNA strands diverge.

Restriction endonuclease An enzyme hydrolysing DNA on recognition of specific nucleotide sequences therein.

Reverse transcriptase RNA-directed DNA polymerase.

RNA processing Cleavage, rejoining and modification of primary RNA transcripts (*hnRNA*).

rRNA Ribosomal RNA, apparently not *translated*.

Semi-conservative replication Replication of double-stranded DNA such that they separate and allow synthesis of a new strand *complementary* to each.

Sense strand The strand of the DNA *duplex* against which mRNA is synthesized.

40S subunit Smaller subunit of the eukaryotic ribosome.

60S subunit Larger subunit of the eukaryotic ribosome.

Template The pattern of bases, in sequence, which serve to express information in other sequences, for example proteins.

Terminal transferase An enzyme catalysing the addition of nucleotides to the *3'-end* of DNA.

Termination signals Nucleotide triplets which specify the termination of a growing peptide chain rather than the addition of further amino acids.

Transcription Process by which one strand of *duplex* DNA serves as a template for mRNA synthesis.

Translation Process by which sequence of *codons* in mRNA is translated to an amino acid sequence in a polypeptide.

Translocation The movement of peptidyl-tRNA from the *aminoacyl site* to the *peptidyl site*.

Triplet A group of three continuous bases.

tRNA An RNA containing an *anticodon* (mRNA recognition site) and an amino acid attachment site.

Unwinding proteins Proteins which promote disruption of the DNA double helix prior to its *replication*.

Vector Any system (for example, *plasmids* and viruses) by which DNA can be incorporated into a host cell and autonomously replicated.

Virion A virus particle as it exists outside the host cell.

'Wobble' Reference to the steric freedom in the pairing of the third base in a codon to tRNA. Thus tRNA for phenylalanine has anticodon AAG and this recognizes either UUU or UUC.

11.14 One-carbon units in biosynthesis

It should be no surprise that one-carbon units also serve as building blocks in biosyntheses. In order of increasing state of oxidation, we may list single-carbon compounds as:

CH_4	CH_3OH	HCHO	HCOOH	CO_2
Methane	Methanol	Formaldehyde	Formic acid	Carbon dioxide

Except possibly for CO_2 these substances are toxic to mammalian systems and mechanisms have been evolved for their assimilation into metabolism as radicals, that is

$—CH_3$	$—CH_2OH$	$—CHO$	$—COOH$
Methyl	Hydroxymethyl	Formyl	Carboxyl

Carbon dioxide is formed in body fluids combined with water (as carbonic acid and biocarbonate) and its assimilation involves the coenzyme biotin. This is treated as a separate subject (11.15). For the other one-carbon units, there is a special set of coenzymes to ensure transfer from donor compounds to biosynthetic pathways (Fig. 11.14.1).

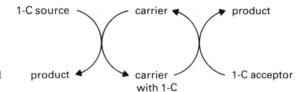

Fig. 11.14.1

The coenzymes are as follows.

1 *Tetrahydrofolate* (FH_4) is derived by reduction of the B-vitamin folic acid. In folic acid itself the pteridine ring is fully aromatic. Substitution occurs at N5 or N10 or both (with cyclization). The forms of FH_4 bearing one-carbon units are shown in Fig. 11.14.2 in order of increasing oxidation state. The forms N^{10}-formyl, $N^{5,10}$-methenyl and N^5-formimido are of the same oxidation state. They are formally interconvertible at the same oxidation level.

Fig. 11.14.2 Aspects of tetrahydrofolic acid (FH_4).

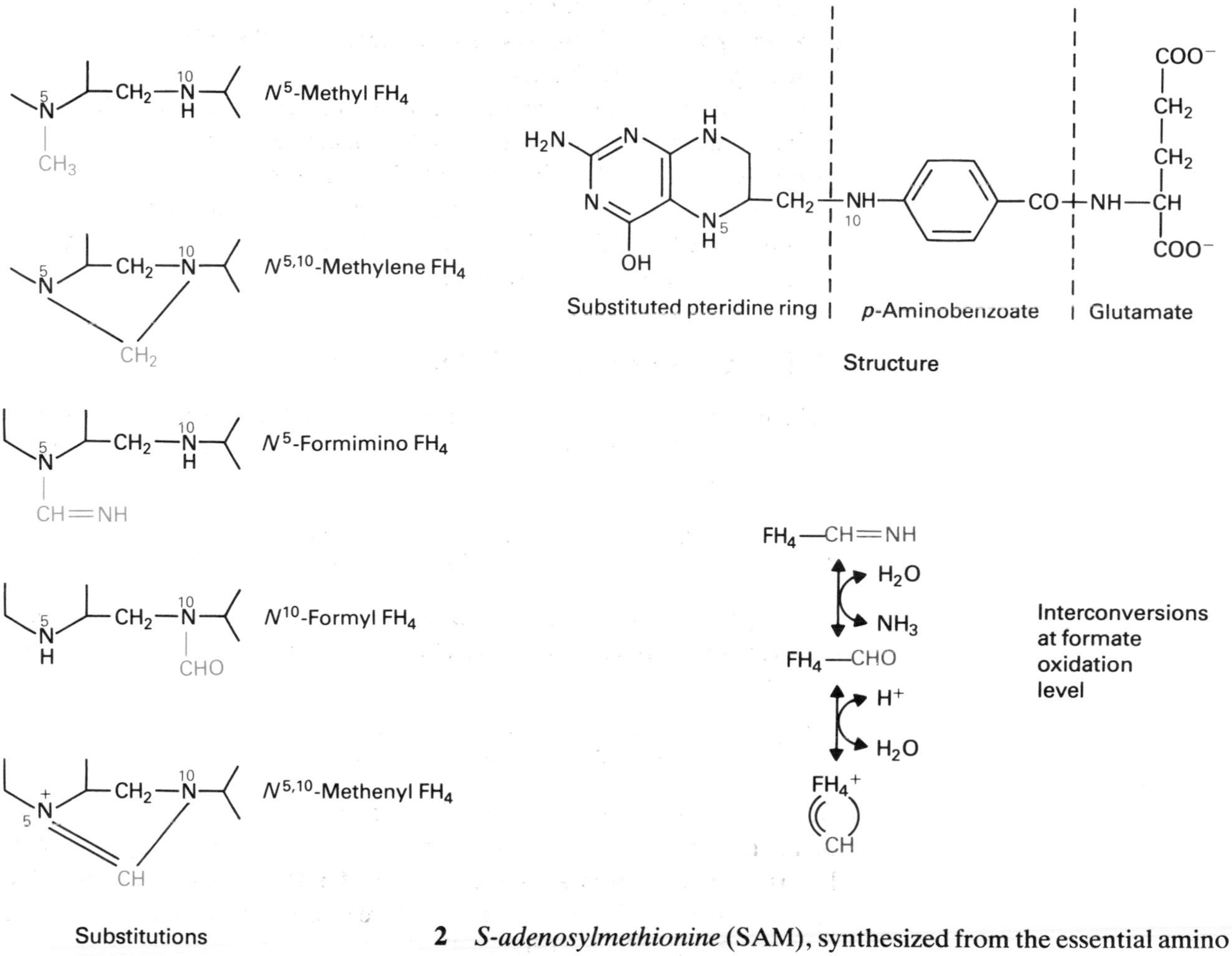

N^5-Methyl FH_4

$N^{5,10}$-Methylene FH_4

N^5-Formimino FH_4

N^{10}-Formyl FH_4

$N^{5,10}$-Methenyl FH_4

Substitutions

Substituted pteridine ring | *p*-Aminobenzoate | Glutamate

Structure

$$FH_4 \text{—} CH \text{=} NH$$
H_2O
NH_3
$$FH_4 \text{—} CHO$$
H^+
H_2O
$$FH_4^+$$
CH

Interconversions at formate oxidation level

2 *S-adenosylmethionine* (SAM), synthesized from the essential amino acid methionine, and ATP (Fig. 11.14.3). This is evidently a potential donor of methyl groups, like N^5-methyl FH_4.

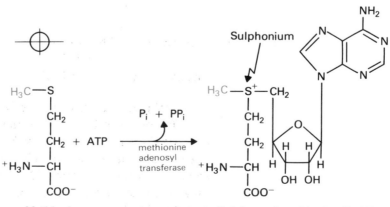

Fig. 11.14.3 Methionine   S-Adenosyl methionine (SAM)

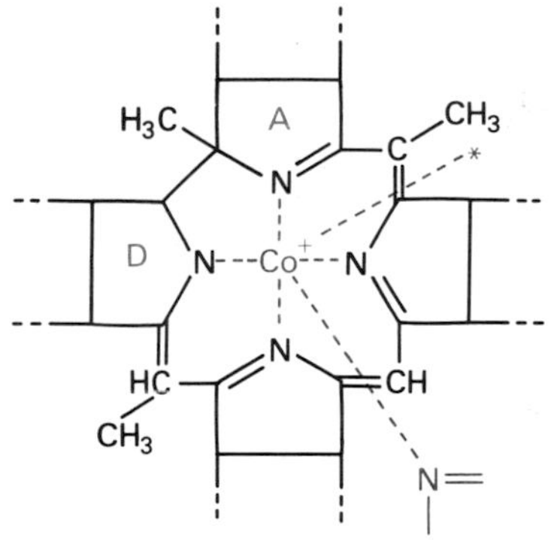

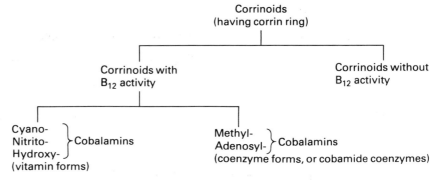

H₃C, A, CH₃, C, *, N, Co⁺, N, D, N, N, HC, CH, CH₃, CH, N, are labels within the corrin ring structure.

* ligand (OH⁻, CN⁻, H₂O in vitamins; methyl, adenosyl in coenzymes)

Fig. 11.14.4 Corrin ring.

Fig. 11.14.5 Nomenclature of the B₁₂ vitamins. One might infer that the corrinoids with vitamin activity are converted, in the body, to the methyl- and adenosyl cobalamin coenzyme forms.

3 *Cobamide coenzymes* These are derived from vitamin B_{12}. Vitamins B_{12} contain a corrin ring, a representation of which is (in black) in Fig. 11.14.4. Obviously it is similar to a reduced porphyrin ring, but there is no methenyl bridge between rings A and D. In the B_{12} vitamins, cobalt is chelated in a manner similar to iron in haemoglobin, and in addition binds cyanide, nitrite, water, or hydroxyl. These are named cyanocobalamin, nitritocobalamin, aquacobalamin, and hydroxycobalamin respectively. Thus the B_{12} vitamins are a series of cobalamins. The coenzymes from these vitamins are methylcobalamin and 5'-deoxyadenosyl cobalamin. The coenzyme forms are often generically described as cobamides. In Fig. 11.14.5 a guide to this confusing nomenclature is provided.

Corrinoids (having corrin ring)

Corrinoids with B_{12} activity

Corrinoids without B_{12} activity

Cyano-
Nitrito-
Hydroxy-
(vitamin forms)
} Cobalamins

Methyl-
Adenosyl-
} Cobalamins
(coenzyme forms, or cobamide coenzymes)

The input of one-carbon units is principally derived from the following compounds (Fig. 11.14.6):
1 glycine
2 serine
3 histidine
4 choline
5 methionine.

Compounds finally synthesized by acceptor systems (Fig. 11.14.7) are:
1 creatine
2 adrenaline
3 choline
4 thymine
5 serine
6 methionine
7 purine bases positions 2 and 8.

This affords us the summary in Fig 11.14.8. If you can memorize this you could for most practical (examination) purposes dispense with the foregoing fairly tedious list of transformations. Some compounds appear both as donors and acceptors. This is of course quite reasonable, since if substances donating 1-C units such as serine and choline are not plentiful (perhaps lacking in the diet), they must be synthesized by the use of 1-C units from other donors.

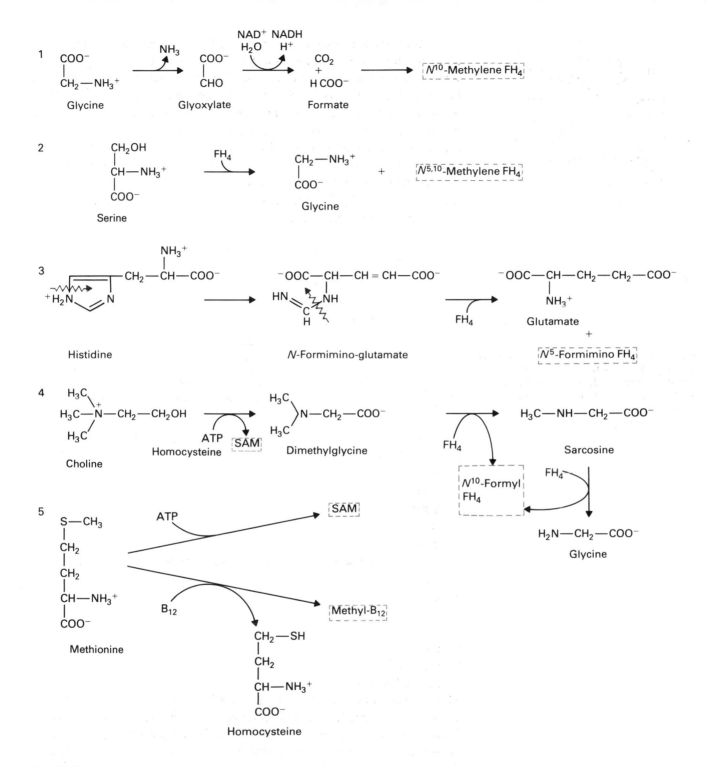

Fig. 11.14.6 Sources of 1-C units.
The arrows obviously represent multiple
reactions (derived coenzymes are boxed).

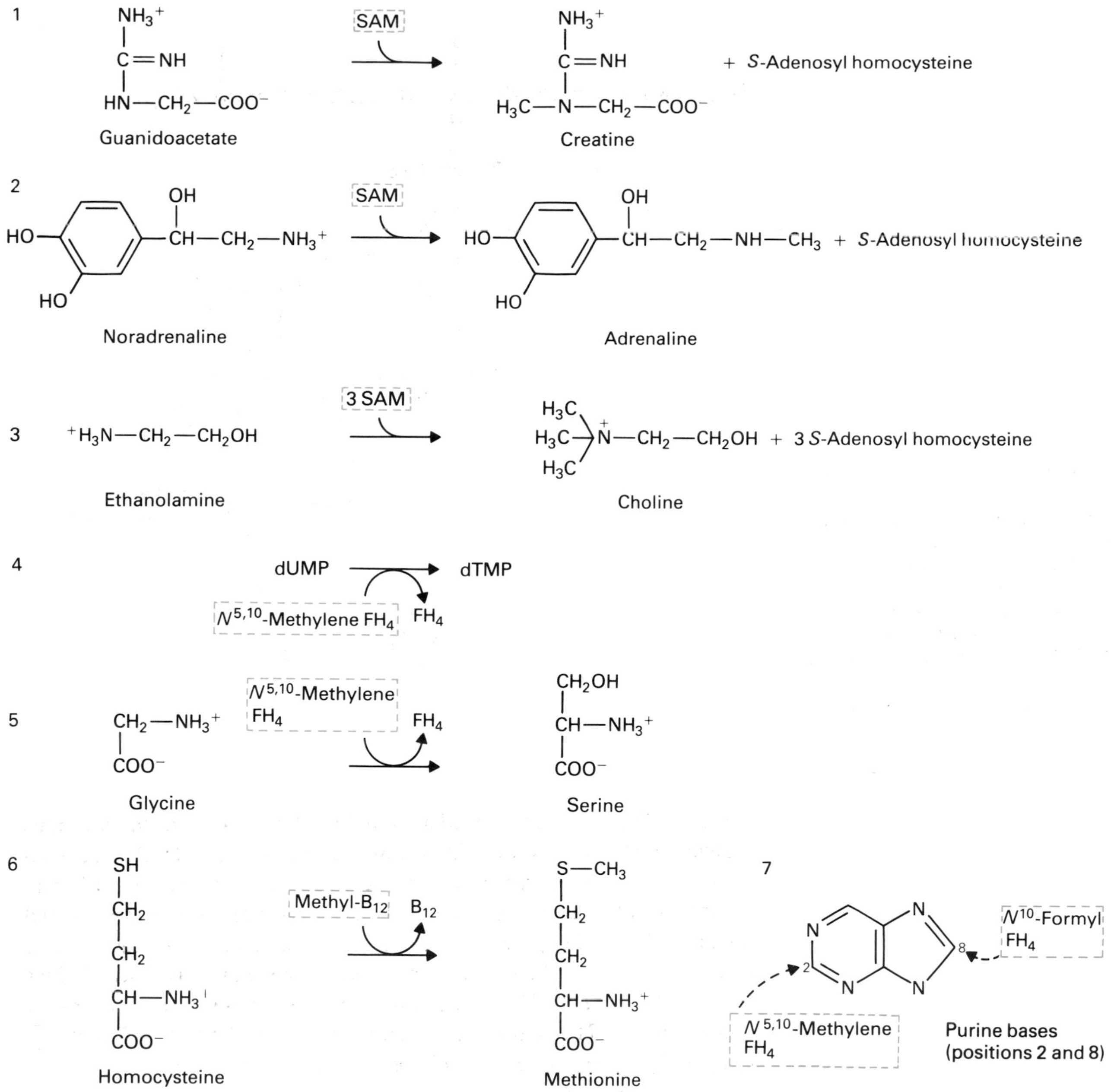

1 Guanidoacetate $\xrightarrow{\text{SAM}}$ Creatine $+$ S-Adenosyl homocysteine

2 Noradrenaline $\xrightarrow{\text{SAM}}$ Adrenaline $+$ S-Adenosyl homocysteine

3 Ethanolamine $\xrightarrow{\text{3 SAM}}$ Choline $+$ 3 S-Adenosyl homocysteine

4 dUMP $\rightarrow$ dTMP ; $N^{5,10}$-Methylene FH$_4$ $\rightarrow$ FH$_4$

5 Glycine $\xrightarrow[\text{FH}_4]{N^{5,10}\text{-Methylene FH}_4}$ Serine

6 Homocysteine $\xrightarrow[\text{B}_{12}]{\text{Methyl-B}_{12}}$ Methionine

7 Purine bases (positions 2 and 8); $N^{5,10}$-Methylene FH$_4$ → position 2; N^{10}-Formyl FH$_4$ → position 8

Fig. 11.14.7

164

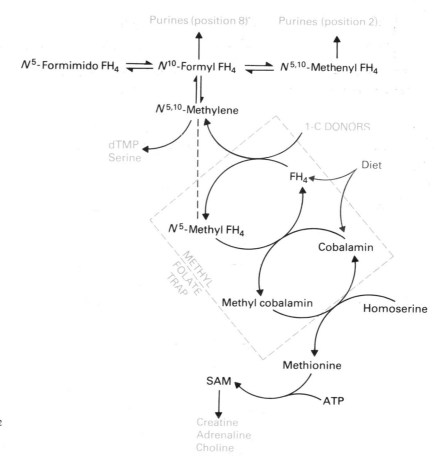

N^5-Formimido FH$_4$ ⇌ N^{10}-Formyl FH$_4$ ⇌ $N^{5,10}$-Methenyl FH$_4$

Purines (position 8)'   Purines (position 2)

$N^{5,10}$-Methylene

dTMP Serine

1-C DONORS

Diet

FH$_4$

N^5-Methyl FH$_4$

Cobalamin

METHYL FOLATE TRAP

Methyl cobalamin

Homoserine

Methionine

SAM

ATP

Creatine Adrenaline Choline

Fig. 11.14.8 Interaction between one-carbon-bearing coenzymes.

The green-boxed area represents the methyl-folate trap; if B$_{12}$ is deficient in the diet then N^5-methyl FH$_4$ cannot be regenerated to FH$_4$ in order that the other important coenzyme forms of FH$_4$ may be synthesized. Clearly the difficulty would be obviated by oxidation of N^5-methyl FH$_4$ to N^5-methylene FH$_4$ (dotted line) but if this reaction occurs, then it is too slow to avoid the consequences of long term B$_{12}$ deficiency.

Methyl-folate trap The boxed area in Fig. 11.14.8 contains the reactions involved in what has become known as the 'methyl-folate trap'. This illustrates the role of the cobamide coenzyme, methyl cobalamin, in the scheme. It is necessary to ensure the flow of methyl groups from donors to SAM and thence synthesis of important products such as adrenaline. If vitamin B$_{12}$ is deficient in the diet then formation of methyl cobalamin is blocked. This means that N^5-methyl FH$_4$ accumulates and there is a shortage of FH$_4$ for other reactions. The implication is that the enzyme for conversion of N^5-methyl FH$_4$ to $N^{5,10}$-methylene FH$_4$ is not very active.

Folate antagonists A discussion of one-carbon metabolism might be sensibly rounded off by a note on the folate antagonists. Thus aminopterin (4-aminofolic acid) and methotrexate (4-amino-10-methyl folic acid) (Fig. 11.14.9) interfere, as competitive inhibitors, with the enzymatic reduction of folic acid to FH$_4$. Since FH$_4$ is necessary as a carrier of one-carbon units, for the synthesis of thymine and the purine bases, a reduction in its optimal concentration might lead to reduced DNA synthesis, especially in rapidly dividing cells. Thus these compounds have been used in the therapy of leukaemias and choriocarcinomas. As might be imagined, the problem is

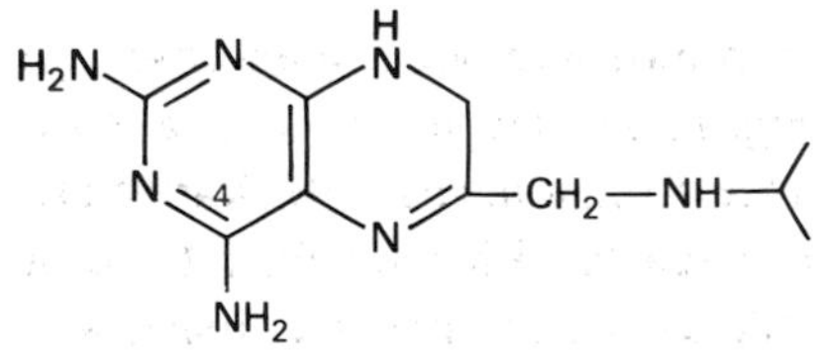

Aminopterin

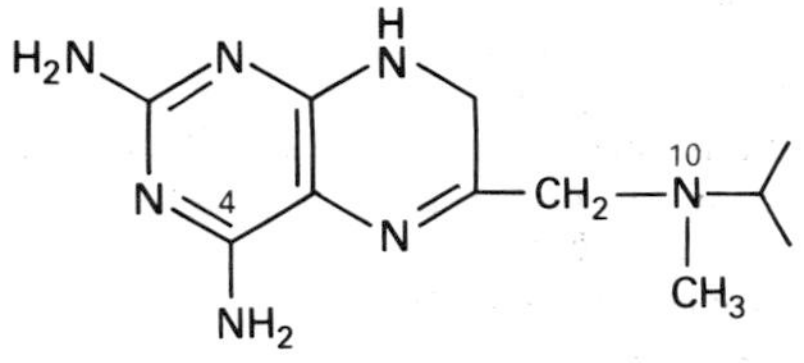

Methotrexate

Fig. 11.14.9 Folate antagonists.
Compare to tetrahydrofolic acid in Fig. 11.4.2.

that there is no specificity for cancer cells; all rapidly dividing cells are susceptible.

Postscript In spite of its importance, one-carbon metabolism does not seem to be a very popular subject for examinations. Is it possible that examiners themselves are uneasy about the range of oxidation states and the variety of carriers but, in view of its relationship to nucleic acid and amino acid metabolism, it cannot be discounted.

11.15   Biotin in carboxylation reactions

Experimental animals fed large amounts of raw egg-white develop a paralysis and dermatitis, called 'egg-white injury'. This is due to the protein avidin in egg-white; it can tightly bind the B-vitamin biotin, rendering it non-absorbable by the gut. Egg-white injury is thus a biotin deficiency syndrome. The exterior manifestations of the syndrome must have some relationship to the biological role of biotin, which is to act as coenzyme for enzymes which 'fix' carbon dioxide. The connection has not been discovered, however. (This is a generalization for the B-vitamins. They have well-established coenzyme roles; if lacking in the diet, cutaneous and neurological symptoms develop.

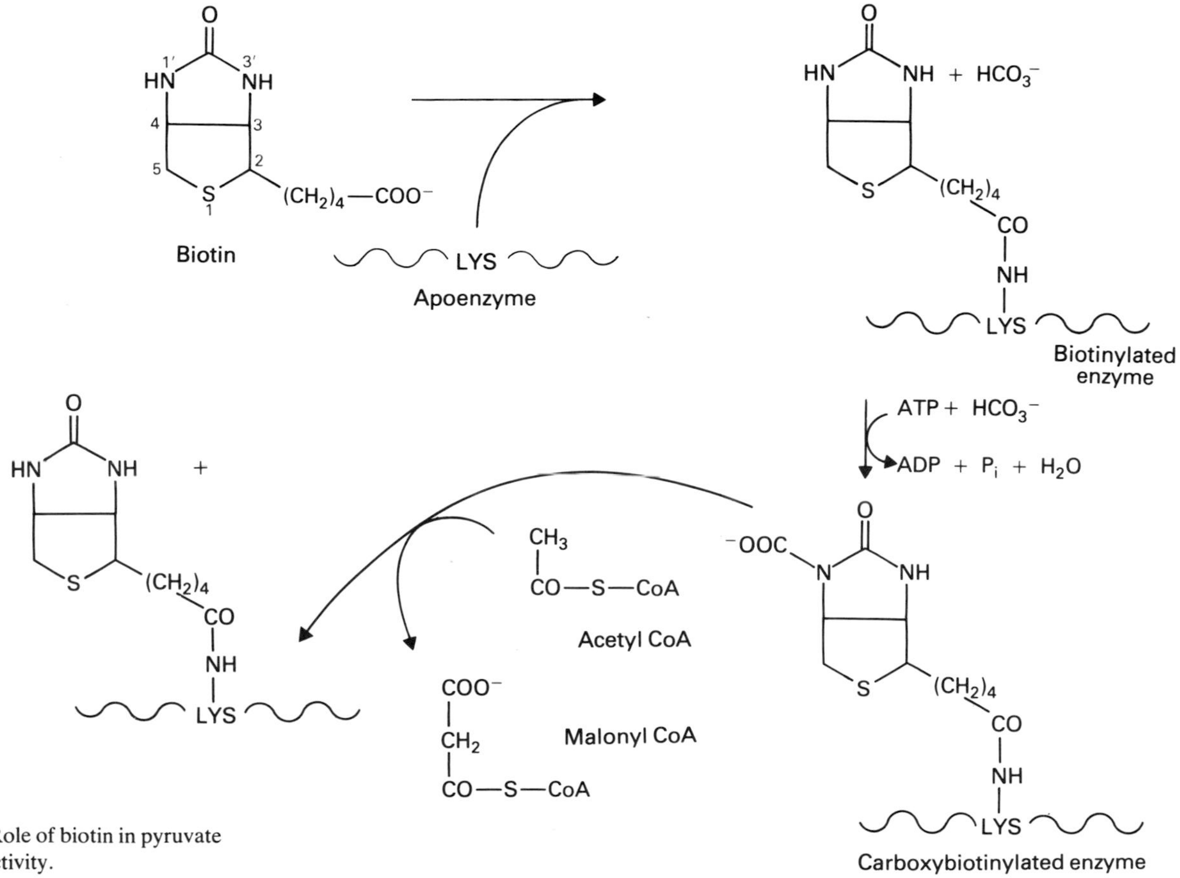

Fig. 11.15.1 Role of biotin in pyruvate carboxylase activity.

The connections between coenzyme function and symptoms remain obscure.)

Biotin has a peculiar ring structure (Fig. 11.15.1) which can attach to the lysine side-chains of enzyme proteins. In this form it reacts with bicarbonate to give (bound) carboxybiotin. The latter reacts with the substrate to be carboxylated, for example with acetyl CoA. This is an important reaction in fatty acid synthesis. Other important carboxylation reactions are shown in Table 11.15.1.

Table 11.15.1 Carboxylation reactions involving biotin

Substrate	Product	Significance
Acetyl CoA	Malonyl CoA	Rate-limiting step for palmitate synthesis
Propionyl CoA	Methylmalonyl CoA	Reaction in threonine and isoleucine catabolism
Pyruvate	Oxaloacetate	Key gluconeogenesis reaction

Note that there are two important CO_2 fixing enzymes which do *not* have biotin as a coenzyme. These are phosphoenolpyruvate carboxykinase (p. 114) and carbamyl phosphate synthetase (p. 96).

12   Regulation of Metabolism

12.0   Preamble

The survival of a cell or organism depends upon its ability to alter metabolic flux in response to changing conditions, although the result may be a maintenance of the *status quo* (called homeostasis in whole animals). Examples are: the switch from glucose to fatty acid metabolism in the starving mammalian liver; the induction of enzymes to metabolize lactose in *E. coli* when that sugar is included in the medium; and the inhibition of cholesterol synthesis in fibroblasts when they are presented with cholesterol in low density lipoproteins. Chapters in books and examination questions often approach the phenomena of regulation in general terms, and are adorned with words such as 'regulation', 'control', 'modulation', and so on. The precise definition of these words and their implications for teleology* need not concern us at this level since we are again aiming at simplification for ease of regurgitation. As ever, the approach is to identify the essentials of the topic (in this case, regulatory factors) to try to classify them and to discuss them briefly. Then, it is hoped, we sum up overall with a clear memorizable diagram. But two preliminary points can be made.

1   It is at present difficult to classify regulatory factors, since primary effects cannot always be distinguished from secondary ones. Thus cAMP and glucagon both cause lipid breakdown in liver, but glucagon's role is as an effector for cAMP synthesis. In this case the relationship is known; there are many other cases where the hierarchy of effectors remains obscure.

2   It is the catalytic activity of enzymes which is the basis of metabolism, so that regulation of metabolism ought to be explicable, ultimately, in terms of rates of transformation due to enzyme catalysis.

Following up the latter point, the rate of flow of atoms through an enzyme-catalysed reaction is dependent upon:

1 the amount of substrate or product present;
2 the amount of enzyme available;
3 the catalytic efficiency of the enzyme.

All dietary, hypothalamic, endocrine, and nervous controls should be reducible, perhaps through many levels of effector, to 1, 2, and 3, either singly or in combination. We set down what we know about each.

---

*Teleology—'the suggestion that something is shaped by a purpose or directed towards an end. . . .'

12.1 Regulation of the rates of enzymic reactions

1 Availability of substrate and product

In the glucokinase reaction in the liver

Glucose + ATP $\rightarrow$ glucose 6-phosphate + ADP

the nature of the enzyme is such (high K_m for glucose) that at low glucose concentrations the reaction cannot proceed. Thus glycolysis is slowed down because there is just not enough glucose available. Although this is a coarse (= insensitive) control, it is fast because in itself it involves no synthesis or transport mechanisms.

If a substrate is confined by a membrane then it is not accessible to catalytic activity by an enzyme on the other side of the membrane. Thus, acetyl CoA cannot cross the mitochondrial membrane, so that when it is generated in the mitochondria it cannot leak out to be converted to fat by the action of acetyl CoA carboxylase. When energy is required, presumably the primary consideration, it is oxidized within the mitochondria. Only where ATP is plentiful is it transported outside as citrate; limitation of the rate of lipogenesis, then, is due to compartmentation of acetyl CoA.

2 Enzyme availability

When an enzyme is synthesized in response to chemical stimulus the phenomenon is known as *induction*. When normally-proceeding enzyme synthesis is slowed, there is *repression*. Induction is normally associated with control of catabolism, repression with control of anabolism. An example of induction is the synthesis of penicillinase by *Penicillium notatum* in response to appearance of penicillin in growth medium. An example of repression is when *Salmonella typhimurium*, synthesizing leucine because it is lacking in the growth medium, ceases to synthesize appropriate enzymes when sufficient leucine has accumulated. Inducible enzymes in mammalian tissues include lactase (synthesized by adult intestine in response to a milk diet) and liver arginase (synthesized in response to a high protein diet; it is necessary for urea production). Other terms used in this context are as follows.
Constitutive enzyme: an enzyme present in a cell in fairly constant amounts, that is non-inducible.
Coordinate induction: a single inducer may bring about synthesis of a number of inducible enzymes. The inducer is normally the first substrate in the pathway. Complimentary to this is *coordinate repression*.

Derepression is the cancelling of repression, so that an induction is apparently observed—a negative cancelling a negative appears as a 'positive'.

Induction and repression are thought to be the result of modification of gene expression. The smallest unit of genetic expression is the *cistron*, a structural gene coding for the synthesis of an enzyme subunit. The cistrons are functionally combined in a unit called the *operon*, which may be described as a regulated gene cluster.

There is no escaping from the *lac* operon in this context, observation of which first suggested the concept. *Escherichia coli* grown on a lactose-free medium has very low activities of lactase (β-galactosidase), but the enzyme appears (is induced) on introduction of lactose. Not only that, but a protein termed a permease is detectable. The permease is a carrier to effect penetration of the inducer lactose into the cells. Genetic mapping techniques indicated that the operon or gene cluster for this coordinated protein synthesis consists of: (a) structural genes for the permease and lactose proteins; (b) an operator gene; and (c) a regulator gene (Fig. 12.1.1). The regulator gene transcribes RNA for repressor at a constant rate, so the repressor is constitutive. The repressor binds with high affinity to the operator gene, which prevents the initiation of transcription at the structural genes. When allolactose (the physiological inducer, with glucose and galactose 1,6 linked rather than 1,4 as in lactose) is present, it binds to the repressor, detaching it from the operator gene. This allows initiation of transcription on the structural genes. Lactase permease then allows utilization of lactose in the medium.

In eukaryotes, genes coding for metabolically-related sequences of enzymes are not in general closely linked to form an operon. Protein synthesis is often associated with endocrine stimulation instead of the direct presence

Fig. 12.1.1 General description of the *lac* operon.

It is descriptive of the genetic material necessary for synthesis of the enzymes to assimilate lactose. In the upper part of the figure the situation in the absence of lactose is depicted. The operator gene is blocked by repressor. In the lower part the repressor is bound by the inducer and the information for enzyme synthesis is expressed. In reality it is an analogue of lactose which is the inducer and more than the two enzymes shown are synthesized. The analogue (allolactose) is synthesized by small amounts of pre-existing enzymes in the bacterium.

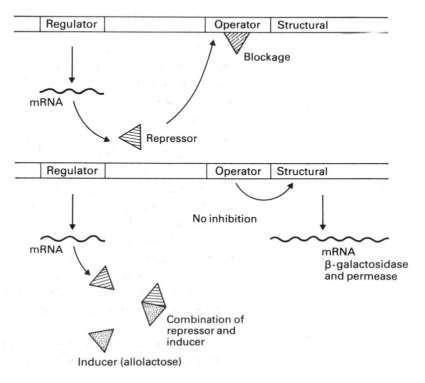

171

of the metabolizable inducer. Hormones which alter the rate of protein synthesis include insulin, thyroxine, somatotrophin, testosterone, and cortisol.

3 Alteration of catalytic efficiency

(i) *By allosteric modification*: the allosteric effector modulates catalytic activity by binding to a site other than the catalytic site. It can thus increase or decrease activity through some conformational change induced in the enzyme. The phenomenon is described by the sigmoid reaction rate–substrate concentration curve (Fig. 5.6.5), the implications of which are critical. Small concentrations of an effector are without significant influence. At intermediate concentrations a small change in effector concentration produces a proportionately large effect upon the enzyme. There is then a saturation region. The control effect resides in the large change produced by a small build-up of effector. An example is the allosteric inhibition of phosphofructokinase by ATP and citrate (see below).

(ii) *Covalent modification*: some enzymes such as glycogen synthetase, phosphorylase, pyruvate dehydrogenase, and hormone-sensitive lipase possess one or more seryl or tyrosyl residues which can be phosphorylated at the expense of ATP. The phosphorylated form has a different catalytic activity from the dephosphorylated form. The phosphorylated form is reconvertible to the dephosphorylated form by the action of a phosphatase. Table 12.1.1 summarizes some of the knowledge about covalent modification of enzyme activity.

Table 12.1.1 Enzymes whose catalytic activity is modified by phosphorylation

Pathway	Enzyme	Effect of phosphorylation
Glycogenolysis	Glycogen phosphorylase	Activation
Lipid mobilization	Lipase	Activation
Glycogenesis	Glycogen synthetase	Inactivation
Carbohydrate catabolism	Pyruvate dehydrogenase	Inactivation
Palmitate synthesis	Acetyl CoA carboxylase	Inactivation
Cholesterol synthesis	β-Hydroxy-β-methyl glutaryl CoA reductase	Inactivation

12.2 Interaction of enzyme systems in control

This title has the merit of being self-explanatory. There are perhaps four mechanisms of interaction, to which we might add 'the use of messengers' which is taken as a separate topic (12.4).

1 *Feedback loops* are a means by which inductions, repressions or allosteric modifications are manifested. Feedback control in itself is not a mechanism of enzyme regulation—it is applicable, as a description, to any system whereby a component of the output affects the rate of the output.

Feedback circuits in electronics are of course very common, being used in automatic control systems. In cells the feedback loop allows one enzyme or enzymes, producing a product, to interact with another enzyme, earlier in the synthetic sequence. The classical example (Fig. 11.11.1) is the action of CTP, an end product of pyrimidine synthesis, in allosterically inhibiting the enzyme aspartyl transcarbamylase. Overproduction of the end-product is avoided. Typically, the feedback loop:

(1) involves, as effector, the last small molecule before a polymer, for example CTP; (the next step in the synthetic process is polymerization of nucleotides to nucleic acids);

(2) is exerted on an early 'committed step' (see below) in the synthetic pathway.

2   *Cascade phenomena* are observable when an initial stimulus is amplified by a series of linked reactions. Again, this is applicable to many systems; amplification in electronics is well-known to all. Examples are set out in Table 12.2.1. The most oft-quoted one in elementary biochemistry is that for glycogen phosphorylase (8.1); here the signal for glycogen breakdown is the arrival of some molecules of adrenaline at receptors on the surfaces of the muscle cells. Amplification through the cascade mechanism ensures the breakdown of large quantities of glycogen relative to the small quantities of messenger arriving. To put it another way, sensitivity is increased. Also, the multiple steps offer more points of control.

Table 12.2.1 'Cascade' phenomena

System	Amplification of signal for
Glycogen phosphorylase	Glycogenolysis
Glycogen synthetase	Glycogenesis
Complement activation	Bacteriolysis
Haemostasis	Blood clotting

3   *Compartmentation* Confinement of enzymes within the space enclosed by a membrane (or within the structure of the membrane itself, for that matter) necessarily enhances control potential by concerted factors. The enzymes of the tricarboxylic acid cycle, including pyruvate dehydrogenase, are in the mitochondria. Gluconeogenesis is largely by a cytosolic pathway, but the enzyme of the initial step, pyruvate carboxylase, is confined in many species within the mitochondria. Thus in general, mitochondrial pyruvate can yield oxaloacetate (pyruvate carboxylase) or acetyl CoA (pyruvate dehydrogenase). When ATP is plentiful it inhibits the dehydrogenase, as does excess of acetyl CoA. But acetyl CoA also activates the carboxylase, while ADP inhibits it. Thus pyruvate transformation can be switched one way or the other, since both the enzymes are in the mitochondria, yielding oxaloacetate when energy and acetyl CoA are plentiful, and acetyl CoA for the tricarboxylic acid cycle when ATP is short (Fig. 12.2.1).

Fig. 12.2.1

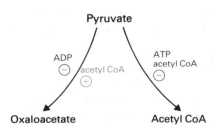

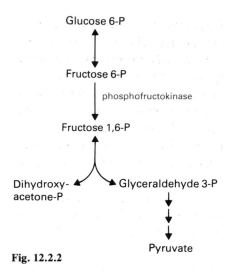

Glucose 6-P

Fructose 6-P

phosphofructokinase

Fructose 1,6-P

Dihydroxy-acetone-P   Glyceraldehyde 3-P

Pyruvate

Fig. 12.2.2

4 *Employment of a 'committed step'* A committed step, often used synonymously with 'rate-limiting step', is typically the first reaction in a pathway yielding a metabolite which has no role other than as an intermediate in that particular pathway. It is obvious that the committed step should occur early in a long pathway. If the amount of an end-product is to be controlled, then it should be controlled before most of the later intermediates accumulate and side-reactions also increase. Thus in the pathway in Fig. 12.2.2, the committed step is that catalysed by phosphofructokinase. It requires ATP and so proceeds with a large loss of free energy; in other words, it is irreversible. You would of course expect this to be true of a committed step. It is inhibited allosterically by citrate and ATP, and it is activated, also allosterically, by AMP, phosphate, and both substrate and product (fructose 6-phosphate and fructose 1,6-diphosphate). Thus the enzyme is subject to various effectors; it can be likened to a valve with fine controls.

The committed step is another mechanism of universal application,* applied to enzyme activity. In summary, in metabolism a committed step:
(1) is early in a pathway,
(2) produces an intermediate without a role in other pathways,
(3) is associated with a loss of free energy (usually),
(4) is controlled by multiple effectors (usually)'
(5) is rate limiting.

Table 12.2.2 Some 'committed steps' of metabolism

Enzyme	'Committed' end-product
Phosphofructokinase	Pyruvate
Acetyl CoA carboxylase	Palmitate
β-Hydroxy-β-methyl glutaryl CoA reductase	Cholesterol
7α-Hydroxylase	Bile salts
Aspartyl transcarbamylase	Pyrimidine nucleotides

12.3   What metabolic control avoids

Evidently the mechanisms outlined above do three things.
1 They prevent wasteful synthesis. CTP inhibition of aspartyl transcarbamylase tends to prevent oversynthesis of nucleic acids. ATP inhibition of pyruvate dehydrogenase and isocitrate dehydrogenase prevents wasteful oversynthesis of ATP itself in the tricarboxylic acid cycle.
2 They prevent wasteful catabolism. Obviously excesses of both synthesis and catabolism are wasteful, but it is conceivable that they could operate simultaneously, and this is linked to the concept of the *futile cycle*. The conversion of fructose 6-phosphate to fructose 1,6-diphosphate and that of

*The classical expression of the committed step is 'crossing the Rubicon'. Less classically, Mussolini said: 'We cannot change our policy now. After all, we are not political whores.'

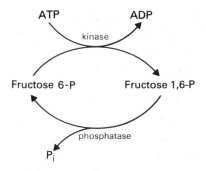

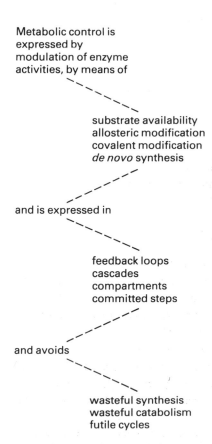

Fig. 12.3.1

Metabolic control is
expressed by
modulation of enzyme
activities, by means of

substrate availability
allosteric modification
covalent modification
de novo synthesis

and is expressed in

feedback loops
cascades
compartments
committed steps

and avoids

wasteful synthesis
wasteful catabolism
futile cycles

Fig. 12.3.2

fructose 1,6 diphosphate to fructose 6-phosphate, as has been stressed, involve two quite different enzymes, the former requiring ATP (Fig. 12.3.1). Were these two enzymes to be active simultaneously, the result would be the consumption of ATP to no apparent purpose. It is therefore supposed that in the ordinary way metabolic control avoids such a futile cycle; for example AMP activates the kinase and inhibits the phosphatase. Thus the kinase reaction is favoured when ATP is in short supply, and this channels carbon to the tricarboxylic acid cycle. Some have cast doubt on this interpretation, maintaining that both enzymes are active simultaneously, with the consumption of ATP; it is proposed that this provides heat and ensures that some fructose 6-phosphate for conversion to glucose is always available. When heat has to be provided very rapidly, for example in the insect flight muscle which must reach 30 °C for normal functioning, futile cycles may be very useful indeed.

3 They promote homeostasis, but metabolically this is saying the same as 1 and 2 above.

For a summary of metabolic control, see Figs 12.3.2 and 12.3.3.

12.4 Hormones and other messengers

The foregoing control mechanisms, theoretically, could all take place in a single cell, or at least in a unicellular organism such as a bacterium. Where communication takes place between cells or between cells organized into tissues, control is mediated by messengers operating through either a nervous system, with its neurotransmitters, or an endocrine system ('ductless glands') with its hormones, or both.

There is no doubt that in the modern biochemistry textbook the section on hormones is one of the most formidable, and that is saying something. There are fifty or so recognized hormones in mammals, of diverse structure and physiology, so how can the topic be hammered into manageable form ('manageable' for examinations, that is)? Firstly, a number of general principles have now emerged and you would be expected to know these. Secondly, it is probable that only the most important hormones are to be asked about at this stage, and then primarily in connection with intermediary metabolism. Then, for the purposes of elementary biochemistry, anatomy can be ignored, so there is no need to worry about the sites and structures of the glands.

To begin with principles, it is established that the signal for the release of a hormone can be chemical or neural. The chemical stimulus can be a rise in the blood concentration of a substance, for example the surge of glucose after a meal triggers off insulin secretion. The end-result can be a consequence of both neural and chemical stimuli—thus neural events cause a release of thyrotropin releasing factor in the hypothalamus; this is transported through the pituitary stalk to the anterior pituitary to effect the secretion into the blood of thyrotropin, also known as thyroid stimulating

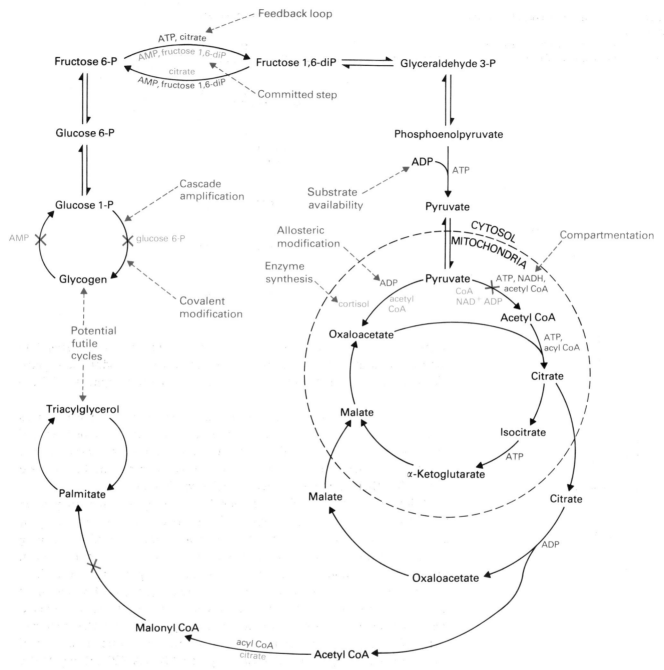

Fig. 12.3.3 A 'control' diagram.
Some positive (green) and negative (red) effectors are shown for this segment of intermediary metabolism (glycolysis, the tricarboxylic acid cycle, and triacylglycerol synthesis). The names of the enzymes have been omitted, but careful attention has been given to the direction of the arrows. The four ways by which enzyme activity is modulated (substrate availability, allosteric modification, covalent modification, and *de novo* synthesis) are grouped together in the centre of the diagram. The expression of these (in cascades, compartments, feedback loops, and committed steps) is also indicated. The potential futile cycles in the diagram are triacylglycerol ⟳ palmitate, glycogen ⟳ glucose 1-P and fructose 6-P ⟳ fructose, 1,6-P. Some of these may not involve ATP directly, but they require energy for one direction, so it comes to the same thing. Another potential futile cycle, glucose ⟳ glucose 6-P is not shown. It is fairly easy to rationalize the point of action of most of the effectors. An exception may be fructose 1,6-diphosphate, in activating phosphofructokinase. Its action seems to be amplifying the effect of AMP, as a product of the reaction itself. Its inhibition of the reverse, phosphatase reaction is then entirely comprehensible. Reactions known to be modulated by phosphorylation of the enzyme concerned are marked with a blue cross.

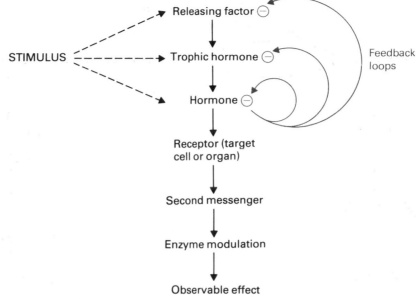

Fig. 12.4.1 Some generalizations about endocrine action.

Not all hormones can be demonstrated to display all the steps and loops noted here.

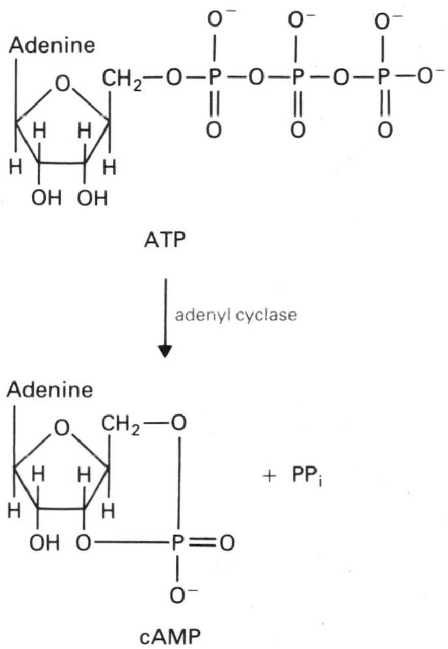

Adenine

$$CH_2-O-\overset{\overset{\displaystyle O^-}{|}}{\underset{\underset{\displaystyle O}{||}}{P}}-O-\overset{\overset{\displaystyle O^-}{|}}{\underset{\underset{\displaystyle O}{||}}{P}}-O-\overset{\overset{\displaystyle O^-}{|}}{\underset{\underset{\displaystyle O}{||}}{P}}-O^-$$

ATP

adenyl cyclase

Adenine

+ PPᵢ

cAMP

Fig. 12.4.2

hormone. *Trophic hormones* are those from the pituitary which stimulate the synthesis and/or release of other hormones from endocrine organs. However some trophic hormones, such as adrenocorticotropin (ACTH), in themselves directly affect metabolism.

Secretion of releasing factors in the hypothalamus and trophic hormones in the pituitary may be inhibited by a rise in concentration, in the blood, of the end-product hormone. For that matter, a rise in concentration of the end-product hormone sometimes inhibits further secretion of the hormone itself. Thus there are 'multiple feedback loops' (Fig. 12.4.1).

A hormone interacts with a target cell via a receptor, which may be on the surface of the plasma membrane (as in the case of insulin, glucagon, adrenaline, calcitonin, parathyroid hormone, and thyroid stimulating hormone) or in the interior of the cell (as for the thyroid hormones and all the steroid hormones such as the oestrogens, androgens, and corticoids). This is one of the few valuable distinctions between groups of hormones. As a result of hormone-receptor interaction some other substance, the 'second messenger', may operate and whereas few if any of these have been characterized for hormones interacting with intracellular receptors, in the case of the group interacting at the cell surface, cyclic AMP (cAMP) is thought to be involved. The binding of hormone by receptor causes activation of adenyl cyclase (Fig. 12.4.2).

cAMP then alters the rate of one or more reactions; it is known to activate protein kinases which catalyse the phosphorylation of enzymes. Thus phosphorylase kinase is activated by cAMP-dependent protein kinase resulting in glycogenolysis (8.1). This then is the basis of the glucose releasing action of glucagon and adrenaline. Acetyl CoA carboxylase, the key enzyme in palmitate synthesis, is phosphorylated and inactivated following exposure of adipose tissue cells to adrenaline, so the end result is inhibition of lipogenesis.

The following points summarize features of the principal hormones.

Insulin

Structure: polypeptide.
Stimulus for secretion: rise in blood glucose; gastrointestinal hormones.
Feedback phenomena: fall in blood glucose, consequent on insulin activity, inhibits its further secretion.
Target organs: Primarily adipose tissue, liver and muscle, not kidney or brain.
Receptor: protein in plasma membrane.
Second messenger: not definitively identified, but something which effects the phosphorylation and dephosphorylation of enzymes.
Principal effects: glycogenesis, lipogenesis, protein synthesis, depletion of blood glucose (hypoglycaemia).

Glucagon

Structure: polypeptide.
Stimulus for secretion: exercise, starvation, autonomic nervous system.
Feedback phenomena: fatty acids released as a result of glucagon action, and insulin in response to hyperglycaemia, inhibit further secretion.
Target organs: adipose tissue, liver, muscle.
Receptor: protein on plasma membrane.
Second messenger: cAMP.
Principal effects: glycogenolysis, lipolysis, rise in blood glucose (hyperglycaemia).

Adrenaline

Structure: a catecholamine (Fig. 12.4.3).
Feedback phenomena: not prominent, possibly because of the transitory effects of the hormone.
Target organs: most.
Receptor: on plasma membrane, but of several types, classified into 'α-adrenergic, β-adrenergic, and 'dopamine' receptors. The α and β receptors

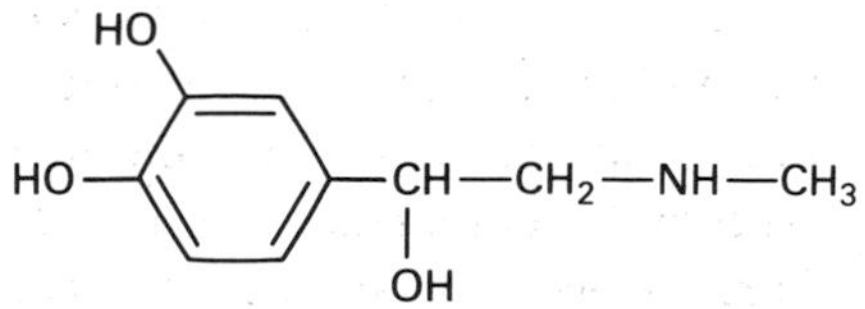

Fig. 12.4.3 Adrenaline.

are usually associated with opposite effects, for example, the former mediate smooth muscle contraction in the blood vessels and the latter inhibit it.

Second messenger: cAMP, calcium.

Principal effects: mobilization of substrates (glycogenolysis and lipolysis); blood pressure rise; in general, antagonism to insulin and reinforcement of glucagon.

Cortisol

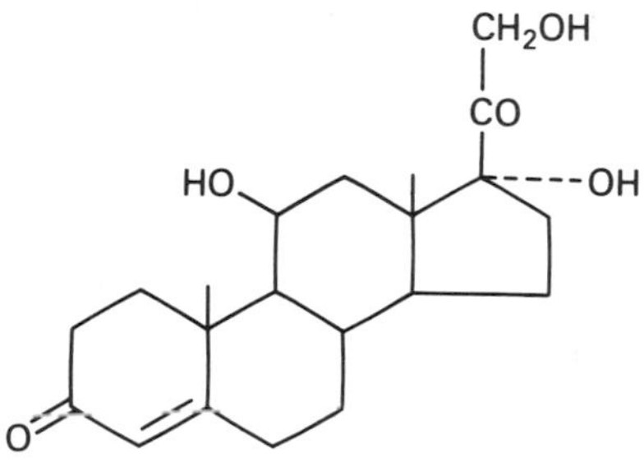

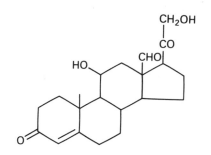

Fig. 12.4.4 Cortisol.

Structure: a steroid (Fig. 12.4.4).

Stimulus for secretion: corticotropin-releasing factor stimulates ACTH secretion.

Feedback phenomena: cortisol and ACTH inhibit secretion of releasing factor; cortisol inhibits ACTH secretion.

Target organs: most.

Receptor: cytosolic.

Principal effects: gluconeogenesis, lipolysis, anti-inflammatory.

Aldosterone

Fig. 12.4.5 Aldosterone.

Structure: a steroid (Fig. 12.4.5).

Stimulus for secretion: angiotensin II, in turn synthesized from renin from the kidney; also elevated serum potassium.

Feedback phenomena: sodium retention caused by secretion of the hormone causes a rise in blood pressure and a fall in renin secretion; also angiotensin II inhibits further renin secretion.

Target organs: kidney, sweat glands, intestinal mucosa.

Cell receptor: cytosolic.

Principal effects: increased resorption of sodium in the distal tubules of the kidney, and increased excretion of potassium and ammonia; elevation of blood pressure.

Thyroxine

Structure: amino acid derivative, iodinated (Fig. 12.4.6).

Stimulus for secretion: thyroid stimulating hormone from the pituitary, in turn induced by thyroid releasing factor of the pituitary.

Feedback phenomena: inhibition, by thyroxine and triiodothyronine, of the pituitary releasing factor and TSH.

Target organs: most, but not spleen and testis.

Cell receptor: receptor in nuclear chromatin; binding causes synthesis of hnRNA and mRNA, subsequently synthesis of various enzymes.

Second messenger: none identified.

Principal effects: stimulation of growth; enhancement of oxidation reactions; increase in basal metabolic rate.

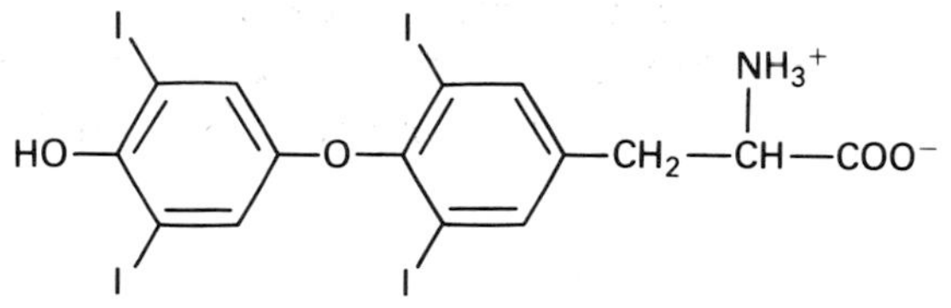

Fig. 12.4.6 Thyroxine.

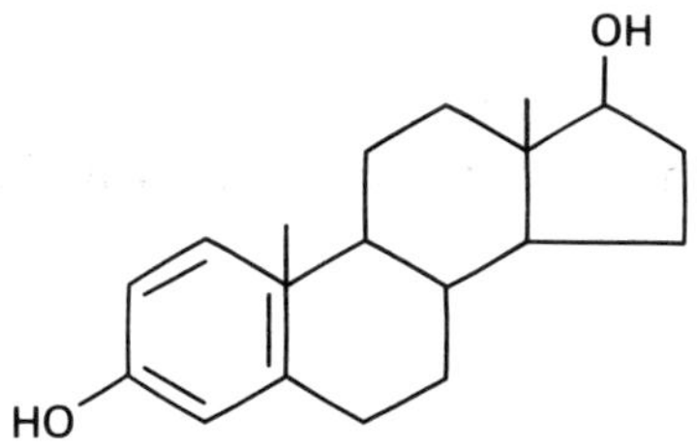

Testosterone

Structure: a steroid (Fig. 12.4.7).

Stimulus for secretion: luteinizing hormone (LH) of anterior pituitary.

Feedback phenomena: suppresses LH secretion.

Target organs: genitalia, skin, skeleton, muscles.

Cell receptor: cytosolic.

Principal effects: development of male phenotype; anabolism and growth.

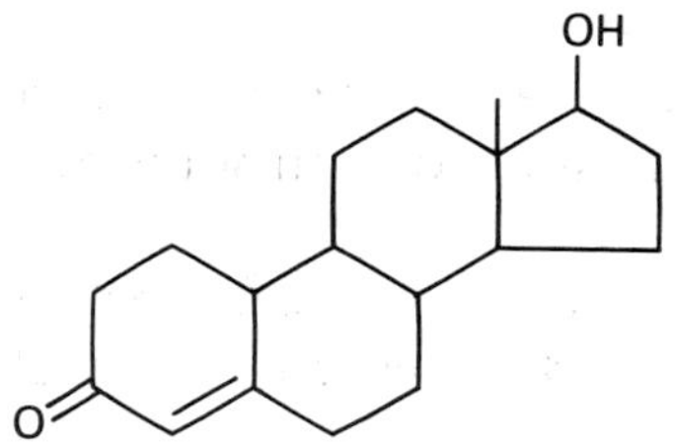

Fig. 12.4.7 Testosterone.

17β-oestradiol

Structure: a steroid (Fig. 12.4.8).

Stimulus for secretion: follicle stimulating hormone (FSH) and luteinizing hormone (LH) of the anterior pituitary.

Feedback phenomena: inhibition of release of FSH and LH, as well as hypothalmic LH releasing factor.

Target organs: uterus, mammary gland, bones, skin.

Cell receptor: cytosolic.

Principal effects: increase in RNA for protein synthesis; bone remodelling.

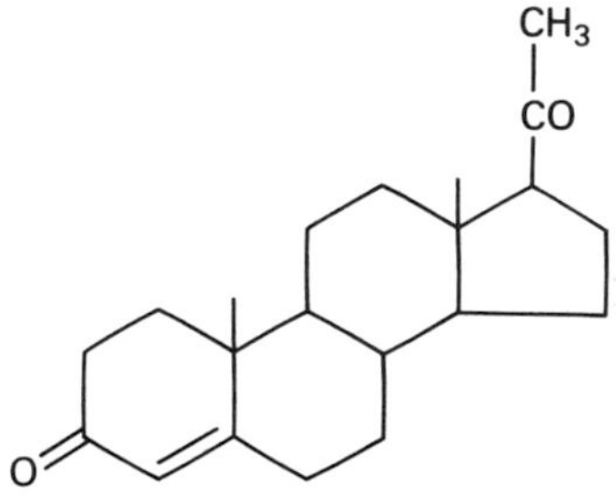

Fig. 12.4.8 17β-Oestradiol.

Progesterone

Structure: a steroid (Fig. 12.4.9).

Stimulus for secretion: LH.

Feedback phenomena: inhibits release of LH and hypothalamic LH releasing factor.

Target organs: endometrium, mammary gland.

Cell receptor: cytosolic.

Principal effects: preparation of endometrium for ovum implantation; growth of mammary tissue.

Fig. 12.4.9 Progesterone.

Parathyroid hormone

Structure: polypeptide.
Stimulus for secretion: drop in plasma ionized calcium.
Feedback phenomena: by drop in plasma ionized calcium.
Target organs: kidney, bone, and gastrointestinal tract.
Cell receptor: on plasma membrane.
Second messenger: cAMP.
Principal effects: maintenance of plasma ionized calcium; renal excretion of calcium and conservation of phosphate.

Calcitonin

Structure: polypeptide.
Stimulus for secretion: rise in plasma ionized calcium.
Feedback phenomena: fall in plasma calcium due to hormone action inhibits its further secretion.
Target organs: bone, kidney.
Receptor: on plasma membrane.
Second messenger: cAMP.
Principal effects: inhibition of bone resorption, lowering of blood calcium.

Vitamin D

Structure: Secosteroid (a steroid-derived structure), the active form of which is 1,25-dihydroxycholecalciferol (Fig. 12.4.10).
Stimulus for synthesis: sunlight impinging on skin (cholecalciferol from 7-dehydrocholesterol); parathyroid hormone (hydroxylation of cholecalciferol).
Feedback phenomena: rising blood calcium and phosphate due to action of 1,25-dihydroxycholecalciferol inhibits further hydroxylation of cholecalciferol in kidney.
Target organs: bone, intestine.
Cell receptor: cytosolic.
Principal effects: synthesis of calcium binding protein, and, therefore, enhanced calcium absorption.

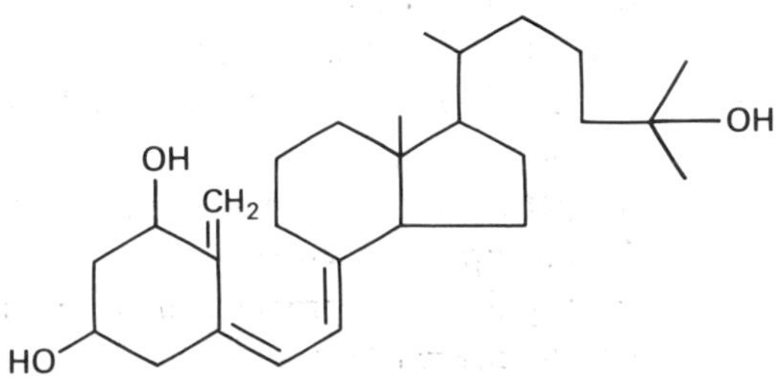

Fig. 12.4.10 1,25-Dihydroxycholecalciferol.

12.5 Glucokinase and hexokinase

This is a sort of addendum to the discussion of control because the roles of glucokinase and hexokinase constitute a favourite exam topic, illustrating as they do a very neat form of regulation of blood glucose and nutrient supply to organs.

Glucose is phosphorylated by ATP on assimilation by tissues
In most tissues, including brain, there are hexokinases to catalyse this reaction, and hexokinases have a low K_m for glucose, about 50 μM; they are also inhibited by the product of the reaction, glucose 6-phosphate. Glucokinase on the other hand is a liver enzyme, with a high K_m, about 20 mM; it is not inhibited by glucose 6-phosphate. In summary:

Liver	Glucokinase	High K_m	Not inhibited by product
Brain	Hexokinase I	Low K_m	Inhibited by product
Other tissues	Other hexokinases	Low K_m	Inhibited by product

This situation has several consequences. Bearing in mind that a low K_m means a high affinity for substrate, the brain will be able to abstract glucose from the circulation even when the glucose concentration is at its lowest (about 2 mM). This is a protective mechanism for the brain, which is absolutely dependent upon glucose as substrate. On the other hand the liver enzyme has little affinity for glucose when blood glucose concentration is low and it is not able to phosphorylate it to the detriment of the supply to the brain. When the amount of glucose in the blood is large, as it is in the portal system after a carbohydrate meal, it is rapidly taken up by the liver; moreover, storage as glycogen is not limited by inhibition by the product of the reaction, glucose 6-phosphate.

So this is an adaptation to the fasting-feeding cycle, and is complemented by endocrine activity (12.4). Adrenaline and glucagon release glucose in the former state, whereas insulin enhances assimilation of it in the latter state.

Hexokinase and glucokinase are able to exert this control because they are compartmented, as described in 12.2. However, it is an organ compartmentation, not an intracellular organelle one.

13 Distilled Topics

13.0 Preamble

The sequence chemistry/biology → structure → catabolism → biosynthesis → control takes the student to a point of departure, departure from the basics. Unfortunately there is still a great deal to cover. Examinations are full of questions on membranes, on inborn errors of metabolism, on muscle contraction, on whatever you care to name. This section covers these, or most of them.

They are presented in a fairly economic way, though not as *reductiones ad absurdum*. All knowledge is subject to a multiple distillation process. First of all, in science, the results of experiments are published in journals devoted to particular specialities. Authorities in their own fields read these research papers and prepare reviews, when invited. These may be published in journals but they sometimes also appear in books. The reviews are then scoured for yet more generalizations for chapters in rather more discursive books and also for the purpose of putting together student textbooks. The college lecturer, unless he also happens to be a specialist in the field in which he has been told to give lectures, makes a further digest from these texts for the purposes of his own presentation in class. The student may manage to get down in his notes a summary of what the lecturer says, writes on the board, and shows on slides; if he is a good student, he will recall most of what is in these notes and the final condensate will appear on the examination script.

Here we make an attempt to come in during the final stage or two of this tortured process. This is not to say that all the preliminary distillations are unnecessary, but they may not be necessary for you.

13.1 Steroid nomenclature and structure

Steroids are important in many areas of biochemistry; leaving aside the bile salts and the steroid hormones and the vitamins D, there remains the role of cholesterol itself in the pathogenesis of the great killer of the twentieth century: ischaemic heart disease. If you are asked specifically about steroids you have to know a little more about their structure than has been given hitherto.

Writing cholesterol as in Fig. 13.1.1 is a complete formal description of its structure, including its steric properties, but this rests on several conventions and to understand these you have to regress to the cyclohexane molecule. Although cyclohexane is written on the page as though its carbons are all coplanar, this is not possible, as the strain on the carbon-carbon bonds

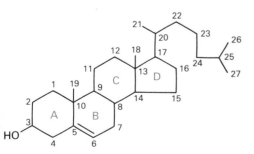

Fig. 13.1.1 Cholesterol.

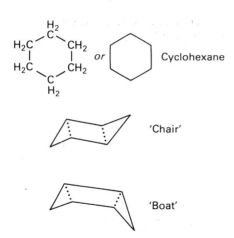

Cyclohexane

'Chair'

'Boat'

Fig. 13.1.2

Fig. 13.1.4

trans-Decalin

cis-Decalin

would be too great. The planar structure just cannot geometrically accommodate the 105° angle. Thus cyclohexane is puckered, and there are two possible puckered forms, appositely termed 'chair' and 'boat' (Fig. 13.1.2).

Of the two, the chair form is the more stable. If two cyclohexane rings are fused together (as in rings A and B of the steroid nucleus) they can be either *trans* or *cis* to each other. These are rather easily imagined and can be drawn in an attempt to render perspective, Fig. (13.1.3). At positions 5 and 10 there is the possibility of geometrical isomerism, that is substituents, as well as the hydrogen atom in the absence of substituents, could project above and below the plane of the ring. But the steric constraints on the two forms above are such that in the *cis* form the substituents or hydrogen atoms are *cis*, and in the *trans* form they are *trans*.

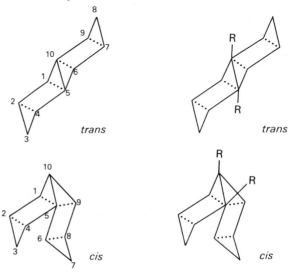

Fig. 13.1.3 *Cis*-and *trans*-Decalin

In most but not all of the natural steroids, the rings A and B are *trans*, so the substituents or hydrogen atoms at C-5 and C-10 are *trans*. As drawn, the substituent at C-10 is above the plane of the ring system, and conventionally groups or atoms on the same side are 'connected' with a solid line. They are said to be β-orientated. Conversely a group below the plane is connected with a broken line and said to be α-orientated. Thus *cis*- and *trans*-decalin may be written as in Fig. 13.1.4, although they are identical with the perspective forms described above. When third and fourth rings are fused as in the steroids they are generally also *trans*. Thus a perspective drawing of cholesterol is as shown in Fig. 13.1.5. This is somewhat easily translated into the planar form as written at the start of the section. Note that the C-3 hydroxyl is written with a solid 'bond', being β-orientated. The two reduction products of cholesterol are shown in Fig. 13.1.6. In coprostanol, a bacterial degradation product of cholesterol, the A and B rings are *cis* to each other, so that the hydrogen at C-5 is above the plane of the ring, namely β-orientated.

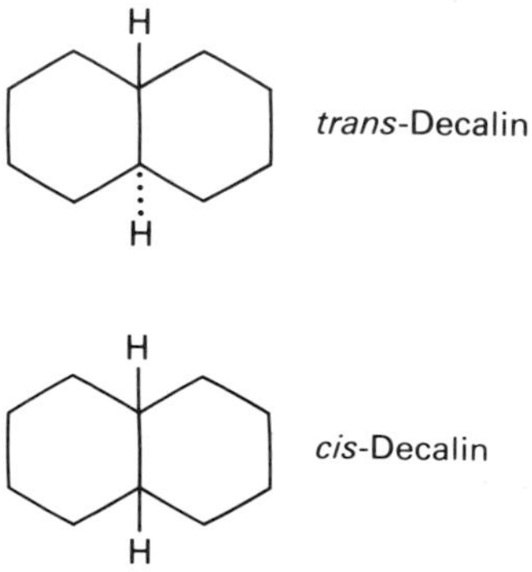

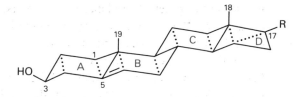

Fig. 13.1.5 Cholesterol (a perspective drawing).

Fig. 13.1.6 Reduction products of cholesterol.

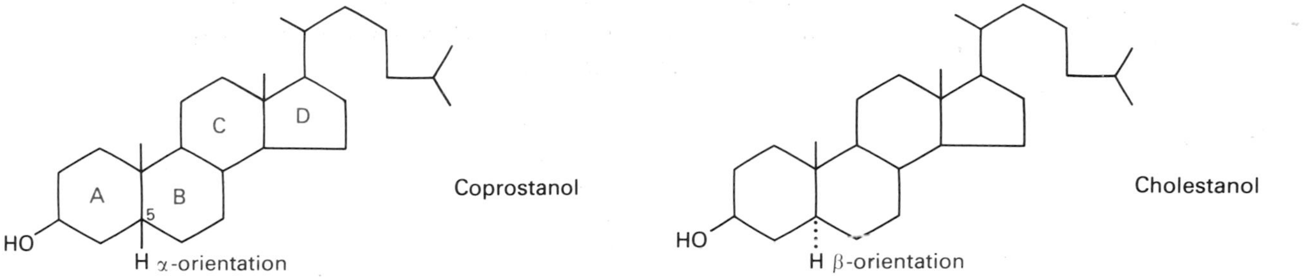

Coprostanol

H α-orientation

Cholestanol

H β-orientation

Three more representative steroids may be chemically dissected as shown in Fig. 13.1.7.

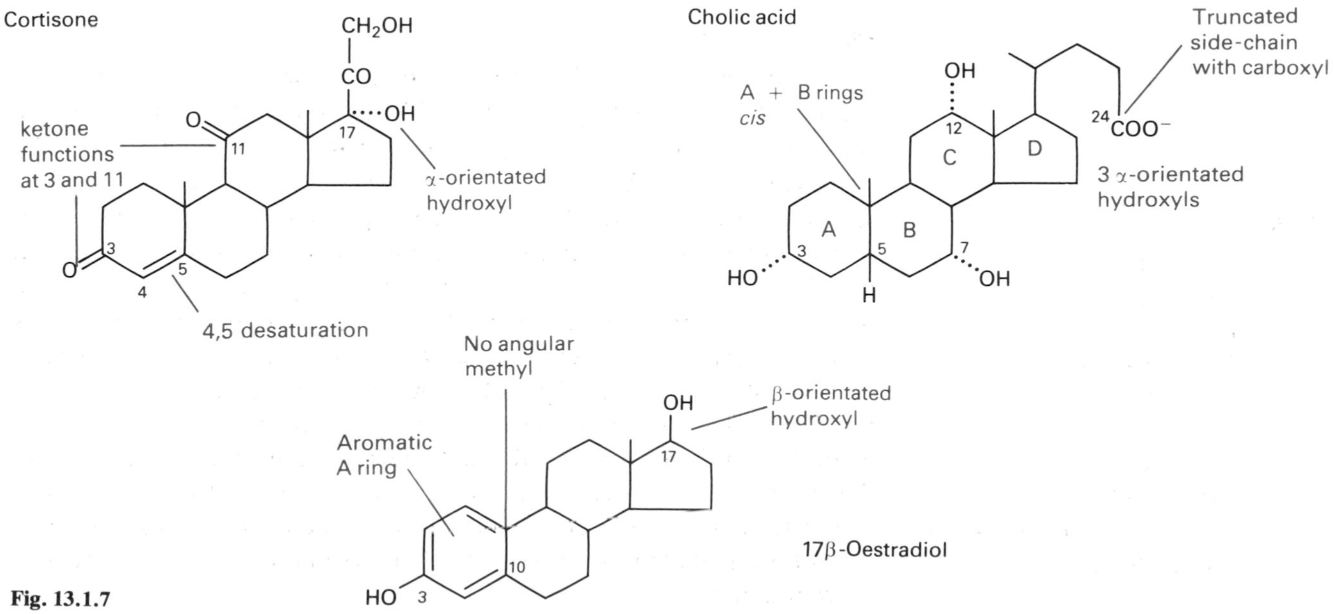

Cortisone

ketone functions at 3 and 11

α-orientated hydroxyl

4,5 desaturation

Cholic acid

A + B rings
cis

Truncated side-chain with carboxyl

3 α-orientated hydroxyls

No angular methyl

β-orientated hydroxyl

Aromatic A ring

17β-Oestradiol

Fig. 13.1.7

13.2 Globular and fibrous proteins

One broad classification of proteins merely distinguishes the globular from the fibrous. *Globular proteins* are more or less spheroid and include such families as the haemoglobins and the lipoproteins as well as the enzymes. The *fibrous proteins* exhibit a variety of structures but their molecules are all lengthy (in technical language, of *high axial ratio*). They occur in tissues when some property of molecular elongation is necessary for mechanical function; for example the long threads of fibrin are able to clump together to form a mat (the clot) which stops extravascular blood flow.

Fig. 13.2.1 The peptide bond.

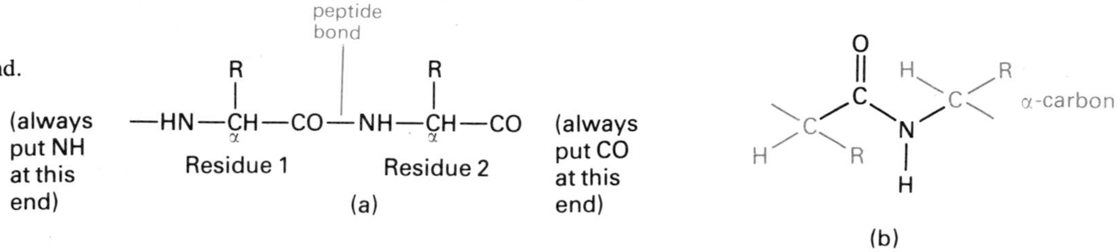

(always put NH at this end)

Residue 1   Residue 2

(always put CO at this end)

(a)

α-carbon

(b)

To explain protein structure or, hopefully, function in terms of structure, it is necessary to take a hierarchical view (Table 5.1.1). But the starting-off point is naturally the nature of the bond joining two amino acids, that is the *peptide bond*. It can be written simply as in Fig. 13.2.1(a) but if you write it out as in Fig. 13.2.1 (b), its main properties are at once suggested:

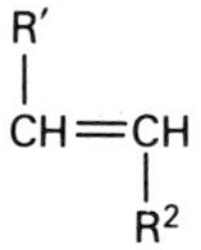

(i) it is a *trans* structure (O and H on opposite sides) — refer to the *trans* alkenyl bond for reference (Fig. 13.2.2);

(ii) it is non-rotatable, unlike the

Fig. 13.2.2

bonds adjacent to it (here again there is an analogy to the non-rotatable nature of the alkene double bond);

(iii) it is a coplanar (N, C, O, and H are in one plane);

(iv) it can form hydrogen bonds with other peptide bonds, —NH with $CO=O$ and vice versa. These are usually written

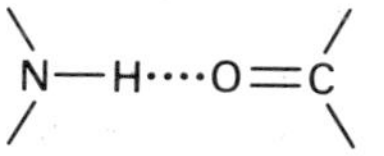

Each and every protein has a multiplicity of hydrogen bonds, because in general they allow the formation of the lowest energy state of the molecule, that is the most stable state.

The *α helix* is a conformation of the peptide chain stabilized by hydrogen bonds within the chain itself. If the criteria above are used, then a model can be constructed which has 3.6 amino acid residues per turn of a right-handed helix (Fig. 13.2.3) and such a structure is also deducible from the results of X-ray diffraction studies. (There is no reason why there should be an

Fig. 13.2.3 How to draw a passable α helix. 'Passable' refers to exams. A random squiggle can be eschewed in favour of the simple but pictorially correct structure shown.

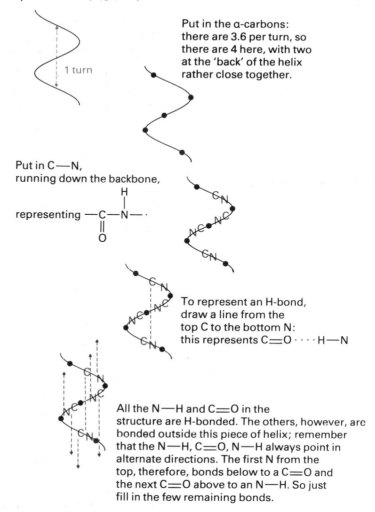

Draw a loose piece of helix (1½ turns).

1 turn

Put in the α-carbons: there are 3.6 per turn, so there are 4 here, with two at the 'back' of the helix rather close together.

Put in C—N, running down the backbone,

representing —C—N—·
with H on the N and O double-bonded to the C

To represent an H-bond, draw a line from the top C to the bottom N: this represents C=O · · · · H—N

All the N—H and C=O in the structure are H-bonded. The others, however, are bonded outside this piece of helix; remember that the N—H, C=O, N—H always point in alternate directions. The first N from the top, therefore, bonds below to a C=O and the next C=O above to an N—H. So just fill in the few remaining bonds.

integral number of amino acids per turn, as students often suppose.) A left-handed helix is quite possible, but is somewhat less stable.

The *β-pleated sheet* is a non-helical conformation stabilized by hydrogen bonds between two peptide chains, or between two segments of the same peptide chain folded back on each other (Fig. 13.2.4). This is the anti-parallel variety, that is the chains 'go'

and   not

and is the most common variety of pleated sheet. The parallel form is, however, perfectly possible. Segments of β-pleated sheets are quite frequent in globular proteins.

Random coils, as implied, are regions of polypeptide chain with no particular folding. They tend to occur when there are amino acids which do not easily accommodate the α-helix structure due to some peculiarity of the

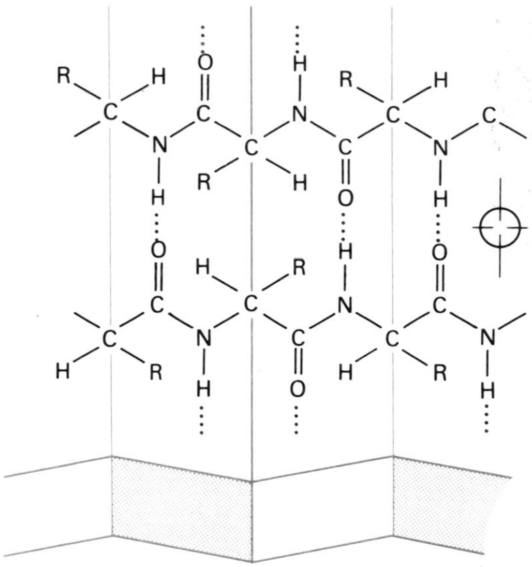

Fig. 13.2.4 The β-pleated sheet.
Since the peptide bonds are planar, the structure has a pleated aspect. It looks, in the representation, as though the side chains (—R) have little room and must impinge upon each other; of course they project out of the sheet, so that there is no steric hindrance.

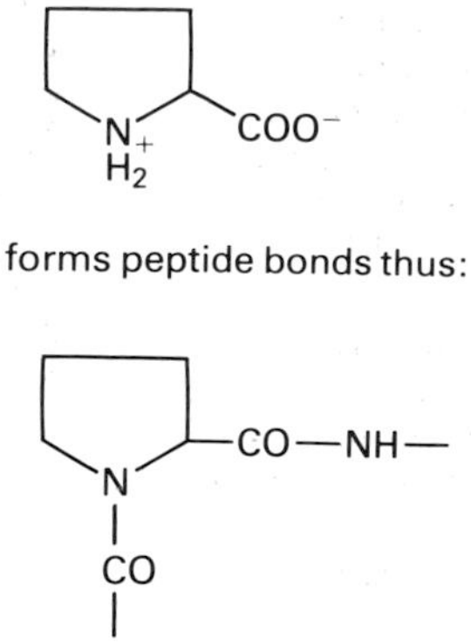

forms peptide bonds thus:

Fig. 13.2.5 Proline in peptide chains.

side-chain. The amino acid proline forms peptide bonds as shown in Fig. 13.2.5, and it is obvious enough that this is an 'abnormal' peptide bond. Prolyl residues tend to occur at corners in the folded structures of proteins. Denatured proteins, that is proteins whose folded structures are destroyed by heating, high pressures, concentrated salts, etc. are also made up of random coils.

Idiosyncratic foldings

Not all proteins exhibit the features just outlined: many have foldings or helices peculiar to themselves. For example the connective tissue protein, collagen, has a triple helix (Fig. 13.2.6) which is the result of the peculiar amino acid composition of its subunits. These subunits are of several types but they all contain large amounts of glycine and proline; they line up end to end and form the triple helix, locked together by the twist of the pyrrolidine (prolyl residue) peptide bonds. Glycine bears only a single hydrogen as a side-chain and so does not disrupt the structure; bulky side chains of other amino acids project outwards from the helix. The three chains constituting the helix form *tropocollagen*, about 3000 nm long and 15 nm in diameter—it is evident that as a biological molecule tropocollagen has one of the largest known axial ratios, for were it the width of a lead pencil it would be a metre and a half long! The tropocollagen can aggregate to form insoluble collagen fibrils, but the association too is of a rather special sort (Fig. 13.2.6).

The cohesion of the collagen fibril is only possible because there is a large amount of *post-translational modification* of the collagen polypeptides. The general term 'post-translational modification' means any modification to a protein subsequent to the formation, on the ribosome, of the peptide bonds themselves. It may involve scission of parts of the polypeptide, attachment

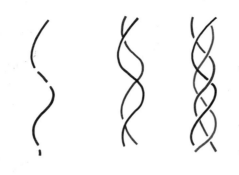

Fig. 13.2.6 The structure of collagen. The triple helical structure is easy to draw, as shown, if you sketch in the first loose helix and then coil the other two chains round about it. One cannot easily depict the great length of the resulting (tropocollagen) molecule, which is asymmetric in the sense that it has a differentiated 'head' and 'tail'.

Below, the tropocollagen molecules are shown, again unrealisticlly shortened, as asymmetric cylinders which aggregate head to tail in a quarter-staggered fashion, with 'holes' in between. The effect of this, as the dotted lines indicate, is to give the collagen fibril a striated appearance, with a periodicity of about one quarter the length of the tropocollagen.

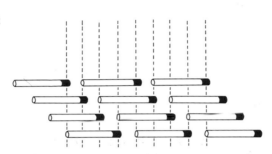

of sugars, modification of amino acid side-chains and others. In the nascent collagen polypeptides, the first of the many post-translational modifications is the hydroxylation of a proportion of the prolyl and lysyl side-chains. In the case of hydroxyproline the reaction proceeds as in Fig. 13.2.7. Ascorbate (vitamin C) is required for the reaction but at the present moment it is not established whether it acts as a specific coenzyme or serves to keep the completely necessary Fe^{2+} in reduced form. The lysyl side-chains are hydroxylated in a similar manner, and many of them are glycosylated. Meanwhile the triple helical structure is forming; the form called procollagen is secreted from the cell (collagen in its mature form is an extracellular substance) and peptides are removed from its N-terminal and C-terminal regions. After such processing the tropocollagen, as the procollagen has now become, associates in the staggered manner described in Fig. 13.2.6.

In the mature form collagen is an insoluble matrix of fibrils (or fibres, as they become when visible in the light microscope). The insolubility is the

Fig. 13.2.7

Prolyl residue + α-Ketoglutarate $\xrightarrow[\text{Prolyl hydroxylase}]{O_2 \quad Fe^{2+} \quad CO_2 \quad H^+}$ Hydroxyprolyl residue + Succinate

189

result of reactions causing cross-links of the covalent sort. These reactions are of a complex series involving lysyl, hydroxylysyl and histidyl residues as well as cysteinyl residues, oxidation of which yields disulphide (—S—S—) cross-links.

The pattern and orientation of the mature fibres reflect the characteristics of specific tissues. Thus in the cornea of the eye they are in regular layers so that they transmit rather than scatter light. In skin or bone they are more randomly orientated. In tendon they exist in strong parallel bundles, illustrating the salient feature of collagen, that is its inextensibility, enabling it to transmit mechanical stress. Tendons transmit the force exerted by contracting muscles to the bones. In contrast elastin, the protein of the ligaments and the arterial wall, is extensible and absorbs mechanical stress.

There are some interesting disorders of collagen synthesis, the various types of Ehlers-Dahnlos syndrome. Those subjects with the most marked joint hyperextensibility, because the tendons did not constrict their rotation, were formerly exhibited in circuses as 'indiarubber men'.

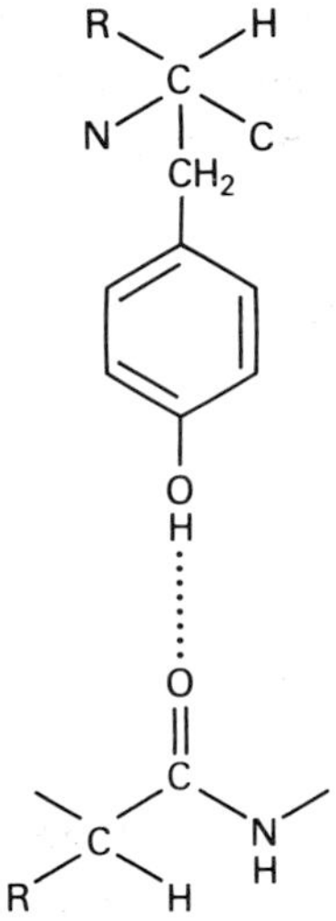

Fig. 13.2.8

Cross-links

Interactions between polypeptide chains can be listed as follows.

1 *Hydrogen bonds* As noted, it is the hydrogen bonds linking the carbonyl and imino groups of the peptide chains which stabilize the α-helical and β-pleated sheet structures. It was the guiding principle of those who first constructed the models that there should be the maximum number of hydrogen bonds in the model to achieve the most stable structures. There are also possibilities for hydrogen bonding in the structures of side chains, for example a phenolic side chain (tyrosine) with a peptide bond carbonyl (Fig. 13.2.8).

2 *Disulphide bonds* A protein containing cysteinyl residues has the possibility of cross-linking by oxidation, thus:

$$\text{CH—CH}_2\text{—SH} + \text{HS—CH}_2\text{—CH} \xrightarrow[\;2H\;]{} \text{CH—CH}_2\text{—S—S—CH}_2\text{—CH}$$

Such cross-links are very common in proteins: keratin, the protein of hair and hooves, is one-fifth cystine, and it is very stable and insoluble. Insulin is maintained in its specific three-dimensional structure by three disulphide cross-links. However there are none in haemoglobin.

3 *Salt linkages* At physiological pH there are two basic amino acid side chains (they have accepted hydrogen ions) and two acidic side chains (they have lost hydrogen ions). Thus if, say, a lysyl and a glutamyl are juxtaposed in a folded peptide chain, the attraction of their opposite charges may promote cohesion:

$$\underbrace{\text{CH—CH}_2\text{—CH}_2\text{—CH}_2\text{—CH}_2\text{—NH}_3^{+}}_{\text{Lysyl}}\;{}^{-}\underbrace{\text{OOC—CH}_2\text{—CH}_2\text{—CH}}_{\text{Glutamyl}}$$

Obviously there is the same potential for salt linkages between N-terminal aminos and C-terminal carboxyls of separate polypeptide chains. This occurs in deoxygenated haemoglobin where the N-terminal valyl of one α-chain associates with the C-terminal arginyl of the other α-chain (13.3).

4 *Hydrophobic bonds* Substances which exclude water tend to associate one with another, like drops of different kinds of oils dispersed in water. Some amino acids have distinctly hydrophobic side chains (Fig. 13.2.9) and they are thought to associate with each other in the interior of globular proteins, such that the total folding is given a cohesion it would not have otherwise. This occurs in the haemoglobin monomer. The concept is reinforced by considering that if there were large numbers of hydrophobic side chains on the surface of a globular protein, it would be insoluble.

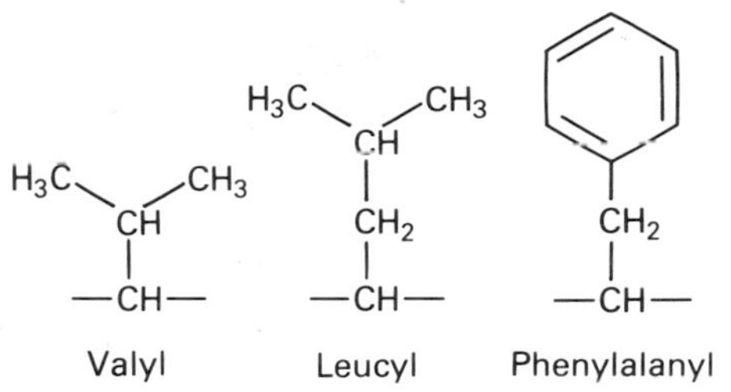

Fig. 13.2.9

```
                              H3C     CH3
                                 \   /
                                  CH                  ⬡
                                  |                   |
         H3C     CH3              CH2                 CH2
            \   /                 |                   |
             CH                 — CH —             — CH —
             |
           — CH —
             Valyl              Leucyl           Phenylalanyl
```

5 *Van der Waals forces* These are also thought to stabilize the complex folding arrangements within protein molecules. They are the weak attractive forces between all atoms and molecules, though they operate at short distances and are important only when side chains are brought into close contact by other cohesive forces or cross-links. Simplistically, they can be imagined merely as side chains getting tangled up with each other.

Hierarchy of structural levels

Peptide bond formation at the translation stage results in a polypeptide chain of specific amino acid sequence. This polypeptide has a *primary structure* only: the sequence of the amino acid residues. On separating from the ribosome, it folds into either an α helix, a β-pleated sheet, or some other folding, or into a sequential combination of these. Primary structure folded into an α helix exhibits a *secondary structure*, which term refers to the way segments of primary structure are arranged in space. The *tertiary structure* refers to the way the sequential segments of helix, sheet or random coil are arranged to form an entire protein molecule. It is this level of structure which is stabilized by the variety of cross-links mentioned above. Many proteins, for example haemoglobin, consist of two or more associated but separate polypeptide chains, that is protein monomers. The way these monomers are arranged with respect to each other is termed the *quaternary structure*.

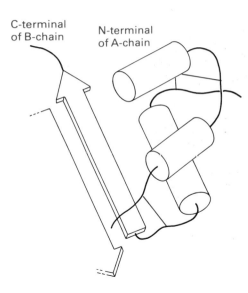

C-terminal of B-chain

N-terminal of A-chain

Fig. 13.2.10 Structure of insulin. Segments of α helix are represented by cylinders, and β-sheet by a solid arrow. Of course the β-sheet has another polypeptide chain, from another insulin molecule, running alongside it. This is only suggested in the diagram. The three disulphide bonds, which are obviously responsible for much of the integrity of the tertiary structure, are represented by red lines.

Within a protein molecule, a *domain* is a cluster of residues in a protein having some function comparable with a similar or identical cluster in another protein. Typically this is the binding of some ligand. Thus the enzymes lactate dehydrogenase, malate dehydrogenase, and alcohol dehydrogenase all contain a section of polypeptide chain, of more or less identical amino acid sequence and shape, which binds NAD^+. Similarly there can be an identical antigen-binding domain on different immunoglobulins. The domain is, therefore, descriptive of a sort of autonomous structure within the primary, secondary, and tertiary hierarchy.

Our aim as always is to summarize the essentials and then produce a clear diagram. However, proteins are complex substances and there are severe difficulties in devising a reproducible picture of the hierarchy of protein structure of any one of them. Nonetheless it might be attempted for insulin, which is a very small protein (Fig. 13.2.10) and for haemoglobin (Fig. 13.3.1).

13.3 Haemoglobin

Even in an elementary biochemistry examination 'haemoglobin' is too extensive a topic to encompass a single essay question. Perhaps three main 'chunks' can be discerned:

1 haemoglobin structure;
2 the role of haemoglobin in gas exchange and acid-base balance;
3 species and pathological variants.

It is the most thoroughly studied of all proteins and there is a wealth of information on all of these topics.

Haemoglobin structure

The 'typical' haemoglobin, for purposes of exposition, is normal adult haemoglobin from human beings, abbreviation HbA_1. It contains two pairs of subunits: a pair each member of which is designated α; and a pair designated β. Thus it is commonly represented as $\alpha_2 \beta_2$ or $\alpha_2^A \beta_2^A$ to show that the chains are the normal adult types. The arrangement of the four chains can be depicted as in Fig. 13.3.1. In view of the number of examination contexts when one may reproduce such a diagram (in its shorter, simplified form) it is probably worthwhile to try to memorize it. This might seem to be a problem (although to put it in perspective, it pales when compared with learning merely to write the *name* of haemoglobin in Chinese: 红 血 球 素, or 'red blood cell substance') but as a basis for discussion of such diverse topics as porphyrins, allostery, cooperativity, acid-base balance, and the biochemical basis of inherited disease, it is surely well worth the effort.

Other subunits exist in other haemoglobins. A minor component of adult human erythrocytes is haemoglobin A_2 which does not have β chains;

Fig. 13.3.1 Structure of haemoglobin.
The top version gives a roughly perspective
view of the four chains packed together. For
clarity the haem is omitted but the position of
the iron atom therein is marked with a red
cross. This iron is associated with a histidyl in
the adjacent chain, marked H. The straight
segments of the chains tend to be of α-helical
secondary structure, shown in the α_1 subunit by
a red spiral. Haemoglobin thus well exhibits the
levels of protein structure, the shape of each
subunit being synonymous with 'tertiary structure' and the way in which the four subunits are
associated being the 'quaternary structure'.

In the simplified version the haem is shown
as a red disc.

(These representations are inspired by the
best-illustrated of all books in protein
chemistry—*The Structure and Function of
Proteins* (1969), by R.E. Dickerson and I. Geis,
Harper & Row Ltd.)

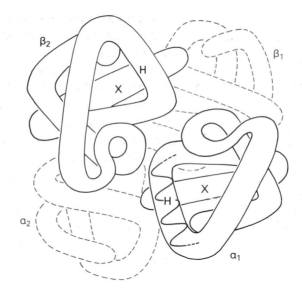

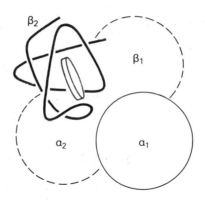

Simplified version

instead it has chains designated δ and is abbreviated $\alpha_2 \delta_2$. Fetal haemoglobin or HbF, which disappears from the blood soon after birth, has chains
designated γ instead of β and so is $\alpha_2 \gamma_2$.

Each chain appears to be so arranged that its hydrophobic residues are
buried in the interior of the folded structure; correspondingly, the bulk of
the hydrophilic residues are on the outside. This makes the assembly
aqueous and also makes possible the existence of a large number of non-
polar bonds in the interior and these are probably the main factors in
stabilizing the tertiary structure.

In addition there are pockets for the haem prosthetic groups. The haem
is so orientated that its iron ion on one face is complexed to a histidyl
residue, leaving the other face accessible for oxygen binding. Naturally
there are also interactions between the chains, but these vary with the state
of oxygenation. The chains are tightly bound to each other when the molecule

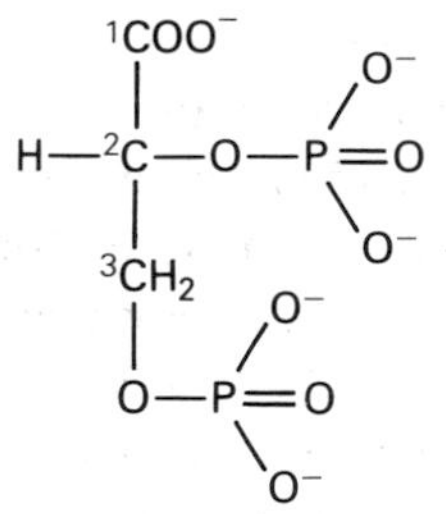

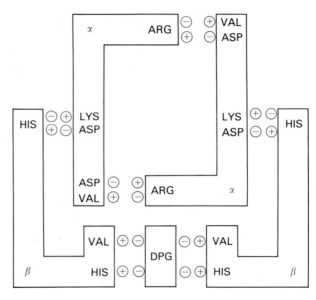

Fig. 13.3.2 2,3-Diphosphoglycerate.

Fig. 13.3.3 Schematic rendering of cross-linkages in deoxygenated ('tense') haemoglobin. Some might say that this is grossly, rather than highly, schematic. Aspartate is not really adjacent to valine as shown, although they are both associated with the C-terminal arginine in the α-chains. Another aspect of the tense form, not shown, is that tyrosyl residues, penultimate to the C-carboxyls of all the chains, are held in pockets distal to them on the same chains; they are ejected on oxygenation of haemoglobin. Note that if histidyls in the β chains become less positive (by releasing H^+ in the lungs), DPG (2,3-diphosphoglycerate) is less tightly bound, and destabilizes the tense form. It is associated with positively-charged residues (e.g. lysyl) other than the valyl and histidyl shown.

is in the deoxygenated state (described as 'tense') but have less cohesion when oxygen is bound ('relaxed').

The cohesion in the tense form is due largely to ionic or salt bonds, by the association of opposite charges in amino acids. In addition, 2,3-diphosphoglycerate (Fig. 13.3.2) a product of erythrocyte glycolysis with four negative charges, is wedged between the β-chains, binding them together by means of its association with the positive charges of a number of amino acid side chains, particularly histidine. There are various ways of representing the tense form on paper, but a simple way is given in Fig. 13.3.3.

The essential point is that the stabilized tense form reacts reluctantly with oxygen. But as pO_2 builds up, an oxygen molecule will eventually be bound by an iron in one of the four haem groups of the haemoglobin tetramer. Oxygenation causes the iron atom to retract more nearly into the plane of the porphyrin ring. This is only the first step in the destabilization of the whole system; the small displacement is transmitted to the histidyl residue on the other side, bringing a segment of subunit helix nearer to the haem. The ionic bonds of the tense form begin to break up. It is thus easier for another three oxygens to react with the remaining three haem irons, which also accelerate displacement and a complete conversion to the relaxed form. While this is occurring, 2,3-diphosphoglycerate is released, and it cannot bind to the haemoglobin so tightly in the lungs anyway as oxygen reacts and hydrogen ions are lost

$$Hb + O_2 \rightleftharpoons HbO_2 + 0.7H^+$$

The hydrogen ions are partially derived from the ionized histidyls in the β-chains which bind 2,3-diphosphoglycerate in the tense form.

So oxygen reacts slowly at first, as shown by the flat part of the oxygen dissociation curve (Fig. 13.3.4) but then rapidly as the tense form relaxes. The sigmoid shape of the curve indicates the cooperativity between the chains—a simple monomeric oxygen binder such as myoglobin (the haem-containing protein of muscles) or indeed an isolated α- or β-chain, yields a saturation curve in the form of a rectangular hyperbola. This is analogous to the simple adsorption isotherm of physical chemistry, for example a gas adsorbing on to charcoal. Haemoglobin F shows even more affinity for oxygen that HbA, but only in the presence of 2,3-diphosphoglycerate, because it binds this substance poorly. Recall that 2,3-diphosphoglycerate binds to the tense, non-oxygenated form of haemoglobin. A haemoglobin which binds it poorly then is liable to become more oxygenated. It is this which ensures that the fetus can safeguard its respiration—HbF depletes HbA of oxygen in the placental circulation.

Fig. 13.3.4 Oxygenation curves of haemoglobin.
When oxygen saturation is plotted against increasing pO_2, normal haemoglobin exhibits a sigmoid curve. This indicates cooperativity between the chains of the tetramer; when one takes up oxygen, it induces the others to do the same more readily. Myoglobin, a monomer, as well as isolated α- or β-chains, exhibit a simple rectangular hyperbolic curve. In a more acid medium the haemoglobin curve is shifted to the right (the Bohr effect). This indicates that acid tends to make oxygen dissociate more readily from haemoglobin as it contributes to the switch to the tense form.

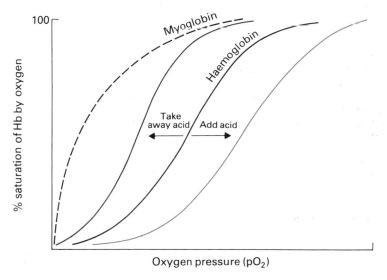

Thus the structure of haemoglobin is an oscillating one: it is almost like the reverse of the lungs, for it tenses or contracts when oxygen is taken up and relaxes or expands when oxygen is released. In the capillaries, where pO_2 is low and hydrogen ions from carbonic acid are plentiful, oxygen tends to be released, but again slowly at first. Then, when a molecule of oxygen at last dissociates, there is cooperativity among the chains to achieve the tense form; hydrogen ions become bound once again to the histidyls which take up 2,3-diphosphoglycerate. The other oxygens dissociate rapidly and reproduce the sigmoid curve in the other, dissociation direction.

Role in gas-exchange and acid-base balance

The Bohr effect is the increase in oxygen dissociation from haemoglobin as a result of increasing concentration of hydrogen ions (Fig. 13.3.4). It was first

195

noticed as an increase in dissociation in relation to increasing concentrations of carbon dioxide. That these are equivalent is evident from

$$CO_2 + H_2O \rightleftharpoons H_2CO_3 \rightleftharpoons H^+ + HCO_3^-$$

The basis of the Bohr effect lies in the conformational changes described above. When the relaxed, oxygenated form of haemoglobin arrives in the capillaries, where pO_2 is low and pCO_2 (and therefore H^+) is high, the hydrogen ions tend to be taken up by the histidyl residues of the β-chains (Fig. 13.3.3). This causes binding of 2,3-diphosphoglycerate and a switch to the tense form, stabilized by ionic bonds, as oxygen is given up to the tissues. One can regard the Bohr effect as due to H^+ ions driving the reaction

$$HbO_2 + 0.7\ H^+ \rightleftharpoons HHb + O_2$$

The Bohr effect makes haemoglobin a much more suitable vehicle for oxygen carriage than, say, myoglobin. Haemoglobin in the capillaries gives up its oxygen more quickly than a simple oxygen carrier would do, and before the milieu is of dangerously low oxygen tension.

Concomitantly, the haemoglobin acts as a buffer. It removes hydrogen ions in the capillaries as above and carries them to the lungs, where the reversion to the relaxed form of haemoglobin causes their release. The equilibrium

$$HCO_3^- + H^+ \rightleftharpoons H_2CO_3 \rightleftharpoons CO_2 + H_2O$$

which is accelerated by carbonic anhydrase, is driven to the right and carbon dioxide blown off.

In the tissues, when H^+ is taken up by haemoglobin during the conversion of oxygenated to deoxygenated haemoglobin, an excess of HCO_3^- accumulates in the erythrocytes. Some diffuses out into the serum as the erythrocyte membrane is quite permeable to it. This would create an electrical charge across the membrane were it not compensated for by an equal amount of chloride ions diffusing from the serum to the erythrocytes. This is known as the *chloride shift*. There is of course a chloride shift outwards in the lungs.

Species and pathological variants of haemoglobin

Many of the haemoglobins of invertebrates, as well as that of the lamprey, consist of single chains. Thus the haemoglobins of these species do not exhibit a sigmoid oxygen dissociation curve or a Bohr effect. All bony vertebrates, however, have the $\alpha_2 \beta_2$ pattern in their principal haemoglobin fraction and show cooperative oxygen binding. However there is considerable interspecies variation of amino acid sequence in both the α- and β-chains. Charting these differences gave rise to the substudy of molecular paleogenetics, whereby the relationships between species can be explored in parallel with traditional morphological techniques. Thus the gorilla β-chain differs from the human β-chain in one site only; the pig β-chain differs in 17 sites. As well as the physiological variants HbF and HbA$_2$, human red blood cells may contain variants which confer disadvantages in certain circumstances, or are frankly pathological. The most famous is sickle-cell haemoglobin or HbS. Homozygous sickle cell subjects, who have inherited the gene for HbS from both parents, are prone to crises during which deformed or sickled cells become entangled with each other in the small blood vessels, thus occluding them. The cells become distorted because the haemoglobin within them crystallizes, forming long fibre-like polymers. Since the distorted cells are constantly being removed from the circulation, the patient is anaemic and, in addition, the hypertrophied bone marrow becomes exhausted. The HbS crystallizes because it is of low solubility at reduced oxygen tensions. The nature of the peculiarity which causes the lowered solubility was explored by splitting the HbS globin chains with enzymes and subjecting the fragments to electrophoresis and chromatography. It was found that HbS differs from HbA solely in having valyl instead of glutamyl at position 6 of the β-chains. Thus it is designated $\alpha_2^A \beta_2^{6\,VAL}$ meaning that the α-chains are the normal ones of HbA and in the β-chains position 6 has valyl instead of the glutamyl found in HbA. This small change is sufficient to lower the solubility of the entire molecule with dire effects on the erythrocyte at low oxygen tensions, as described above. Sickle cell anaemia is an autosomal recessive conditon (see glossary on p. 234) and so subjects with sickle cell trait only, that is with about half their haemoglobin in the variant form (they may be designated HbAS), live normal lives and are free from crises as long as they are not subjected to anoxia. The sickle cell gene occurs in populations in various parts of the tropics but principally in Central and West Africa (and therefore in modern times in the Americas) and it has been established that its survival is due to the advantages possessed by heterozygotes (sickle trait) in regions where falciparum malaria is endemic. Erythrocytes with HbAS sickle when infected with the parasite and are removed from the circulation; in HbA erythrocytes the parasites are not at risk of removal and do very nicely.

Other important haemoglobinopathies/haemoglobin variants are as follows.

Thalasaemias are due to mutations in the regulator genes which control peptide chain synthesis. In β-thalasaemia, wherein the normal synthesis of the β-chains is repressed, there is a compensatory increase in the synthesis of other chains, and the excess of α-chains combine with the δ- and γ-chains which are the normal constituents of HbA$_2$ and HbF respectively. However, much of the haemoglobin is in the form of α_4, which does not have the normal oxygen dissociation curve and fails to exhibit a Bohr effect. This of course has dire effects on proper oxygen loading and unloading.

HbC is a variant in which the life span of the red cell is slightly shortened, and in which the haemoglobin can be designated $\alpha_2^A \beta_2^{6\ LYS}$. In haemoglobin SC disease there is sickling and it has the severity of homozygotic sickle cell disease.

HbD (Punjab) is a haemoglobin in which the β-chain 121 glutamate is merely amidated. A charged amino acid side chain is therefore replaced by a less soluble neutral one. Heterozygotes (HbAD) are essentially normal but HbDS is associated with a pathological condition which is, however, less severe than homozygotic sickle cell anaemia.

13.4 Peptide sequences

Once isolated and purified, the amino acid sequence of a polypeptide can be determined by a combination of techniques. (There is no clear-cut distinction between proteins and polypeptides—a polypeptide is merely a small protein.) The first step is usually to determine the N-terminal, that is the amino acid with a free amino group. Classically, the Sanger reagent has been employed. In alkali it reacts with the amino group to yield a stable dinitrophenyl peptide (Fig. 13.4.1). When hydrolysed with strong acid in the normal way the dinitrophenyl peptide gives a series of amino acids, plus the yellow dinitrophenyl amino acid representing the N-terminal. It is identified by chromatographic means using a series of known dinitrophenyl amino acids for comparison.

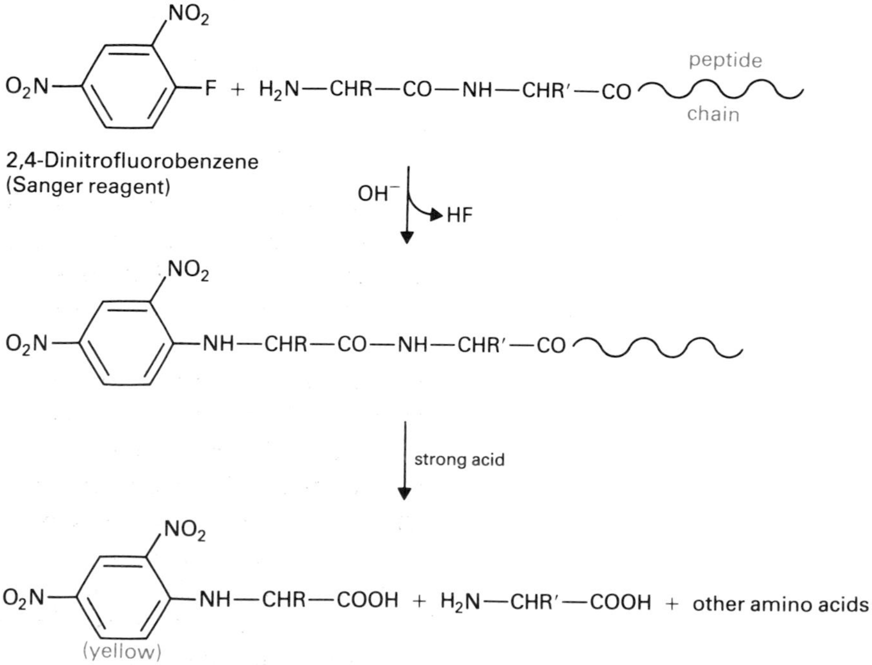

Fig. 13.4.1

Obviously, this method cannot determine the sequence at the terminal, only the N-terminal itself. Peptides can also react in alkali with phenylisothiocyanate to yield a phenylthiocarbamoyl peptide which when treated with weak acid gives a phenylthiohydantoin of the terminal amino acid as well as the remaining, intact sequence. The new N-terminal can then be reacted and the new phenylthiohydantoin isolated for identification (Fig. 13.4.2). This is evidently a sequential method, and has been powerfully adapted in the manufacture of automatic 'sequencers'.

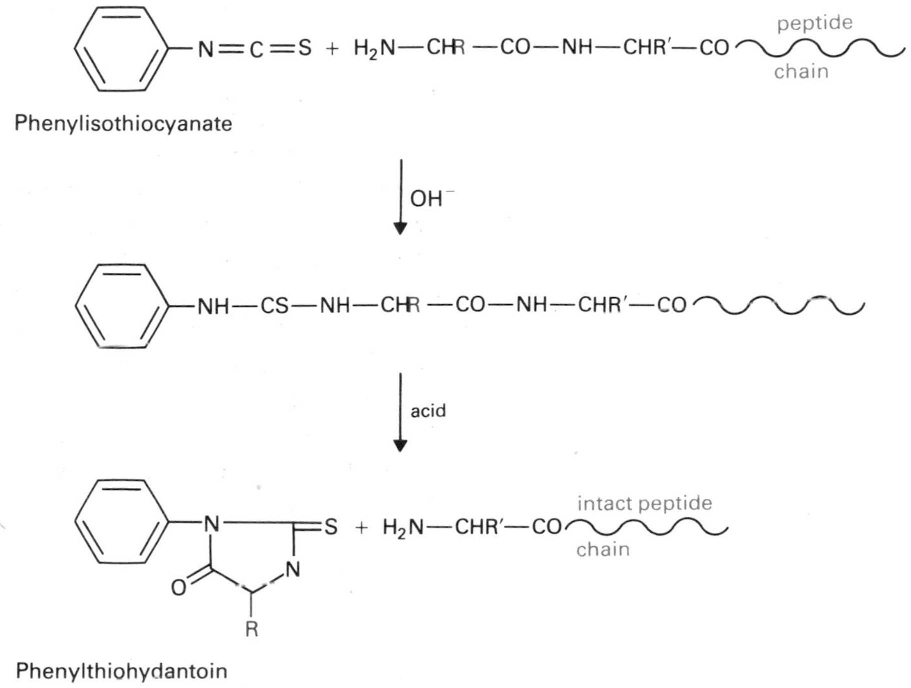

Fig. 13.4.2 The Edman method for sequencing peptides.

There is also an enzymic method for the N-terminal, utilizing the enzyme leucine aminopeptidase. This is a pancreatic endopeptidase, that is it hydrolyses the peptide bond at the end of the peptide chain. The description 'leucine' is inaccurate, for the enzyme splits a variety of amino acids from the N-terminal end. When a peptide is incubated with the enzyme and the amino acids released in turn by the enzyme are identified, then, under ideal conditions the N-terminal sequence can be determined.

Methods for the C-terminal are not so plentiful. There is a pancreatic exopeptidase which hydrolyses the peptide bond attaching the C-terminal amino acid. This enzyme, carboxypeptidase, is employed in the same way as leucine aminopeptidase.

Sequence determination has elucidated the primary structures of a large number of effector peptides from the brain. Interrelationships have been found among the following.

ACTH (adrenocorticotrophic hormone), is synthesized in the adenohypophysis. It is the trophic hormone for the adrenal cortex. It has 39 amino acids, of which only the first 23 are required for biological activity.

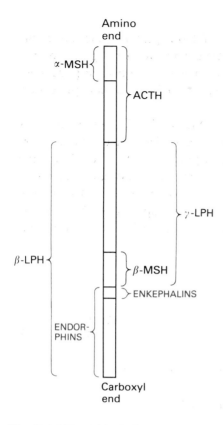

Amino
end

α-MSH

ACTH

γ-LPH

β-LPH

β-MSH

ENKEPHALINS

ENDOR-
PHINS

Carboxyl
end

Fig. 13.4.3 Economies in the synthesis of peptide sequences.
The original precursor (transcribed from the mRNA) is in fact a much bigger peptide. The abbreviations are explained in the text. The segments are drawn in rough proportion only.

It increases the synthesis of corticoids by the adrenal and has a small melanocyte-stimulating activity.

α- and *β-MSH* (melanocyte stimulating hormones) cause dispersion of pigment in the melanophore cells of cold-blooded animals and have some effect on the deposition of melanin in human skin. Cortisone inhibits MSH secretion, as well as ACTH secretion. That is why there is often pigmentation of the skin in adrenocortical insufficiency when cortisone is lacking.

β-lipotropin (LPH) causes lipolysis in experimental animals but is thought to have little endocrine function in humans; it has 1/20th the ACTH activity of ACTH itself.

Endorphins are a series of peptides of 31 amino acids, isolated from the brain, which bind to the same receptors as opium. Both the endorphins and opium, therefore, are anodynes (pain relievers).

Enkephalins are another series of peptides, but of five amino acids, also with morphine-like activity. They are less potent than the endorphins however.

It was observed in relation to these peptides that:

1 to some extent they have overlapping biological activity, as above;
2 they have some amino sequences in common;
3 they tend to be synthesized simultaneously.

The explanation is that they all have a common precursor, and all represent pieces, in some cases pieces of pieces, of that precursor (Fig. 13.4.3) which has been isolated from a mouse pituitary tumour cell line. Thus α- and β-MSH are 'segments' of ACTH and β-lipotropin respectively. The endorphins are segments of β-lipotropin, and the enkephalins are segments of the endorphins.

Such homologies are found in other peptide systems. Thus insulin of the β-cells of the Islets of Langerhans and relaxin from the corpus luteum (relaxing the symphysis pubis) both consist of an A-chain and a B-chain, linked by disulphide bonds, with a number of sequences in common. The pituitary hormones TSH (thyroid-stimulating hormone or thyrotropin), luteinizing hormone (LH), follicle-stimulating hormone (FSH or follitropin) and human chorionic gonadotrophin (HCG) all consist of α and β subunits, the subunits being almost identical in each hormone. Evidently these homologies allow economies in the maintenance of the pool of structural genes.

13.5 Complex lipids

In the context of energy supply and of the storage of energy-yielding substances, the only lipid to be considered is triacylglycerol (triglyceride). In the context of lipoproteins and membranes cholesterol is also very relevant, but along with its derivatives also constitutes a fairly simple class of lipid structures. However all tissues, quantitatively the most notable of which are brain and nerve, contain a range of very complex lipids. Although it is tedious to plough through the structures of these lipids, they are too important to neglect even in an elementary treatment: questions purely on the chemistry of tissue constituents are for some reason rare in essay papers but you are likely to have to deal with complex lipids in the multiple choice section.

A scheme of classification is given in Fig. 13.5.1. The problem arises because a given lipid or class of lipid can be viewed and therefore named in different ways. Thus lipids containing phosphate and the base sphingosine are described equally as phospholipids or sphingolipids. As can be seen from the diagram, sphingomyelins are a subgroup of both. Gangliosides, sulpholipids and cerebrosides are at the same time glycolipids and sphingolipids. These interconnections dealt with, it remains to list and describe the several types.

Fig. 13.5.1 A classification of lipids. Many of the terms are overlapping, so that many 'glycolipids' are also 'sphingolipids'. In the older terminology a 'neutral fat' was one without ionizing groups, for example triacylglycerols and waxes. Phosphatides are derivatives of phosphatidic acid.

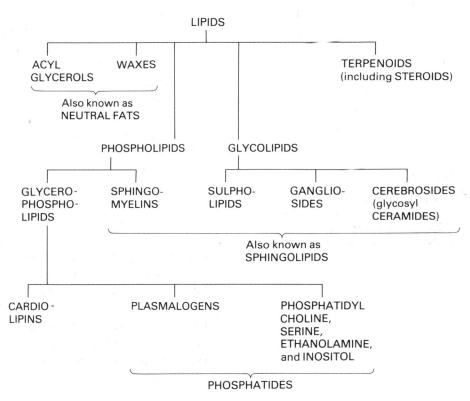

Fig. 13.5.2 Lecithin.

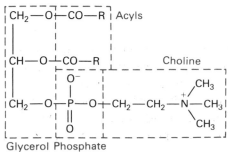

Glycerol Phosphate

Phosphatides are a class of glycerol-containing phospholipids (glycerophospholipids). They contain, in addition, fatty acids and an organic base. By far the best known is lecithin, which is present in all mammalian cells and is a constituent of such foodstuffs as soya beans and eggs. Its structure is shown in Fig. 13.5.2. Lecithin is evidently not a single substance, since the acyl groups (—CO·R) derived from fatty acids may vary. A more systematic name for lecithin is phosphatidylcholine and this shows it to be a derivative of phosphatidic acid, like the related phosphatides (Fig. 13.5.3). If the acyl

201

Fig. 13.5.3 Phosphatides.

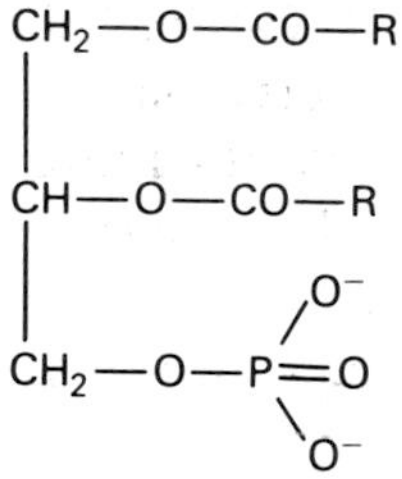

$$CH_2-O-CO-R$$
$$CH-O-CO-R$$
$$CH_2-O-P{=}O \quad (O^-, O^-)$$

Phosphatidic acid

$$CH_2-O-CO-R$$
$$CH-O-CO-R$$
$$CH_2-O-P(O)(O^-)-O-CH_2-CH_2-NH_3^+$$

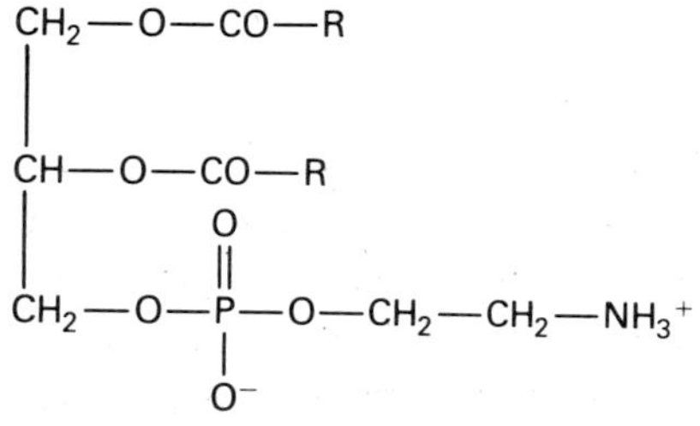

Phosphatidyl ethanolamine

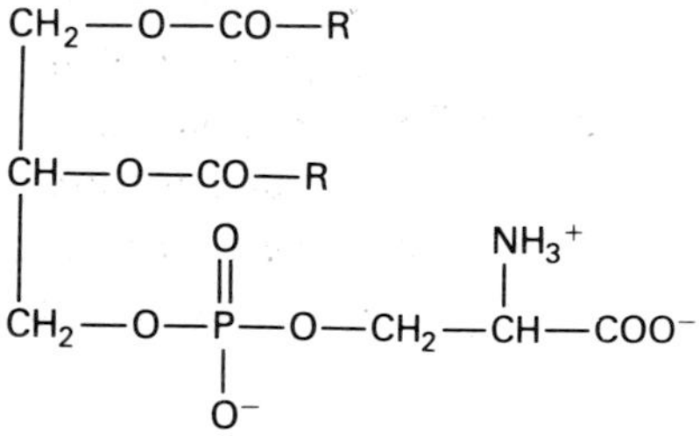

$$CH_2-O-CO-R$$
$$CH-O-CO-R$$
$$CH_2-O-P(O)(O^-)-O-CH_2-CH(NH_3^+)-COO^-$$

Phosphatidyl serine

$$CH_2-O-CO-R$$
$$CHO-CO-R$$
$$CH_2-O-P(O)(O^-)-O-\text{(inositol ring)}$$

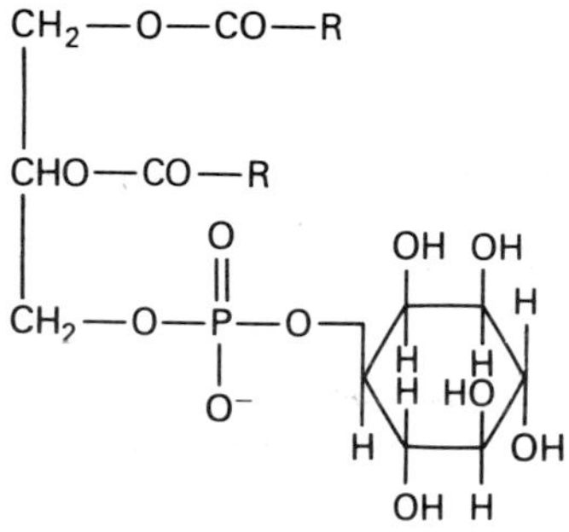

Phosphatidyl inositol

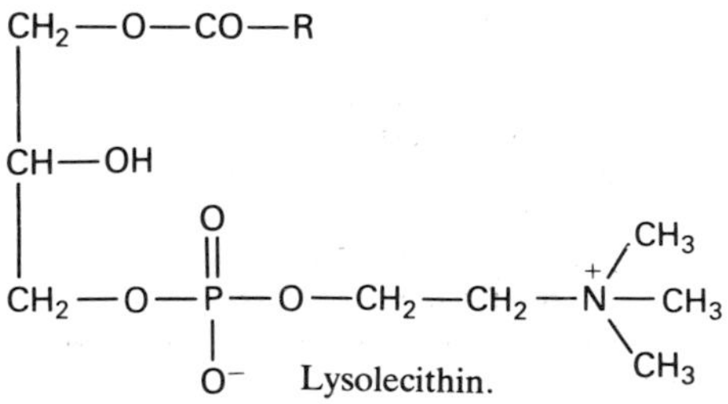

$$CH_2-O-CO-R$$
$$CH-OH$$
$$CH_2-O-P(O)(O^-)-O-CH_2-CH_2-N^+(CH_3)_3 \quad \text{Lysolecithin.}$$

group on C-2 is released by hydrolysis the resulting compounds have powerful detergent activity, to the extent that they can rupture or lyse red blood cells. These are the lysophospholipids, for example lysolecithin (Fig. 13.5.3).

Plasmalogens are similar to lecithins but have an unsaturated alcohol in place of a fatty acid. Their general structure is shown in Fig. 13.5.4. The vinyl group being $-CH_2{=}CH_2-$, the C-1 has a vinyl ether attachment. Hydrolysis yields $HO-CH_2{=}CH_2R$ which is an unsaturated alcohol.

$$CH_2-O-CH{=}CH-R$$
$$CH-O-CO-R$$
$$CH_2-O-P(O)(O^-)-O-CH_2-CH_2-N^+(CH_3)_3$$

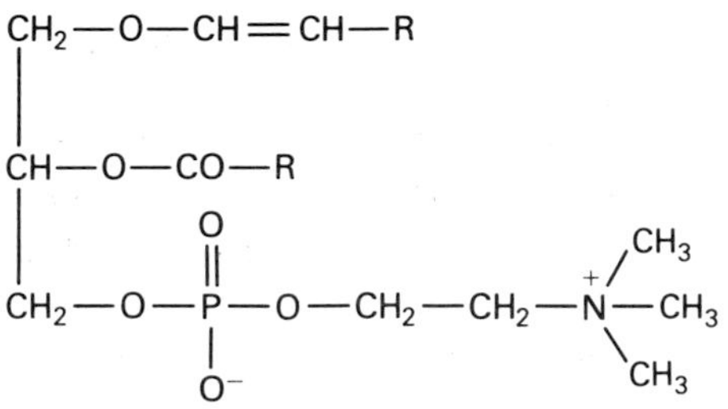

Fig. 13.5.4 Plasmalogen structure.

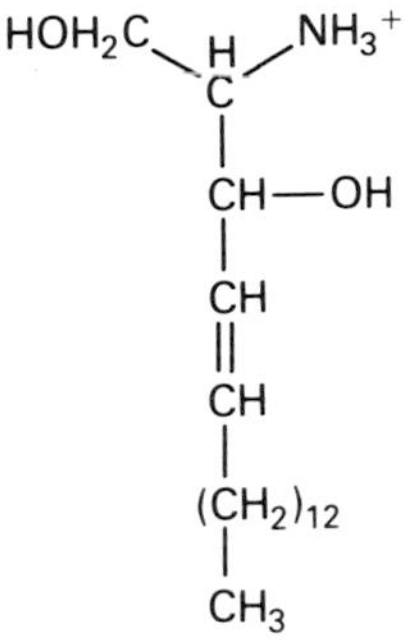

$$HOH_2C-\underset{H}{C}-NH_3^+$$
$$CH-OH$$
$$CH$$
$$CH$$
$$(CH_2)_{12}$$
$$CH_3$$

Fig. 13.5.5 Sphingosine.

Cardiolipins are diphosphatidylglycerols. Thus their summary formula is 1,2-diacylglycerol – phosphate – glycerol – phosphate – 1,2-diacylglycerol.

Sphingolipids all contain the complex base, sphingosine (Fig. 13.5.5). The terminal hydroxyl can be esterified to phosphate and the amino group may be *N*-acylated. A *sphingomyelin* in summary structure is

Sphingosine—Fatty acid
|
Phosphate—Choline

Ceramides are similar, containing *N*-acylated sphingosine.

Gangliosides consist of *N*-acylated sphingosine (ceramide), glucose, galactose, acetylgalactosamine and a sugar base, *N*-acetylneuraminic acid (NANA). A typical arrangement is

Acylsphingosine—Glucose—Galactose—Acylgalactose—Galactose
|
NANA

Cerebrosides contain a molecule of glucose or galactose (glucocerebrosides and galactocerebrosides respectively). The monosaccharide is glycosidically linked to *N*-acetylated sphingosine. The general structure, then, is

Sphingosine—Fatty acid
|
Monosaccharide

Sulpholipids are related to cerebrosides; it is generally the galactose residue which is sulphated.

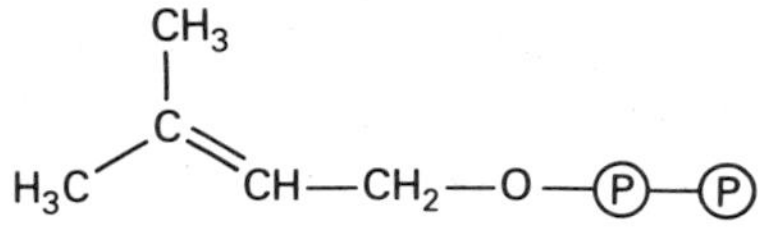

Fig. 13.5.6 Isopentenyl pyrophosphate.

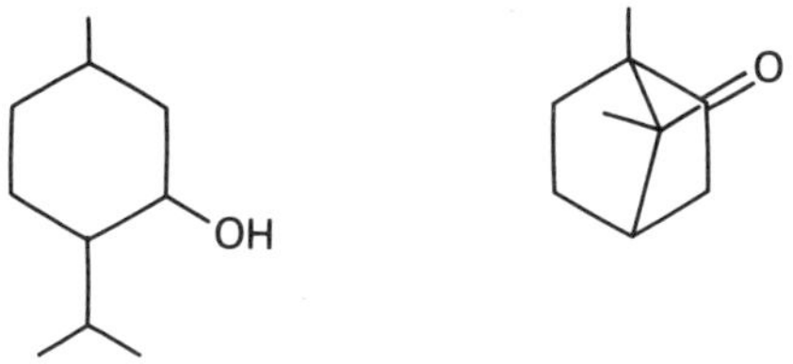

Menthol   Camphor

Fig. 13.5.7

Terpenoids or *isoprenoids* comprise a large group of substances derived in living tissues from isopentenyl pyrophosphate (Fig. 13.5.6). They include the steroids (13.1), rubber, carotenoids (the vitamin A precursors in plants), retinol, tocopherols, and such substances as camphor and menthol (Fig. 13.5.7). Since, after polymerization, the isoprene unit is subject to a great deal of modification, there is no general formula for the group.

Waxes are fatty acids esterified not to glycerol but to long-chain alcohols. Their hydrolysis can thus be represented as

$$R-CO-O-R \rightarrow RCOOH + ROH$$

Wax   Fatty acid   Alcohol

Nothing has been said so far about the occurrence of these lipids, although some of them have cropped up in other contexts. Their distribution is summarized in Table 13.5.1 which also indicates some pathological conditions in which there is accumulation or accretion of lipid in the tissues. Notably, there is derangement of the metabolism of many of the complex lipids in the inherited diseases of metabolism termed *lipidoses*. In Niemann-Pick disease for example, there is an inherited deficiency of lysosomal sphingomyelinase so that the cells of most tissues of the body, but especially spleen and liver, become loaded with sphingomyelin. Under normal conditions, of course, there is a balance between synthesis and degradation. Since the catabolizing enzymes are for the most part situated in the lysosomes, such conditions have often been termed 'lysosomal diseases'. Some of the glycogenoses (13.13) also come into this category.

Table 13.5.1 Distribution and associated pathologies of the lipids

Lipid	Occurence	Associated conditions
Acylglycerols	Animal depot fat; lipoproteins; plant oils	Obesity
Cholesterol and its esters	Membranes and lipoproteins in animals (not plants)	Ischaemic heart disease
Waxes	Plant and insect surfaces	—
Phosphatides	Membranes; lipoproteins; alveolar surface; egg yolk; plants	—
Plasmalogens	Membranes; nervous tissue	—
Spingomyelins	Membranes; lipoproteins	Niemann-Pick disease*
Sulpholipids	Brain	Metachromatic leuco-dystrophy*
Gangliosides	Brain	Tay-Sachs disease*
Cerebrosides	Myelin sheath	Gaucher's disease* Fabry's disease*

*lysosomal disease, with accumulation of relevant lipid

13.6 An organ: the brain

There is perhaps no rigorous *a priori* reason why the brain's metabolism should be peculiar but it is difficult to avoid the impression that it is up there doing something special, even if in humans it might seem to offer more analogies to a tumour than anything else. In fact there are a number of metabolic idiosyncracies to be discussed if you are asked to write about the biochemistry of the brain. They fall under perhaps four heads.

1 *Composition* The brain contains more lipid (11% of 'wet' weight) than any other tissue except adipose tissue. Many of these lipids are found only in the brain as constituents of the myelin. These complex lipids (cerebrosides, sulpholipids, plasmalogens and gangliosides) are classified and described in 13.5. There are few 'simple' lipids (fatty acids and triacylglycerols) in the brain. Also, it has little glycogen and this deficiency no doubt contributes to its heavy dependence on blood glucose (below). Turning to the cells, there are two specialized types, the *neurones* and the *glia*. Each neurone has a long process called an axon which terminates at the *synapse* on the cell which the neurone innervates. The glia may form the blood-brain barrier (below) and support the neurones metabolically.

2 *Dependence on glucose and oxygen* The brain's dependence on oxygen is easily appreciated by visualizing the effects of a few minutes choking or throttling. Although only 3% of the body weight, the brain accounts for 20–25% of the body's oxygen consumption, that is 50 ml oxygen per minute. This stays relatively constant, in contrast for example to the fluxes in muscle. (The oxygen consumption of muscle can vary by a factor of five according to the amount of exercise.) The large amount of oxygen entering the brain is consumed in oxidizing glucose, and energy is obtained for the translocation of ions important for electrical activity and for the very active biosynthetic mechanisms taking place. Even during intense mental effort, for example learning biochemical pathways, there is no increase in the oxygen consumption of the brain. In ordinary sleep oxygen consumption does not drop. However in the type of sleep associated with rapid eye movements, REM, there is an *increased* oxygen consumption.

Since glucose is the principal source of carbon for the reactions leading to oxygen consumption, the brain is equally dependent on blood glucose and when this falls to about 2 mmol/l the brain no longer functions properly, as shown by the lapse into coma. The brain is not sensitive to insulin, so fluctuations in its blood concentration do not affect the glucose uptake. This is undoubtedly an advantage, offsetting the fact that the brain has little or no glycogen store. Uptake of glucose, of course, involves its phosphorylation. In the brain this reaction is catalysed by hexokinase, which has a Km for glucose of about 5×10^{-5} mol/l; this means that hexokinase has a high affinity for glucose, as compared to the phosphorylating enzyme in liver, glucokinase, with $K_m = 5 \times 10^{-2}$ mol/l. So the brain can still take up glucose when it is low in concentration, whereas the liver tends only to take it at the higher concentrations. The catabolism of glucose in brain involves primarily the tricarboxylic acid cycle but some glycolysis with lactate and pyruvate as end-products also occurs. In starvation when glucose in the blood is low and fat is being mobilized from the adipose tissue, yielding ketone bodies, the latter become substrates for the production of energy. They can never replace more than 50% of the glucose normally utilized by the brain, however. It is the cardinal principle of brain biochemistry that the organ is absolutely dependent on blood glucose.

3 *Selectivity permeability to blood components* Certain dyes, when injected into the bloodstream, permeate all tissues except the brain. This led to the concept of the *blood brain barrier*: a mechanism, or series of mechanisms, to select those substances which may pass freely from the circulation into the brain. Those which pass freely are (of course) oxygen, glucose, and ketone bodies; also water and ethanol, mannose and maltose, and most essential amino acids. Those rejected are fructose, the hydrogen ion, glutamate, aspartate, γ-aminobutyrate, and glycine, as well as most antibiotics.

The barrier can of course be seen as another protective mechanism, protecting the brain from fluctuations in the composition of the plasma and from any toxic or foreign substances it may contain.

The blood brain barrier does not function in very young animals. This is why neonatal jaundice is so dangerous; bilirubin can bind irreversibly the brain tissue (kernicterus) and cause mental retardation. In phenylketonuria, phenylalanine cannot be converted to tyrosine and is metabolized to other intermediates such as phenylpyruvic acid. This cannot be rejected by the brain in the very young and similarly causes damage with mental retardation.

4 *Amino acid and neurotransmitter metabolism* Glutamate, one way or the other, is the paramount amino acid in brain metabolism. Although the brain may have a urea cycle (10.2), ammonia is detoxicated by glutamine formation (Fig. 13.6.1). Glutamate is the source of γ-aminobutyrate (GABA) which is probably a neurotransmitter, although it may enter the citric acid cycle.

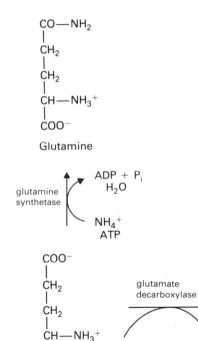

Fig. 13.6.1 Glutamate metabolism in the brain.

Neurotransmitters are the chemical messengers which convey electrical impulses from one neurone to another. Thus when an electrical impulse travels down an axon to the presynaptic region it depolarizes the terminal membrane (Fig. 13.6.2). This results in the release of a neurotransmitter which diffuses across the cleft to the postsynaptic membrane, where it reacts with a specific receptor. The interaction causes a postsynaptic potential on the neurone containing the receptor. The neurotransmitter in the synaptic cleft is meanwhile rapidly inactivated. It is the excitatory neurotransmitters which produce the postsynaptic potential (a depolarization of the membrane due to Na^+ entering and K^+ leaving the cell). Inhibitory neurotransmitters cause return to a resting potential, or stabilize (in the sense of decreasing sensitivity to stimulation) undepolarized membrane segments. The known neurotransmitters are all small molecules and amines, and probably many more will be discovered in due course; so far, it has been determined that a given neurone secretes only one neurotransmitter. A list of neurotransmitters includes the following compounds.

Fig. 13.6.2 A synapse.
This is a representation of a cholinergic synapse. Acetylcholine is synthesized in the synaptic knobs and forms vesicles. The wave of depolarization sweeping down the axon releases the acetylcholine into the synpatic cleft. It combines with receptors in the postsynaptic membrane and this interaction alters the permeability properties of the membrane, such that Na^+ enters and K^+ leaves. Thus the wave of depolarization is propagated to the next cell. Meanwhile the acetylcholine is rapidly hydrolysed (and thus inactivated as a neurotransmitter) by cholinesterase.

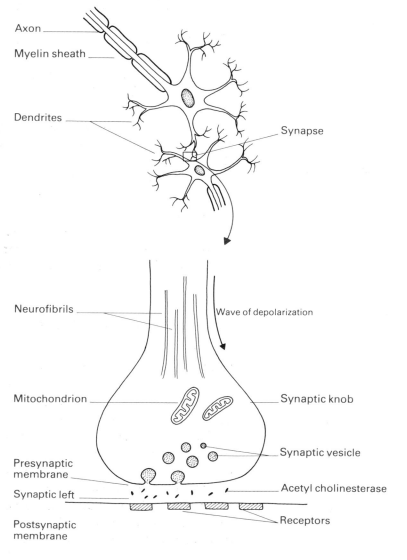

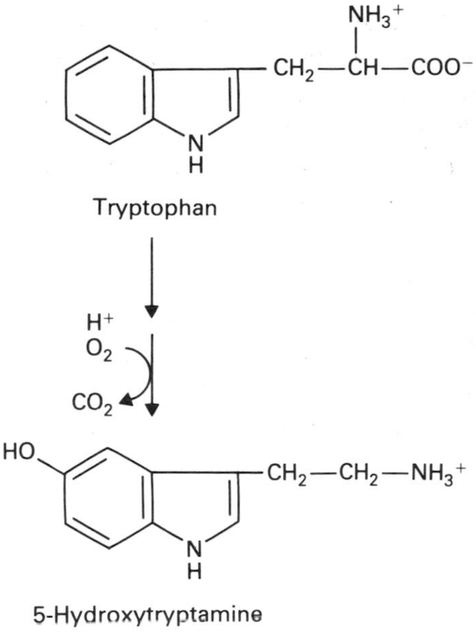

Tryptophan

H^+
O_2

CO_2

5-Hydroxytryptamine

Fig. 13.6.3

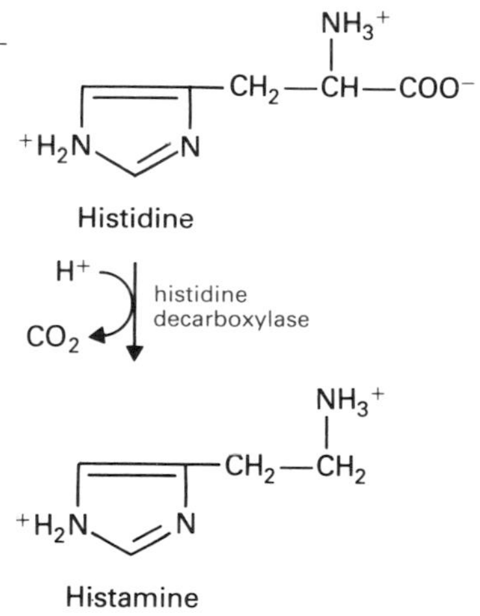

Histidine

H^+ histidine decarboxylase

CO_2

Histamine

Fig. 13.6.4

(1) *Acetylcholine*: in *cholinergic* nerve endings acetylcholine is released. The receptors on its target cells have been shown to be of two types: one stimulated by muscarine as well as acetylcholine; and another stimulated by nicotine as well as acetylcholine. Acetylcholine is destroyed by hydrolysis by cholinesterase in the synaptic cleft.

(2) *Catecholamines* are (loosely) amines based on 1,2-dihydroxylbenzene. Dopamine, adrenaline, and noradrenaline are the catecholamine neurotransmitters. They are all synthesized from tyrosine (Fig. 11.8.10). The receptors for catecholamine are also of two types, designated α and β. They are distinguished by their separate inhibition or blockage by drugs. In the synaptic cleft these transmitters are destroyed by the enzymes monoamine oxidase and catechol-*O*-methyltransferase.

(3) *Serotonin* is synthesized from tryptophan by hydroxylation and decarboxylation (Fig. 13.6.3). Its role is not established.

(4) *Glutamate* is probably an excitatory neurotransmitter in mammals.

(5) γ-*Aminobutyric acid* (GABA) is an inhibitory neurotransmitter in the brain and spinal cord. Parts of the brain in patients with Huntington's chorea, a genetic disease characterized by involuntary movement, are abnormally low in GABA.

(6) *Glycine* is a neurotransmitter in inhibitory neurones in the spinal cord.

(7) *Histamine* is a possible neurotransmitter. It is produced by decarboxylation of histidine (Fig. 13.6.4). Histamine is of course also associated with the mast cells; its release from them is an aspect of the allergic reaction.

Thus, in summary, we have described the dependence of the brain on oxygen and glucose. Some defence mechanisms have been described (the blood brain barrier, the lack of sensitivity to insulin, the low K_m of hexokinase). Some neurotransmitters have been described. You would hardly be expected to know about the biochemistry of memory or consciousness at this stage; indeed little of these subjects is known by anybody.

13.7   A tissue: skeletal muscle

Some problems remain uninvestigated because their very existence is difficult to conceptualize. However the nature of animal locomotion does not come into this category, since the muscle twitch and muscle activity are such prominent features of most forms of life that they have cried out for investigation since the dawn of biology.

Like all tissues, skeletal muscle is composed of cells, but these are highly eccentric, being packed with striated material and having multiple nuclei. On closer inspection the striated material is found to be composed of a number of individual strands called *myofibrils*. Between the myofibrils lie glycogen granules and mitochondria which are evidently involved in providing energy for the concentration of the myofibrils, which is the basis of the contraction of the whole muscle. Of the three histologically-distinct kinds of muscle—skeletal, smooth, and cardiac—it is usual to discuss only skeletal, for the simple reason that most is known about it.

In discussing contraction, you have to fill in some of the details of a story which essentially goes like this: a nerve impulse arriving at the junction between the nerve ending and the muscle (the motor end-plate) depolarizes the outer membrane of the muscle cell, the *sarcolemma*. The surface depolarization is propagated to the interior via the T-tubules (Fig. 13.7.1) which are a series of invaginations of the sarcolemma. In turn they transmit the signal to a highly intricate system of membranes round the myofibrils, the *sarcoplasmic reticulum*. (It is a highly specialized form of the endoplasmic reticulum.) Depolarization of the T-tubules in effect causes a release of calcium from those regions of the sarcoplasmic reticulum which are adjacent.

Fig. 13.7.1 Organization of the skeletal muscle cell.

The cell is surrounded by a membrane, the sarcolemma, which contains discrete fibrils, the myofibrils. The spaces between the myofibrils contain nuclei, glycogen granules, and mitochondria.

At the top of the diagram two myofibrils are represented, one surrounded by T-tubules ('T' for transverse) and sarcoplasmic reticulum. A representation of the complex striated pattern of the myofibril at high magnification is shown. On contraction, since the thick and thin filaments slide across each other, the A-band stays relatively constant while the H-zone disappears and the I-band becomes smaller.

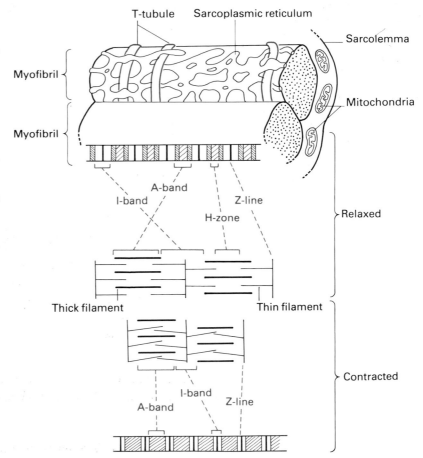

The released calcium binds to the protein troponin in the myofibrils, causing a conformational change such that troponin reacts with another protein, tropomyosin. In relaxed muscle tropomyosin prevents the interaction of myosin and actin, but when it interacts with troponin the constraint is removed, myosin and actin also interact, and ATP is hydrolysed (Fig. 13.7.2). Myosin is the protein of the thick filaments, and actin of the thin filaments, so the latter slides between the former. This is the fundamental energy-consuming, that is ATP-consuming, physical movement. Relaxation involves a regeneration of the whole system, including replenishment of ATP.

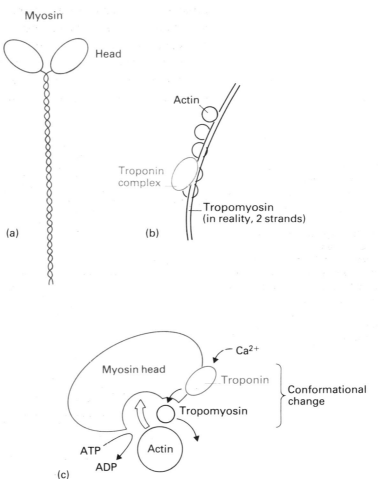

Fig. 13.7.2 The elements of muscular contraction.
The thick filaments are composed of myosin (a) with the individual molecules stacked into a fibre. The thin filaments (b) are composed of long fibres of actin, with which strands of tropomyosin and, intermittently, the troponin complex are associated. The tropomyosin is in fact a double helix, but only a single strand is indicated; also, the actin is really a double strand of bead-like structures. These simplifications are offered in the cause of reproducibility, since considerable artistic skill would be required to draw the full structures—this can be pointed out in an exam. In (c) the functional relationships are indicated. Calcium initiates the contraction through troponin and tropomyosin, which are here imagined in a sort of transverse section across the myofibril. The arrows indicate conformational change more than movement, and the shapes are purely arbitrary.

Details can be filled out under the headings of (1) the proteins involved and their interactions and (2) ATP as the energy supply, and its sources.

Proteins involved in muscular contraction

Myosin in the thick filaments is a large globulin, that is, by definition, it is insoluble in water. It has the property of catalysing ATP hydrolysis, so it is an ATPase. It is a very lengthy molecule, consisting of two polypeptide chains wrapped round each other. Each of the chains has a globular head (Fig. 13.7.2).

Actin, in the thin filaments, is also a globulin, existing in two forms; globular (G-actin) and fibrous (F-actin). G-actin can polymerize to F-actin in the presence of salt. Removal of salt or addition of ATP causes depolymerization.

Troponins are associated with actin in the thin filaments. The troponin complex consists of three subunits, designated T, C, and I. One troponin complex is associated with about seven molecules of actin. The subunits have different functions but in general they interact with tropomyosin in the presence of calcium to permit the mechanics of the all-important actin-myosin interaction.

Tropomysin is a thin, double-stranded, coiled-coil α-helix. In the thin filaments two strands of F-actin are wound round one another and tropomyosin is located in the grooves formed between these strands. In the current theory of muscle contraction, called the 'steric blocking model', it is suggested that tropomyosin blocks the interaction of actin and myosin and this inhibition must be removed before contraction can take place.

Thus there are two contractile/structural proteins (actin and myosin) and two regulatory proteins (troponin and tropomyosin).

On stimulation of the muscle, the head of the myosin molecule binds to actin; before interaction, ADP is bound to myosin, which in that state has a low ATPase activity. As calcium is released from the endoplasmic reticulum, however, interaction begins to take place

$$\text{Myosin---ADP---P}_i + \text{actin} \rightleftharpoons \text{actin---myosin} + \text{ADP} + \text{P}_i$$

ATP can now be taken up

$$\text{Actin---myosin} + \text{ATP} \rightleftharpoons \text{actin} + \text{myosin---ATP}$$

In the absence of ADP myosin has a high ATPase activity

$$\text{Myosin---ATP} \rightleftharpoons \text{myosin ADP---P}_i$$

The hydrolysis of the ATP provides the energy for a change of conformation of the myosin head, allowing it to twist relative to the actin and to pull the thick and thin filaments alongside each other, as summarily represented in the first reaction above. Myosin—ADP—Pi was the starting material for the whole cycle, which can therefore start again when the muscle is re-stimulated, should ATP be available.

ATP as the energy supply for muscular contraction

The ATP in skeletal muscle is sufficient to maintain vigorous contraction for a second or so. It must be replenished, and there are three mechanisms for replenishment.

1 *By utilization of creatine phosphate* Creatine phosphate is a muscle

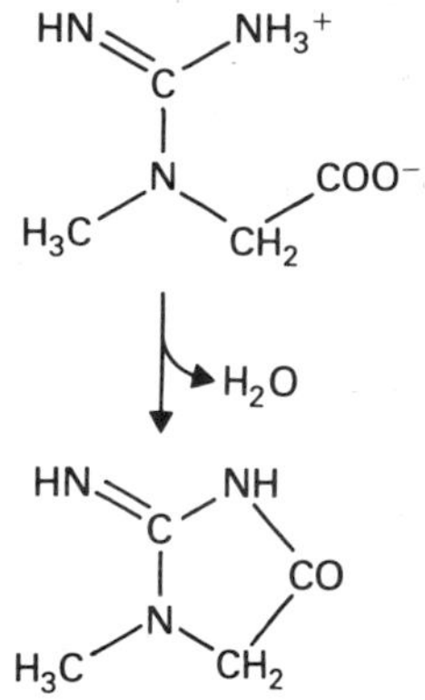

Fig. 13.7.3 Creative cyclization

phosphagen, that is a donor of phosphate to ADP to yield ATP. In actively contracting muscle, creatine kinase catalyses

Creatine ~ P + ADP ⇌ creatine + ATP

When ATP later becomes available in resting muscle, creatine phosphate is synthesized by the reverse reaction. Creatine spontaneously cyclizes to creatinine (Fig. 13.7.3). Creatinine represents an end-product which cannot be utilized by the body, and which is taken up in the bloodstream for excretion by the kidney. The quantum of this excretion indicates the total muscle mass of an individual, and is relatively constant from day to day. It was previously thought that measuring it in a 24 hour sample of urine would indicate the completeness of that sample (patients often conceal the fact that they have failed to make a complete collection) but it is not sufficiently constant, it is now known, for that purpose.

Creatinine is filtered at the glomerulus but is not resorbed. If the concentration in the urine is compared with that in the blood a measure of kidney function (the 'creatinine clearance') is obtained.

2   *By vigorous glycolysis* In contracting skeletal muscle oxygen cannot be supplied at a sufficient rate to allow aerobic oxidation of the glucose from the glycogen stores. Glycolysis, although producing very little ATP compared to the citric acid cycle, does not need oxygen because its end product is lactate. Lactate, as well as some pyruvate, is removed by the blood during exercise, and is resynthesized to glucose by the liver using the enzymes of gluconeogenesis. (This is the *Cori cycle*.) This glucose is then available, in turn, to replenish muscle glycogen. There is concomitant oxidation in the muscle of fatty acids utilizing such oxygen as may be available. Fatty acids are the principal energy supply of resting muscle, which of course is supplied with sufficient oxygen to oxidise them. Fatty acids must also be the principal energy supply in a subject like a marathon runner who has exhausted his carbohydrate stores—he has to oxidize fatty acids rather than depleting his blood glucose, which is necessary for brain function.

3   *By the adenylate kinase or myokinase reaction* Myokinase catalyses transfer of a phosphate from one ADP to another

2ADP   → ATP + AMP

This is obviously limited by the amount of adenine nucleotide in muscle; therefore it is a last resort as regards the provisions of energy for muscular contraction.

The following points must be made, therefore, if you are asked about the biochemistry of skeletal muscle:

(1) the myofibril contains myosin and actin which interact to contract it;

(2) the actin-myosin interaction is modulated by calcium ions, acting on the regulatory proteins troponin and tropomyosin;

(3) the energy is derived from ATP, itself made available by anaerobic glycolysis and aerobic fatty acid oxidation;

(4) muscle does not release glucose, for it lacks a glucose phosphatase (but although it cannot directly maintain blood glucose, it releases lactate and pyruvate into the circulation, and by gluconcogenesis contributes to liver glycogen and thus ultimately to blood glucose);

(5) it is sensitive to insulin (glucose uptake increased) and adrenaline (glycogenolysis increased);

(6) resting muscle largely conducts aerobic oxidation of fatty acids, released from the adipose tissue by glucagon;

(7) muscle in the fasting subject provides a store of amino acids for gluconeogenesis. Branched-chain amino acids in the liver are converted to alanine and glutamine for conversion to glucose in the liver.

13.8 An organ: the liver

An older type of examination question asked students to discuss the liver as a 'chemical factory'. This was to stress the great synthetic and metabolic capacity of the liver, as well as its central role in metabolism. There is no doubt that quantitatively it is an important organ for metabolism, since it is about 1.5 kg in weight. It receives absorbed foodstuffs from the portal system, and amino acids and fatty acids mobilized for utilization, so there is no doubt that it is also 'central'.

Presented with absorbed foodstuffs or metabolites, the liver can carry out a range of synthetic activities, and the synthesized materials can either be stored or released into the bloodstream. Much of the synthetic capacity, however, is committed to processing toxic materials, or making them less toxic. As an extension of this detoxication function, the liver inactivates effector substances, like steroid hormones, to terminate their effector status. Thus bilirubin from lysed erythrocytes is conjugated and excreted in the bile; cholesterol is hydroxylated to bile salts and similarly voided; ammonia is incorporated into urea; drugs are hydroxylated and conjugated to render them more diffusible for excretion; alcohol is oxidized.

A round dozen of products/metabolites can be memorized to illustrate the range of the liver's activities.

1 *Ketone bodies* are normal products of lipid breakdown. Since some of the capacity of the tricarboxylic acid cycle in the liver is pre-empted for urea synthesis, there is limited ability to take up acetyl CoA and it is directed to ketone body synthesis. It is only on overproduction that ketogenesis is pathological (11.6), leading to acidosis and dehydration.

2 *Lipoproteins* The lipoproteins designated VLDL (very low density) and HDL (high density) are synthesized by the liver. LDL (low density) is a product of VLDL catabolism. The hepatocyte endoplasmic reticulum assembles apolipoproteins, cholesterol, triglyceride, and phospholipid into these lipoproteins; in the resulting water-soluble form triglyceride is transported to the muscles and adipose tissue. Cholesterol is transported to the liver in HDL for conversion to bile salts (13.16).

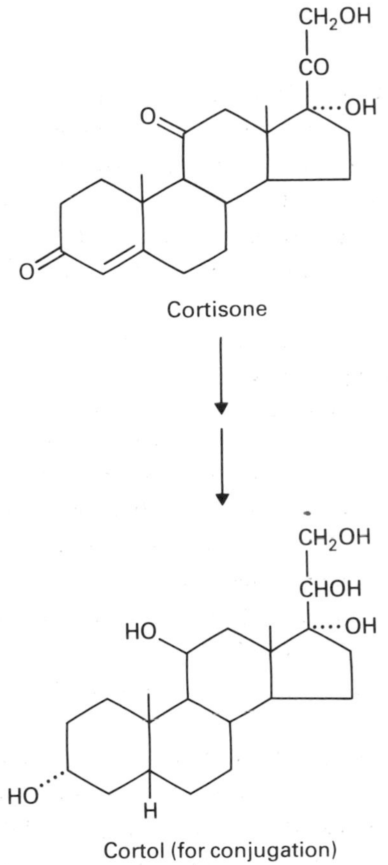

Cortisone

↓

↓

Cortol (for conjugation)

Fig. 13.8.1

3 *Plasma proteins* comprise fibrinogen, albumin, and the globulins. These are old-fashioned terms describing solubility. Albumin maintains the osmotic pressure of the blood and is a transport substance for fatty acids and free bilirubin. Its synthesis depends on the integrity of the liver and a fall in serum albumin is one indication of liver damage. The α- and β-globulins are of hepatic origin, but the γ-globulins are a product of lymphoid tissue.

4 *Glucose* arrives in the liver from the portal blood and, in a subject who has previously been fasting, is used to replenish liver glycogen (11.1). Conversely, at the onset of fasting, liver glycogen stores become depleted as glucose is released to maintain brain function. Liver possesses a glucose 6-phosphatase, unlike muscle which cannot release free glucose. In the absence of dietary glucose, the liver synthesizes glucose from amino acids, principally alanine, by means of the gluconeogenic pathways; after exercise, gluconeogenesis occurs from lactate and pyruvate (11.3). In times of glucose excess, in some subjects, glucose is converted to depot fat by way of acetyl CoA (11.4).

5 *Glycogen* can accumulate to about 5–6% of the fresh weight of liver; glycogenolysis and glycogenesis in the liver are enhanced by glucagon and insulin respectively (11.1).

6 *Bile salts* are synthesized as essential participants in fat digestion, but since a small amount of them are lost in the faeces they represent an excretory pathway for their precursor, cholesterol. They are voided into the gall bladder for storage, and the gut hormone cholecystokinin stimulates gall bladder contraction. The bile salts which are not excreted are partially resorbed by the small intestine and make another passage through the hepato-biliary system; they are said to undergo an enterohepatic circulation (11.7).

7 *Cholesterol* is incorporated into the lipoproteins in the liver and is of both exogenous (from the diet) and endogenous (*de novo* synthesized) origin. The reason for the active synthesis of about 1 g/d cholesterol by the liver is unknown. It is also secreted in the bile and is partially reabsorbed before it undergoes an enterohepatic circulation. Precipitation of cholesterol in the gall bladder is responsible for gallstones.

8 *Bilirubin diglucuronide* is the non-toxic, diffusible derivative of bilirubin, secreted into the bile and excreted in the faeces after transformation by the bacterial flora of the large intestine (10.4).

9 *Drug and hormone derivatives* are synthesized in the smooth endoplasmic reticulum of the hepatocytes. For example, corticosteroids are inactivated by ring and ketone function reduction (Fig. 13.8.1). The reduced derivatives are conjugated with glucuronic acid and excreted in the urine.

10 *Coenzymes* such as TPP, CoA, and FAD are synthesized in the liver from their mainly B-vitamin precursors.

11 *Retinyl esters* are synthesized from incoming vitamins A (13.20) and stored in the liver. It appears that in one form or another vitamins A are plentiful in a mixed diet and continual secretion into the blood of the active forms, retinal, retinol, and retinoic acid, might be potentially harmful in view of their toxicity. They are therefore converted to the alcohol (retinol), esterified, and stored in this detoxicated form.

12 *Ferritin* acts as a reserve of iron in the liver. It consists of a group of apoproteins with different affinities for iron, and therefore constitutes both a labile and a long-term store.

Virtually every metabolic pathway hitherto covered is necessary for these fluxes in the liver: glycolysis in the production of acetyl CoA; the hexose monophosphate shunt for NADPH production; glycogenolysis; glycogenesis; gluconeogenesis; protein synthesis. Liver biochemistry can be profusely illustrated by basic biochemical knowledge.

Figure 13.8.2 summarizes the liver biochemistry dealt with.

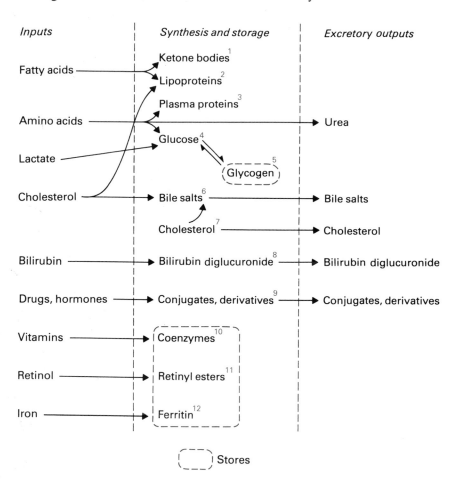

Fig. 13.8.2 Liver biochemistry. The numbers refer to the points in the text.

13.9   Membranes and membrane transport

It is obvious that the boundaries of cells and cell organelles, although serving to isolate their contents from the surrounding environment, must be selectively permeable such that some substances may enter (for example,

nutrients) and some may leave (for example, waste products and se
molecules). Moreover cell membranes must be able to fuse one with a
in the course of cell division, or in phagocytosis (Fig. 13.9.1). This do
however, exhaust the functions of membranes: some contain spe
ceptors for external stimuli such as hormones; some generate chemical or
electrical signals; some incorporate catalysts converting one form of energy
to another, as in photosynthesis and oxidative phosphorylation. Thus the
functions of membranes may be listed:

1   to isolate organized living systems from the environment;
2   to selectively mediate the passage of substances;
3   to respond to stimuli;
4   to generate signals;
5   to house energy conversion systems.

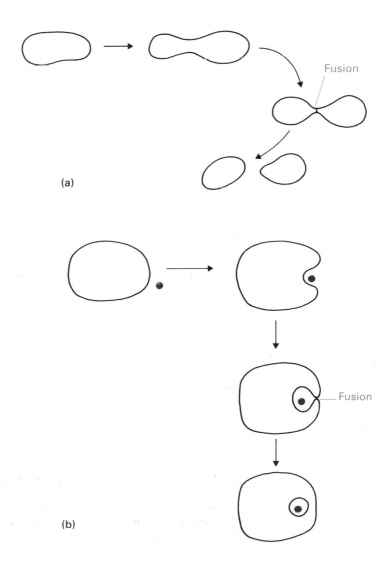

Fig. 13.9.1 Illustrating the necessity of cell membrane fusion in (a) cell division (b) phagocytosis.

215

Composition Membranes are lipid-protein assemblies. The lipids are of the amphipathic type, that is they contain hydrophilic and hydrophobic regions within a single molecule. A typical example is phosphatidylcholine or lecithin. Sphingomyelin is also a prominent component of many membranes, as is cholesterol (for structures, see 13.5). Glycolipids such as cerebrosides may also be present. Amphipathic lipids associate in aqueous media to form bilayers, with the hydrophobic functions associating with each other in the interior and the hydrophilic functions on the outside binding water molecules (Fig. 13.9.2). Many membranes also have as much as 50% protein, reflecting the importance of proteins in membrane function.

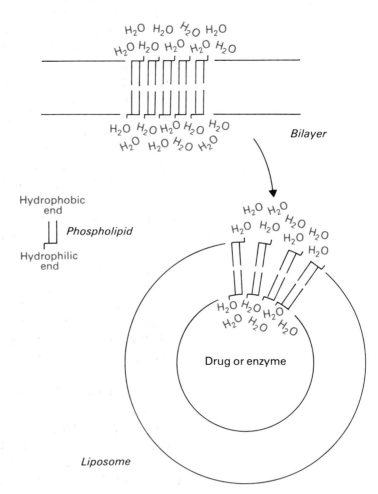

Fig. 13.9.2 Bilayers and liposomes. The bilayer tends to spontaneously form a closed vesicle by self-association of the ends. The symbol π represents a phospholipid. For representational purposes any asymmetric shape would serve, but here the two arms represent the long hydrocarbon chains of the acyl groups, which are hydrophobic, and the little tail represents the hydrophilic phosphorylcholine moiety.

In the middle of the liposome is entered 'drug or enzyme', since these compounds can be so encapsulated. Such loaded liposomes present a potentially effective way of delivering missing enzymes (perhaps in patients with inherited enzyme deficiencies) or drugs poorly taken up from the circulation. Under the right conditions the liposomes fuse with living cells and will deliver their contents thereto.

Structure The current generalized concept of the cell membrane is the Singer-Nicholson or *fluid mosaic* model (Fig. 13.9.3). In this model the membrane is seen as a phospholipid bilayer which is interrupted by protein moieties, some of which are on the exterior, some in the interior, and some running across the entire width. Cholesterol in varying amounts may also be

Fig. 13.9.3 Fluid mosaic model of membrane structure.

There is basically a phospholipid bilayer, with associated proteins which have some mobility in the plane of the membrane. Some of these proteins are on the exterior, some on the interior, and some all the way through. In addition there may be proteins associated with the membrane proteins by weak bonds (peripheral proteins). Many types of membrane also contain cholesterol, orientated such that its hydrophobic region is towards the interior.

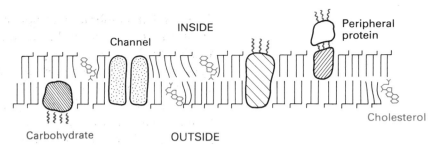

orientated with the hydrophilic C-3 hydroxyl on the outer edge. It is thought to make the membrane structure less fluid; the 'fluid' term in 'fluid mosaic' refers to the idea that the proteins are mobile—they may diffuse in the plane surface of the membrane to some extent. They cannot, however, migrate from the inside to the outside of the membrane, or vice versa. They can also bind yet other proteins on the surface; these are not part of the membrane, and are called *peripheral proteins*.

Liposomes Phospholipids dispersed in water tend to interact to form bilayers. If a bilayer forms it tends to seal itself at the end rather than have exposed hydrophobic chains, so forming a vesicle or liposome (Fig. 13.9.2). Observations such as these have suggested that cell membranes have the capacity for self-assembly. The fusion of living cell membranes evidently utilizes this capacity in so far as two membranes break and reseal themselves spontaneously one to another.

Membrane transport If the fluid mosaic model is correct, it must explain the three types of movement (simple and facilitated diffusion, and active transport) across membranes.

1 *Simple diffusion* might be expected for those compounds readily soluble in the lipid bilayer. The build up of a substance on one side would cause such substances to become dissolved in the bilayer and diffuse to the other side. Such 'transport' occurs as long as there is a concentration gradient across the membrane. When the concentrations on both sides are equal, net flow ceases.

The substances which appear to cross membranes habitually by simple diffusion are water, ethanol, urea, anaesthetics, and some ions. The effect is not necessarily explicable on the basis of solubility in lipids—water is the prime example thereof. However water does seem to be able to pass between the hydrocarbon chains even in the absence of protein.

2 *Facilitated diffusion* also occurs solely in the direction of the concentration gradient, but is accelerated by the presence of some protein, often called a *permease*. It can be imagined that a protein situated right across the membrane (Fig. 13.9.3) might form a gate for hydrophilic substances through the lipid bilayer. This would be especially true if the protein were composed of two or more subunits which could form a channel. In some cells substances observed to be transported by facilitated diffusion include sugars and amino acids.

3 *Active transport* occurs against a concentration gradient. It is easy to prove formally, but easier just to conceptualize, that it requires the expenditure of energy, no doubt involving the hydrolysis of ATP. This constitutes one of the three salient features of active transport, thus:

(1) it is energy-dependent;

(2) part, at least, of the transport system must span the membrane; and

(3) it functions in most cases as a proton or sodium co-transport mechanism.

Co-transport systems are divided into *symports* (systems carrying species in the same direction) and *antiports* (carrying substances in opposite directions). These are habitually illustrated by two systems, namely the maintenance of a high internal potassium concentration by cells, and the uptake of glucose by cells. Cells typically have potassium concentrations of about 100 mmol/l and sodium concentrations of about 10 mmol/l. The figures in extracellular fluid are typically 4 and 140 respectively. The gradient is maintained by hydrolysis of ATP, catalysed by the enzyme Na^+K^+-ATPase. The reaction catalysed by this enzyme can be represented

$$H_2O + ATP \xrightarrow[\text{2K}^+ \text{ in}]{\text{3Na}^+ \text{ out}} ADP + P_i$$

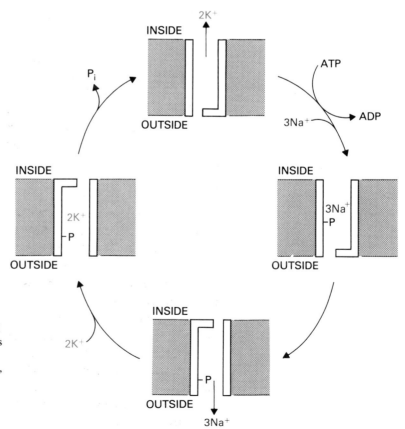

Fig. 13.9.4 The sodium-potassium pump. The form at the top of the diagram has a high affinity for sodium, and when sodium is bound the form is phosphorylated by ATP. This causes a conformational change such that affinity for sodium is decreased and affinity for potassium, taken up from outside the cell, is increased. Uptake of potassium triggers dephosphorylation, and reversion to the form with sodium affinity, at the top.

218

and the mechanism, *in toto*, is known as the sodium-potassium pump. The enzyme cannot catalyse ATP hydrolysis unless sodium and potassium are transported; sodium must be on the inside of the membrane, and potassium on the outside. This immediately points to an orientation of the enzyme protein across the membrane. Indeed it is demonstrably an oligomeric enzyme with some asymmetry. How it catalyses the pump is not exactly known—there are only models. It is known to be phosphorylated by the ATP in a reaction which requires sodium, and dephosphorylated in the presence of potassium. Therefore a simple model can be constructed along the lines of Fig. 13.9.4. Why $3Na^+$ and $2K^+$? This is not clear, but the result is to create a potential gradient or polarization across the cell membrane. The nerve impulse is also a wave of depolarization, with sodium flowing in and potassium out. All such systems are evidently antiports.

Glucose transport, no less than potassium, is coupled to sodium flux. In animal cells sodium and glucose bind to the same transport protein on the outside of the membrane and enter together. Sodium is destined to be transported out again by the action of the sodium-potassium pump. This symport system is described in Fig. 8.0.2 in connection with glucose absorption.

13.10   Metabolism of ethanol

Alcohol can be totally metabolized in the body, its calorific value being 30 kJ/g (7.1 kcal/g), and regular yet moderate drinkers receive about 10% of their energy from this familiar substance. Its production from starch or glucose by yeast in the process known as *fermentation* (defined as the breakdown of substances by microorganisms in the absence of oxygen) is a function of the glycolysis pathway common to almost all forms of life (Fig. 13.10.1). That it is an anaerobic pathway is shown by the stoichiometry from starch or sucrose

$$(C_6H_{12}O_6)_n \rightarrow 2nC_2H_5OH + 2nCO_2$$

The yeasts, either added to the sugar solution (as for beer) or naturally occurring on the surface of fruit (as for wine) are allowed to effect this transformation. The alcohol can be considerably concentrated, to about 40% by volume, by distilling.

The metabolism of alcohol is very peculiar. Its salient factors can be described as follows.

1   It is eliminated at a constant rate of about 100 mg each hour. This is mainly by oxidation, but some appears in the breath and urine.

2   It becomes evenly distributed throughout the body fluids, indicating that there are no membrane barriers to its free diffusion; equally there can be no mechanism for concentrating it.

3   It is preferentially oxidized, before other possible substrates, and mainly in the liver (Fig. 13.10.1).

Fig. 13.10.1 Alcohol synthesis and metabolism. Pyruvate tends to be converted to lactate because there is so much NADH produced from the alcohol. In the well-nourished subject there is ample pyruvate (derived from dietary carbohydrate) and the consequences may be slight. The acetate also derived from the alcohol is lipogenic in the well-fed subject and may well contribute to a slowly-developing obesity. In the malnourished subject pyruvate is in short supply and what there is becomes reduced to lactate. Any glucose becoming available is converted to pyruvate, and there may be a dangerous hypoglycaemia. Meanwhile, fat is being broken down and acetate from this source as well as alcohol contribute to ketosis.

In this diagram the arrows are all unidirectional to show the pressure of carbon and electron flow, but many of them are reversible in the thermodynamic sense.

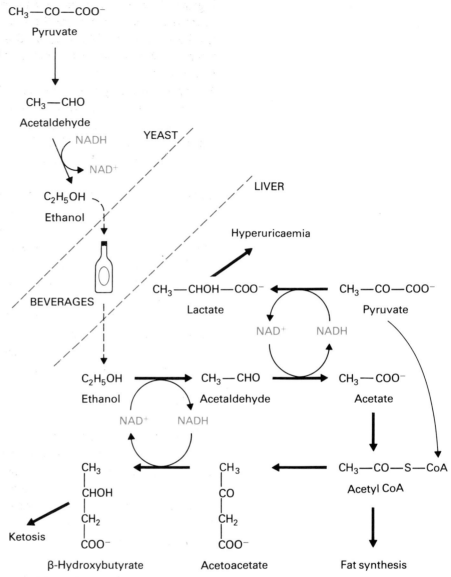

There are two separate oxidation mechanisms operative. The first, quantitatively the most important, involves an enzyme of low specificity, alcohol dehydrogenase, which catalyses the production of acetaldehyde and NADH. Much of the adverse effects of alcohol on the system are ascribed to the former, which in turn is oxidized to acetate by liver aldehyde dehydrogenase, giving more NADH. Acetate is of course converted to acetyl CoA, both in the liver and peripheral tissues. Now, as has been separately emphasized (Fig. 11.3.3), acetyl CoA is not convertible to carbohydrate in human beings. Its fate is to provide carbon atoms for lipid synthesis, or for ketone bodies, or to be oxidized in the citric acid cycle. If a well-nourished subject ingests alcohol there is plentiful pyruvate from dietary carbohydrate, and therefore acetyl CoA from the alcohol is largely converted to lipid. Lipid synthesis requires NADPH, but this is plentiful due to transhydrogenation from NADH, involving the malate shuttle. The excess NADH from ethanol is utilized in three ways.

1 By transhydrogenation to NADPH for fat synthesis.

2 By conversion of pyruvate to lactate. This occurs to the extent of pyruvate becoming in short supply in the alcoholic who is not eating, and thus not supplying glucose to his liver for replenishment of pyruvate. At the same time, alcohol dehydrogenase competes with lactate dehydrogenase for available NAD^+, so that lactate, once accumulated cannot be cleared if alcohol continues to be consumed. Because lactate competes with urates for excretion by the renal tubules, another consequence is hyperuricaemia, which contributes to the pathogenesis of gout; hence the age-old association between good living (a surfeit of port wine, for instance) and that condition.

3 By conversion of acetoacetate to β-hydroxybutyrate. This is likely to be of greater significance in the case of the malnourished alcoholic, who has insufficient pyruvate to maintain oxaloacetate for the citric acid cycle. Whatever acetyl CoA is not converted to fat thus tends to form acetoacetyl CoA and free acetoacetate.

The alternative oxidizing system, known as the microsomal ethanol oxidizing system (MEOS) utilizes a mixed function oxidase for the reaction

$$C_2H_3OH + NADPH + O_2 + H^+ \rightarrow NADP^+ + CH_3CHO + 2H_2O$$

It is thought to be significant only at high tissue concentrations of ethanol. Since this reaction takes place in the microsomes, the site for drug oxidation, ethanol can interfere with the disposal of drugs. This is why it is dangerous to take tranquillizers at the same time as alcoholic drinks.

Finally, methanol is metabolized in the same way, since alcohol dehydrogenase is of low specificity. But the products are formaldehyde and formic acid. These are extremely toxic, and the optic nerve is particularly sensitive to them; adulterated wines and other drinks cause blindness for this reason. Since methanol and ethanol compete with each other for the active site of the alcohol dehydrogenase, the latter is a useful antidote for the former.

This is all summarized in Fig. 13.10.1.

13.11 Starvation

Probably starvation is a favourite topic for examination questions because even examiners have a subjective idea of it and perhaps also because starvation and repletion evoke the drama of a Manichean dualism, the struggle between good and evil, examples of which are otherwise so rare in biochemistry. If you are asked about starvation it is logical to start by considering the stores of nutrients which may be available to the body on cessation of feeding, and what happens to these stores until refeeding or death. Endocrine changes and the eventual cause of death, as well as the fluxes in metabolism if refeeding occurs, must be considered. Since the bulk of such a discussion applies to adults, starvation in children producing marasmus and kwashiorkor should be mentioned as an addendum.

Body stores of substrates

These are listed in Table 13.11.1. For practical purposes vitamin stores need not be considered since if they are adequate death from energy shortage occurs long before the appearance of avitaminoses. Also, it is said that there is a slowdown of metabolism in starvation so that the vitamin requirement is reduced. The table indicates that carbohydrate stores last for one day only. After that there is a pronounced switch in the metabolic pattern. The time course is as follows.

Table 13.11.1 Stores of nutrients in a 70 kg man

	Store (g)	Daily loss in starvation (g)	Exhaustion time
Carbohydrate	150	150	1 day
Fat	6,500	150	6–7 weeks
Protein	2,400	60	6–7 weeks
Water	4,000	1,000*	4 days
Sodium	35	14**	2–3 days
Vitamins A	0.12	0.3×10^{-3}	1–2 years
Thiamine	0.025	0.35×10^{-3}	2–3 months

*Obviously the correct term here is 'water deprivation', not starvation
**Assumes heavy sweating

Day 1   Glycogen is mobilized to maintain blood glucose, needed for brain function and muscular activity. If a starving organism is to survive these two essential functions must be maintained, that is brain function or consciousness to motivate the search for food, and muscular function to gather it. Insulin secretion starts to decrease.

Day 2   Glycogen from liver being exhausted, insulin secretion further diminishes and lipolysis takes the place of lipogenesis as glucagon effects triglyceride breakdown. The released free fatty acids are utilized by resting muscle for energy. Exercising muscle uses its own glucose, maintained as glycogen stores. Lipolysis to such an extent causes ketogenesis but brain and heart muscle can oxidize acetoacetate. Cortisol effects gluconeogenesis, for blood glucose cannot be maintained from lipolysis but is maintained by the conversion of the glucogenic amino acids to carbohydrate. Thus it is maintained at about 60 mg/100 (3 mmol/l) to allow brain function. The protein

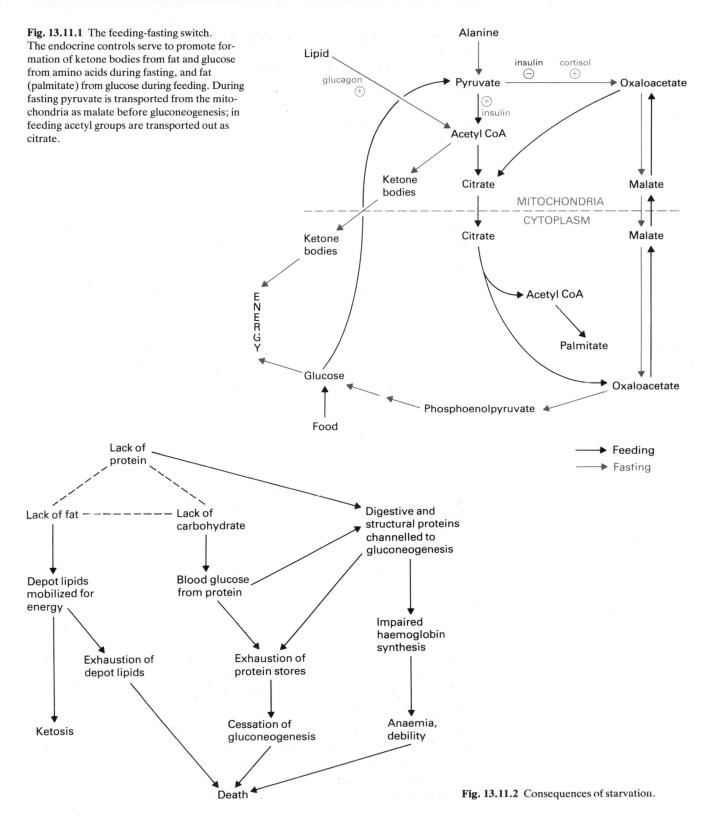

Fig. 13.11.1 The feeding-fasting switch. The endocrine controls serve to promote formation of ketone bodies from fat and glucose from amino acids during fasting, and fat (palmitate) from glucose during feeding. During fasting pyruvate is transported from the mitochondria as malate before gluconeogenesis; in feeding acetyl groups are transported out as citrate.

Alanine

Lipid

glucagon ⊕

Pyruvate

insulin ⊖ cortisol ⊕

Oxaloacetate

⊕ insulin

Acetyl CoA

Ketone bodies

Citrate

Malate

MITOCHONDRIA
CYTOPLASM

Ketone bodies

Citrate

Malate

Acetyl CoA

Palmitate

ENERGY

Glucose

Oxaloacetate

Phosphoenolpyruvate

Food

Feeding
Fasting

Lack of protein

Lack of fat

Lack of carbohydrate

Digestive and structural proteins channelled to gluconeogenesis

Depot lipids mobilized for energy

Blood glucose from protein

Impaired haemoglobin synthesis

Exhaustion of depot lipids

Exhaustion of protein stores

Ketosis

Cessation of gluconeogenesis

Anaemia, debility

Death

Fig. 13.11.2 Consequences of starvation.

223

which is being catabolized comes from the liver, the intestinal tract, and the pancreas, but there is also some conservation of proteins in that the synthesis of digestive enzymes by the pancreas and small intestine ceases. Fat is now being consumed, in the adult man, at the rate of about 150 g/day, and protein at the rate of about 60 g/day. Two analyses of the situation are in Figs 13.11.1 and 13.11.2. The subsidiary controls now operating (the primary metabolic control being the absence of food) are the reduction of circulating insulin, which is lipogenic, and an increase of effector activity of glucagon, which is lipolytic. Insulin represses synthesis of pyruvate carboxylase*, so the reduced level of insulin means that this essential reaction for gluconeo-genesis

$$\text{Pyruvate} \xrightarrow{\;*\;} \text{oxaloacetate} \rightarrow \text{phosphoenolpyruvate} \rightarrow \text{glucose}$$

may proceed. It is also stimulated by cortisol, whose secretion is stimulated by stress and hypoglycaemia.

Day 40   At about this time the average man has used up his body fat stores (obese subjects will last longer and so there are good grounds for viewing obesity as a famine survival trait). All energy needs, plus the requirement for blood glucose, have to be met by gluconeogenesis from the remaining stores of protein. This, however, implies the breakdown of about 250 g/d protein which cannot be maintained for longer than a day or two (Table 13.11.2). Death ensues. The record for a fast with no food consumed at all is 249 days, but this was a grossly obese subject. Scientific data on the total time an individual may survive without eating are naturally rare, but some figures have recently become available under unfortunate circumstances, in that seven Irish republican prisoners this century have starved themselves to death. The range of survival time was 51 to 74 days, with a mean of 61. This suggests that the theoretical estimate of about forty days given above is rather too low and indeed it might seem sensible to modify it to allow for the drop in metabolic rate and basal energy requirement in a terminally starving person lying in bed with a shrinking body mass.

Table 13.11.2 Consequences of total fasting

	Carbohydrate stores	Fat stores	Protein stores
Day 0	150 g in liver	6.5 kg in adipose tissue	24 kg in muscle and other tissues
Day 1	Exhausted	Normal	Normal
Days 2–40 (approx)	Exhausted	Mobilized at 150 g/d	Mobilized at 60 g/d
Day 40	Exhausted	Exhausted	Used up at 250 g/d

Refeeding ordinary foodstuffs is dangerous in the terminal stages because the intestinal wall becomes delicately thin and medical supervision of re-feeding is advisable. Then there is a reversal of the control mechanism.

Insulin release re-occurs and this inhibits pyruvate carboxylase. Pyruvate is channelled to acetyl CoA formation which can be a substitute for resynthesis of fat. Insulin also enhances glycogen formation from the dietary carbohydrate. The effects of glucagon, adrenaline and cortisol are reduced as glucose comes from the diet and itself maintains blood glucose. At the same time dietary proteins rebuild tissue mass, and allow the resynthesis of digestive enzymes and haemoglobin.

Kwashiorkor and marasmus

Children are at greater risk than adults in times of food shortage for a number of reasons, but the most obvious one is the proportionately greater demand on nutrients by growth. Physicians have described the range of conditions associated with starvation as *protein-energy malnutrition* and within this have distinguished *kwashiorkor*, where the greater deficiency is in the protein supply, from *marasmus* where energy-producing foodstuffs are lacking. However there are intermediate forms, and the clinical picture is most often complicated by avitaminoses and infections. Moreover there is a view that marasmus is the end-result of adaptation to protein-energy malnutrition whereas kwashiorkor represents a breakdown in that adaptation process. Be that as it may, the differences between the two are as set out in Table 13.11.3. The reason for the biochemical differences can be explained as follows. In kwashiorkor carbohydrate is almost the only source of energy and is thus rapidly removed from the blood and converted to fat. The demands on what body protein there is, and possibly also liver impairment, reduce serum albumin synthesis and since this protein is primarily responsible for the maintenance of the osmotic pressure of the blood oedema occurs. In marasmus protein synthesis is apparently sufficient to maintain almost normal concentrations of serum albumin. However there is no dietary carbohydrate or fat to spare, little or no adipose tissue is maintained, and the subject is wasted and grossly underweight.

Table 13.11.3 A comparison between marasmus and kwashiorkor

Marasmus, child has	*Kwashiorkor*, child has
low body weight	only slightly lowered body weight
thin muscles, no fat	thin muscles, but fat present
normal hair	hair changes
no oedema	oedema
slightly lowered serum albumin	very low serum albumin
low blood glucose	very low blood glucose
normal serum fatty acids	high serum fatty acids

Malnutrition and the developing brain

In terms of total weight, the greatest spurt of brain growth occurs from around the twentieth week of gestation to eighteen postnatal months. Fears have been expressed that malnutrition during this period might permanently affect intellectual development. From the point of view of intelligence, the area of the brain of greatest interest is the cerebrum and this shows two growth spurts: an eight-week period of neuroblast multiplication before birth, and a later largely postnatal development of glial cells. It is now largely accepted that these growth processes are programmed in such a way that they cannot be altered or prolonged by malnutrition, but that malnutrition can reduce the amount of growth itself. It follows that if malnutrition affects growth in the vulnerable period there is no possibility of catching up later. Animal experiments have demonstrated that during malnourishment there is a reduction in the rate of protein and lipid synthesis in the cerebrum. In humans studies can only be made of the brains of infants who have succumbed to malnutrition, and the sparse results so far indicate a reduction of gangliosides compared to controls, that is to infants who have died for reasons other than starvation. The gangliosides are complex lipids (13.5) located in the dendritic interconnections and since the complexity of the latter must presumably contribute to higher brain functions, this constitutes some evidence of the permanent effect of early childhood malnutrition on these functions. Of course most discussion centres round the long term social damage, but nobody would dispute that this is beyond our present terms of reference.

13.12 Diabetes mellitus

'Mellitus' is derived from a Latin word meaning 'honey' and 'sugar'. Diabetes is the metabolic derangement *par excellence*; possibly more biochemical man-hours have been spent on it than on any other disorder, syndrome or disease (its precise description is still a matter of discussion) and it is a great favourite for essay questions and for vivas. It is usually defined as an absolute or relative deficiency of normally-functioning insulin. Some cases arise in adults and usually they are initially treated by dietary means whereas others are detected in childhood and generally need insulin therapy; these are the non-insulin dependent ('mature onset') and insulin-independent ('juvenile') types respectively. However some clinicians are able to distinguish six separate classes of diabetes.

Any insulin deficiency, from whatever cause, inevitably has these metabolic sequelae:
1 increased release of glucose from the liver (since insulin is glycogenic);
2 decreased uptake of glucose by muscle and adipose tissue (since insulin is necessary for passage of glucose through their cell membranes); and
3 increased lipolysis in liver and adipose tissue (since insulin is lipogenic as well as glycogenic).

1 and 2 together result in a rise in blood glucose; 3 results in a rise in blood lipids and ketone bodies (Fig. 13.12.1). It is obvious from the above that the actions of insulin must be related to the different organs of the body. Indeed most tissues are insulin-sensitive, but the systemic effects of insulin deficiency are mainly due to its effects on skeletal muscle, liver and adipose tissue. Table 13.12.1 outlines some aspects of tissue sensitivity to insulin. The two striking metabolic consequences, namely hyperglycaemia and ketosis, have further consequences of their own (Fig. 13.12.1). Hyperglycaemia results in glucosuria when the concentration reaches about 180 mg/100 ml (10 mmol/l) (the 'renal threshold' for glucose). Glycosuria contributes to loss of water (dehydration) and thus to haemoconcentration, further aggravating the hyperglycaemia. Many authorities think that hyperglycaemia is the primary pathogenic factor, in that it causes progressive glycosylation (p. 233) of proteins such as those in the basement membranes of the blood vessels (resulting in angiopathies) and haemoglobin.

Fig. 13.12.1 Consequences of insulin deficiency.

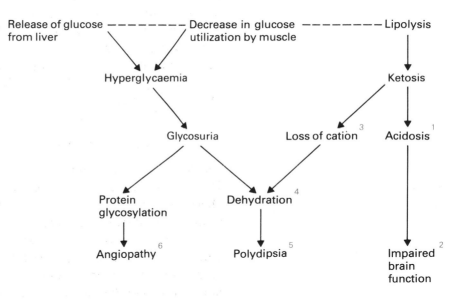

1 since ketone bodies are relatively strong acids
2 since brain is very sensitive to H⁺
3 since Na⁺ and K⁺ are lost to balance acetoacetate⁻
4 water is lost along with the cations, as a solvent
5 excessive drinking due to increased urination
6 glycosylation of proteins in the basement membranes
  causes vascular impairment

The ketosis and its accompanying excretion of anions result in loss of balancing sodium and potassium cations. The water lost in keeping these in solution contributes further to dehydration and thus polydipsia (thirst), a marked symptom of diabetes. The acidosis, if unchecked, can impair brain function, leading to coma ('diabetic coma'). These relationships are summed up in the diagram.

Table 13.12.1 Insulin sensitivity of some tissues

Tissue	Insulin enhances	Insulin deficiency results in
Liver	Pyruvate dehydrogenase, pyruvate kinase, and phosphofructokinase activities	Decreased glucose utilization, therefore breakdown of fat to ketone bodies, also increased gluconeogenesis
Adipose	Glucose uptake	Inhibition of lipogenesis; fatty acids mobilized to liver
Skeletal muscle	Glucose uptake	Lack of glucose for glycogenesis
Cardiac muscle	Glucose uptake	Lack of glucose for contraction
Brain	No specific function	No ill effects, but acidosis due to liver ketogenesis is harmful

The insulin molecule

Insulin is synthesized as a preproinsulin in the β-cells of the islets of Langerhans of the pancreas. Two peptides are split from the precursor molecule to yield insulin itself, which is stored as precipitated protein, in association with zinc. The second peptide to be split off, during the conversion of proinsulin to insulin, is termed the C-peptide and is released into the blood along with insulin (Fig. 13.12.2). The circulating effector molecule, insulin itself, is a two-chain peptide linked by two —S—S— bonds. There is also an internal —S—S— bond. The shape of insulin is such (Fig. 13.12.2) that it interacts with specific receptors on the surface of the cells of the liver, muscle, and adipose tissue. The amino acid sequences of insulins from a large number of species have been determined; the one which is most similar to human insulin, and therefore most suitable for therapeutic use, is pig insulin. It differs from human insulin only in the terminal amino acid of the B-chain (Fig. 13.12.2). Sheep, horse, and cow insulin differ from human insulin in three residues. From such animal sources a large number of injectable insulin preparations have been prepared. Ideally they should not be immunogenic (that is elicit antibodies in the manner of injected foreign proteins) and this has been minimized by sulphating them. 'Monocomponent' insulin is a highly purified variety of low immunogenicity. The action of the hormone can be prolonged by injecting poorly soluble types such as those prepared by crystallization in the presence of zinc. Since it is now known that insulin secretion rises rapidly in response to a meal, it is doubtful whether the continuously-releasing types are of real value.

Insulin can now be produced by recombinant DNA techniques. DNA sequences are chemically synthesized to contain the information for coding the amino acids in the A- or the B-chain of human insulin. These synthetic

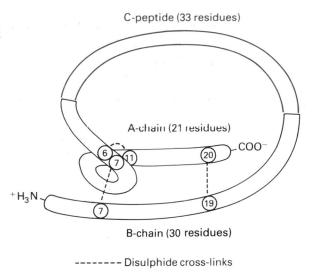

Fig. 13.12.2 A representation of the insulin molecule (as proinsulin, that is before removal of the C-peptide). A rendering of the three-dimensional structure is given in Fig. 13.2.10.

C-peptide (33 residues)

A-chain (21 residues)

B-chain (30 residues)

------- Disulphide cross-links

genes, one for the A-chain and one for the B-chain, are inserted into separate plasmids, in turn separately inserted into *E. coli*. The bacterium synthesizes the A- and B-chains on the basis of the information in the inserted plasmids. The bacterial cells are lysed, and the A- and B-chains purified, reduced, mixed, and oxidized to give functional insulin (13.22).

Mode of action of insulin

The effector status of insulin depends on receptors present in liver, muscle, and adipose tissue. Insulin binds firmly to these receptors, and the extent of binding governs the response of the tissues, that is response is proportional to both the amount of circulating insulin and the number and affinity of receptors. The effect of insulin on tissues is variable (Table 13.12.1) so receptor binding triggers off a response determined by the nature of the tissue and perhaps a second messenger. Despite the intensive research on insulin, the details of its action remain speculative (see 12.4 and 13.19).

Assessment of functional insulin status

(The adjective 'functional' is inserted because insulin molecules may well circulate in the blood without having full activity.) Glycosuria is an overt sign of hyperglycaemia, but indicates nothing about its cause, only that the renal threshold has been exceeded. If a patient is being treated for diabetes mellitus then glycosuria merely shows that control is not effective. Similarly, a single estimation of blood glucose gives little information. Hyperglycaemia may be idiopathic (of unknown cause) or perhaps due to recent ingestion of food despite protestations to the contrary. More informative is the glucose tolerance test (GTT). Here 'tolerance' means 'ability to metabolize'. An oral dose of 75 g, sometimes 50 g of glucose is given in lemon juice, and samples for analysis are taken immediately and at 0.5, 1, 1.5, 2, and 2.5 h. The urine is tested for glucose (glycosuria) at 1 h and 2 h. Representative types of response are shown in Fig. 13.12.3. A patient is considered diabetic, according to a recent report of an international working party, if the fasting blood glucose is greater than 8 mmol/l (140 mg/100 ml) on more than one occasion, and if the GTT exhibits a concentration of greater than 11 mmol/l (200 mg/100 ml) at 2 h and at some other point of sampling. A normal response is that which fails to rise above 11 mmol/l at any time and is less than 8 mmol/l at 2 h; moreover, there should be absence of glycosuria throughout. Subjects with intermediate response are said to have 'impaired glucose tolerance'. Diabetic-type responses are shown in acromegaly (excessive growth hormone secretion), obesity and often, in pregnancy.

Fig. 13.12.3 The glucose tolerance test.

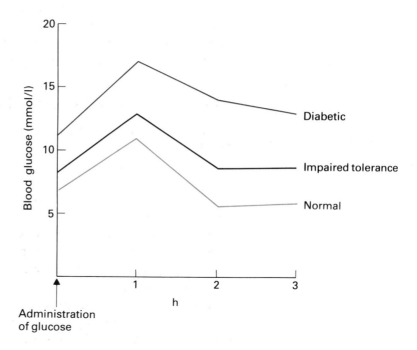

Circulating insulin may be assayed by radioimmunoassay if the clinician wishes to determine whether a deficiency of it is contributing to the patient's condition; it is also assayed for the diagnosis of insulinoma, an insulin-secreting tumour of the β-cells. The problem about estimating circulating insulin is that (a) insulin therapy must be discontinued lest it interfere with the test and (b) prolonged insulin therapy causes the elaboration of insulin antibodies which lower its apparent concentration. In view of this it has been proposed that assay of circulating C-peptide (Fig. 13.12.2) is more useful. C-peptide is released at the same time as insulin and is not immunogenic, so that its concentration is an index of the secretory activity of the pancreas. It is also measured by radioimmunoassay.

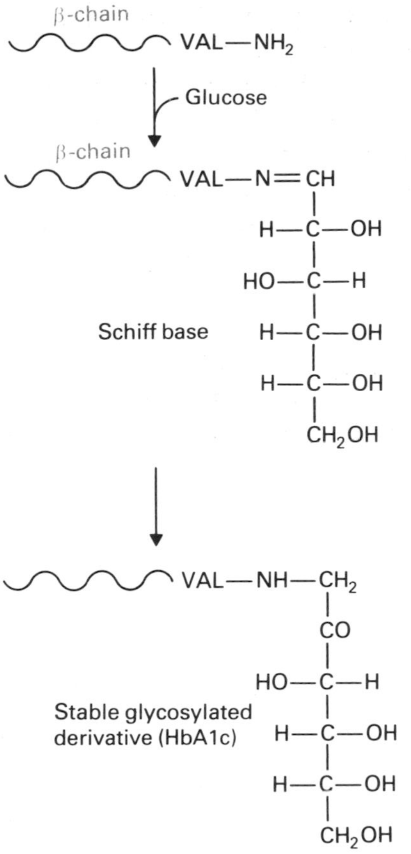

β-chain

VAL—NH₂

Glucose

β-chain

VAL—N═CH
|
H—C—OH
|
HO—C—H
|
H—C—OH
|
H—C—OH
|
CH₂OH

Schiff base

VAL—NH—CH₂
|
CO
|
HO—C—H
|
H—C—OH
|
H—C—OH
|
CH₂OH

Stable glycosylated
derivative (HbA1c)

Fig. 13.12.4

Of great importance now is the estimation of the haemoglobin termed HbA_{1c}. This was long known to be a subcomponent (about 4%) of normal adult haemoglobin (HbA) but in diabetes the proportion can rise to 10% or over. HbA_{1c} was found to be glycosylated, that is it results from a glucose unit attaching to the N-terminal valine of the β-chains of HbA. Since diabetics have chronic hyperglycaemia, the glycosylation is progressive throughout the lifetime of the red cell. It does not go to completion since the red blood cell only survives for about 120 days, after which its haemoglobin is degraded. The significance of the measurement of HbA_{1c} is that it provides an index of hyperglycaemia over the previous 4–6 weeks—whatever the state of diabetic control at the time of examination, progressive glycosylation due to hyper-glycaemia in the preceding period is faithfully recorded. The reaction is thought to proceed as in Fig. 13.12.4. It remains possible that similar glycosylation of basement membrane proteins is the basis of a glucose toxicity in the long-standing diabetic. The changes in the basement membrane are very likely related to the retinopathies, nephropathies, and the circulatory malfunctions leading to amputations.

13.13 Inborn errors of metabolism

The inherited disorders traditionally known as *inborn errors of metabolism* are also termed *inherited metabolic defects* or *transmissible metabolic defects* or *molecular diseases*. Obviously they constitute a group within the overall description 'birth defects'. Originally the idea was that a single faulty gene resulted in a single missing or defective enzyme, but the pathology of this range of conditions is now traceable either to decreased activity of an enzyme or a group of enzymes, or to a specific error in protein synthesis, or to a transport defect, or to a receptor defect. Evidently the concept has widened somewhat, although for a condition to come under the definition, it should be possible to link it to a specific inheritance pattern; nonetheless, sometimes the boot is on the other foot in the sense that the inheritance pattern strongly suggests a specific error in enzyme or protein synthesis or a transport or receptor defect, prior to the demonstration of any of these.

Providing as they do a direct conceptual link between metabolism and disease, the inborn errors are of great interest in biochemistry and are great favourites for examinations, especially within the medical course. This is despite the fact that many are fairly rare, in some instances with only two or three known medical cases. Some are peculiar to certain races. However others, such as sickle cell anaemia, are common enough in many areas to represent major cases of mortality and morbidity.

The generalizations which may be made are as follows:

1 Most enzyme defects are recessive, most protein defects dominant.

2 Heterozygotes generally do not exhibit illness, whether the defect is recessive or dominant. It is important to realize that recessiveness and dominance are attributes of the phenotype. Thus the sickling *phenomenon*

(red cells distorting at low oxygen tensions) is dominant since it can be demonstrated both in homozygotes and in heterozygotes. The *disease* sickle cell anaemia, however, is recessive since it is observed only in homozygotes.

3 Very few are Y-linked, that is they are almost all autosomal or X-linked.

4 Many of the defects in amino acid metabolism lead to accumulation of metabolites toxic to the brain; thus homozygotes may suffer mental retardation.

5 Pauling has pointed out that in a sense we all suffer from inherited metabolic defects in that we cannot synthesize the vitamins or essential amino acids. Thus scurvy is the result of the defect being made manifest through the stress of an ascorbate-deficient diet.

The subject is replete with the language of genetics, so a short glossary of relevant terms is appended.

Allele One of the possible mutational forms of a given gene, demonstrated by some characteristic.

Autosomal Refers to chromosomes other than the *sex chromosomes*. Thus, an autosomal inheritance is not *sex-linked*.

Autosomal dominant A *dominant* gene in a chromosome other than an *X*- or *Y-chromosome*. That is, it can be transmitted by either sex and male and female offspring are equally affected.

Autosomal recessive This is a *recessive* gene on a chromosome other than the *sex chromosome*. See *autosomal dominant*.

Balanced polymorphism The state whereby the adaptive value of the *heterozygote* is greater than that of the *homozygote*; there is the selection to preserve a balanced distribution of the genes involved.

Chromosomes In man there are 22 pairs of non-sex chromosomes, and one pair of sex chromosomes. In the female both are X; in man there is one X and the smaller Y chromosome.

Congenital Present at birth (not necessarily genetic in causation).

Dominance The masking of one *allele* by another in a *heterozygote*. It is an attribute of the *phenotype*, that is it is detected by the outward manifestation of the effects of the so-called dominant gene.

Genotype The genetic constitution of an individual.

Heterozygote An individual carrying different *alleles* for a particular gene.

Homologous chromosomes The chromosomes of a pair in the full (diploid) complement of a cell.

Homozygote An individual having the same *alleles* at loci in homologous chromosomes.

Multifactorial See *polygenic*.

Penetrance The frequency with which a heritable characteristic is observable in individuals known to carry the gene responsible for it.

Phenotype The outward characteristics of an individual, partly determined by *genotype*.

Polygenic Refers to a variable *phenotype* which in turn is due to the complexity of gene interaction. Also called *multifactorial*.

Polymorphism The existence of two or more genetically different groups in an interbreeding population. Can be a *balanced polymorphism*.

Recessiveness That property of a gene causing its effects to be masked in the *heterozygote*. Like *dominance*, it is an attribute of the *phenotype*.

Sex chromosomes The X- and Y-chromosomes.

Sex linked This refers to the mode of inheritance of genes located in the *X- and Y-chromosomes*.

Trait A characteristic determined by genes, but used particularly for the *heterozygous* state of a *recessive* disorder like sickle cell anaemia.

X-chromosome This is the *sex chromosome* located in both sexes. Females are *homozygous* (XX) with respect to it, males *heterozygous* (XY).

X-linked recessive A recessive gene in the X-chromosome, that is *sex-linked* (very few have been allocated to the Y-chromosome). Disorders due to the X-linked recessive affect only males with their XY configuration, but females may be carriers—their XX configuration means that a recessive of an X is not manifested.

Y-chromosome The chromosome found only in the male sex.

Phenylketonuria

Phenylketonuria is sufficiently common (1 in 10,000 births in North America and European populations) for a neonatal screening programme to be economically worthwhile. The defect, which is inherited in an autosomal recessive manner, is due to the absence of activity of a component of phenylalanine hydroxylase which under normal circumstances catalyses the conversion of phenylalanine to tyrosine. There is no risk to the patient due to lack of tyrosine, for it is present in the diet, but when phenylalanine is also present and it cannot be hydroxylated it is converted to other apparently toxic products (Fig. 13.13.1). These affect normal brain development and therapy is aimed at keeping the dietary intake of phenylalanine within strictly defined limits in order to avoid damage. Naturally the dietary intake of phenylalanine cannot be zero since it is an essential amino acid, necessary for normal protein synthesis.

The fact that phenylketonuria is relatively common, at least in European races, suggests that its survival is due to a balanced polymorphism, that is the condition has some survival advantage. It certainly acts to 'conserve' phenylalanine and this might be of some advantage in certain situations of protein deprivation.

Alkaptonuria

Alkaptonuria is an essentially harmless condition, but was generally re-marked upon in the pre-scientific era because the subject's urine turns black on standing. There is a defect in homogentisate oxidase, an enzyme of

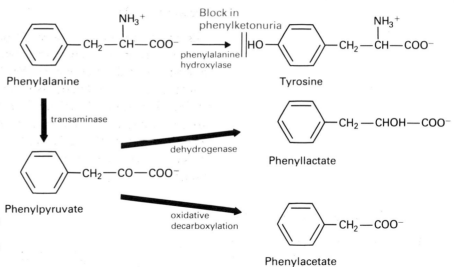

Fig. 13.13.1 Production of the toxic metabolites in phenylketonuria.
The thickened arrows represent the alternative pathways.

tyrosine metabolism. In the absence of adequate enzyme activity homogentisate is excreted and oxidized by air to a dark quinone pigment especially when alkali is present (Fig. 13.13.2). The pigment is as yet uncharacterized.

Fig. 13.13.2 Defect in alkaptonuria.
Arrows represent multiple steps.

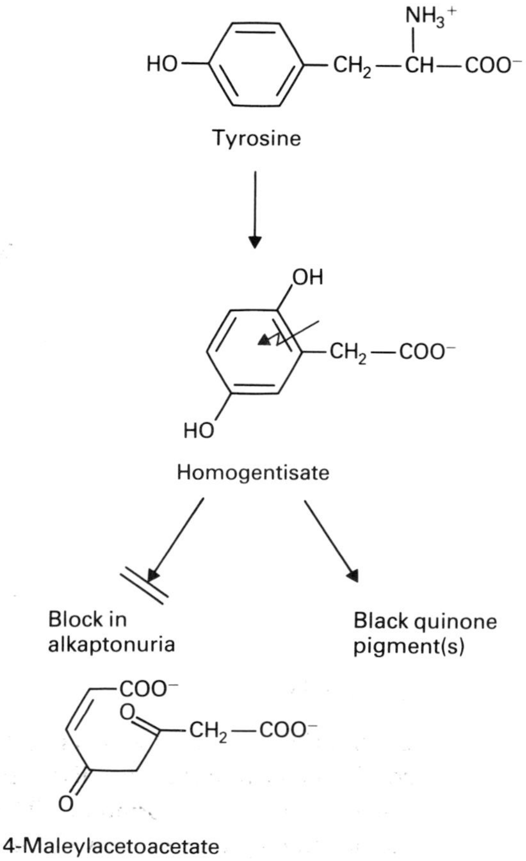

Tyrosine

Homogentisate

Block in
alkaptonuria

Black quinone
pigment(s)

4-Maleylacetoacetate

Homocystinuria

This condition occurs as a result of impaired metabolism of methionine. After methyl group removal the product homocysteine is normally coupled to serine to yield cystathionine (Fig. 13.13.3). When cystathionine synthetase is defective homocysteine may reach toxic levels, in equilibrium with homocystine. Pyridoxal phosphate (vitamin B_6) is required as cofactor by the enzyme and some cases respond to massive dosage with it; this is thought to be explainable on the basis that the failure is not in the synthesis of enzyme protein, rather that for some reason this synthesized apoprotein has a very low affinity for B_6.

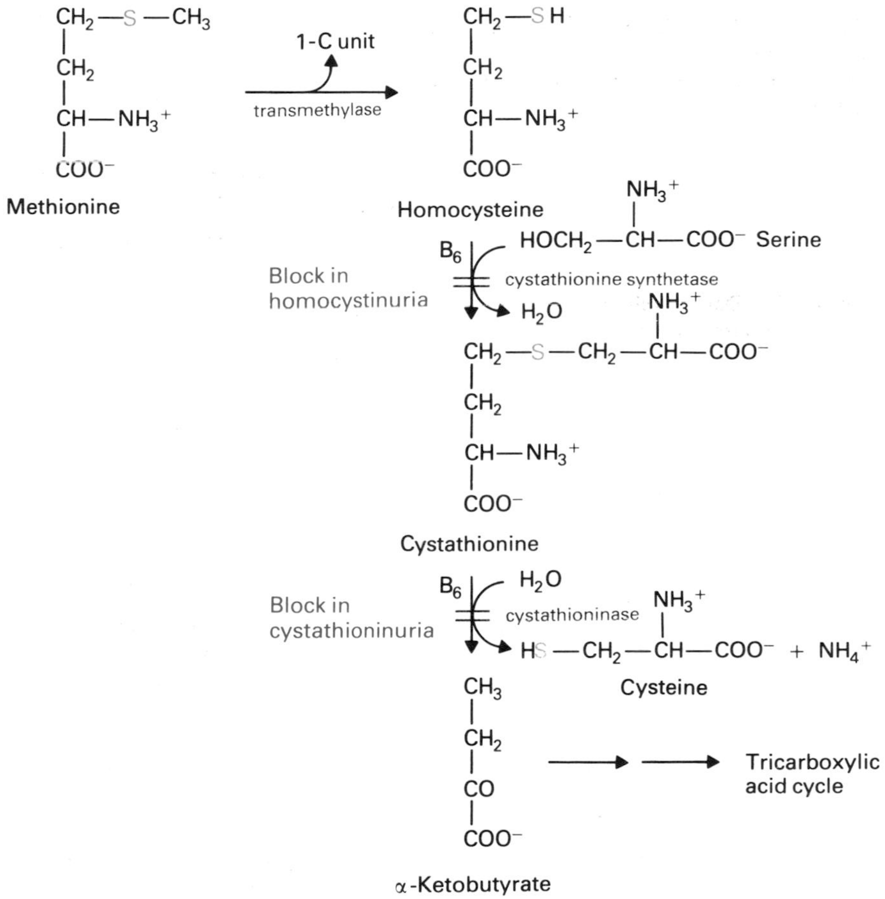

Fig. 13.13.3 Homocystinuria and cystathioninuria.
In homocystinuria the accumulating substance is shown as homocysteine, that is the reduced form. The condition is called homocystinuria because oxidation tends to occur in the urine. The annotation 'B_6' means that pyridoxal phosphate is a cofactor.

Cystinuria

Cystinuria is an example of a genetic defect in membrane transport. In this condition the renal tubule fails to reabsorb four amino acids, namely cystine, lysine, arginine, and ornithine. This would be unimportant were it not for the fact that cystine is relatively insoluble and forms stones.

Glycogenoses

This is a collective term for conditions due to one or other defect in glycogen metabolism, previously known as the 'glycogen storage diseases'. Some seven types are recognized, but probably the two best ones to remember as examples are McArdle's syndrome (type V) and Pompe's disease (type II).

McArdle syndrome patients have: (1) large amounts of glycogen in their skeletal muscles; (2) an inability to undergo strenuous exercise; (3) little lactate in the blood after any sort of exercise; (4) nonetheless the ability to exhibit a rapid rise in blood glucose after glucagon administration. The liver and muscle forms of phosphorylase are distinct proteins and the condition may therefore be explained as a defect in the activity of the muscle enzyme only. On the other hand liver phosphorylase, being normal, responds to glucagon and glucose is released into the blood. It is thought to be an autosomal recessive condition.

Pompe's disease, also autosomal recessive, is both a glycogen storage disease and one of the group referred to as 'lysosomal diseases' because the defect is in lysosomal α-1,4-glucosidase in all tissues. This enzyme is not involved in the normal pathway of glycogen degradation, and so presumably has a scavenging function peculiar to the lysosomes, where it breaks down accumulating glycogen to maltose units. In its absence these bodies become loaded with glycogen and this apparently affects the heart muscle so badly that the condition is normally fatal in infancy.

Lactose intolerance or hypolactasia

Since it is the norm for most of the world's population lactose intolerance is not a pathological condition. It is the inability to hydrolyse lactose in milk due to an absence of intestinal lactase. The lactase activity of the intestinal mucosa, in most races, rises to a maximum about the time of birth but falls away to rather low levels by the third or fourth year. If milk is then ingested, the lactose goes undigested but is fermented by bacteria in the large intestine to give hydrogen gas and organic acids, at a minimum causing discomfort and flatulence. Most of the inhabitants of Northern Europe and their descendants retain the ability to synthesize lactase throughout adult life. It is an autosomal dominant characteristic. Alactasia is rather different, being an inherited inability to synthesize active lactase in infants.

Hyperlipoproteinaemias

These are either primary or secondary: by definition the primary ones are genetic in origin, the secondaries being consequent on such conditions as diabetes or alcoholism. The primaries are classified into five types, with some subdivision. One of the most interesting is type IIa, otherwise known as familial hypercholesterolaemia. It is autosomal dominant, and patients suffering from it have very high concentrations of blood cholesterol, three or four-fold greater than the normal. They tend to die from myocardial infarction before the age of 30, and on autopsy their coronary arteries are occluded with the masses of the cholesterol-containing lipid termed atheromas. The basis of the defect is not in an enzyme, but in a receptor. Receptors for low density lipoprotein are not present on cell surfaces, so that low density lipoprotein cholesterol cannot be removed from the circulation; instead its cholesterol is deposited in the intimas of the arteries and accumulates to develop atheromas.

Lesch-Nyhan syndrome

This is an X-linked recessive, involving a defect in the enzyme guanine phosphoribosyl transferase. This is a salvage pathway enzyme (11.11) whose existence in certain tissues makes possible the phosphoribosylation of guanine to yield guanylic acid (GMP). GMP normally exerts a feedback inhibition on purine synthesis and it appears that lack of the enzyme results in lack of control of the main pathway. Excess purines give rise to excess urates and so Lesch-Nyhan children have the symptoms of gout. The enzyme defect or the accumulating metabolites affect the brain in some way, in that patients are prone to attacks of self-mutilation, for example biting at their own extremities. Later, aggression can be directed at others. This has aroused a great deal of interest, since it indicates some connection between metabolism and psychopathology.

There is a summary in Table 13.13.1. The implications of each condition in the study of biochemistry are given in the fourth column.

Table 13.13.1 Summary of inherited disorders of metabolism

Condition	Inheritance	Survival with no therapy	Relevance in biochemistry
Phenylketonuria	Autosomal recessive	Early adulthood	Illustrates disorder resulting in accumulation of toxic metabolites; ethnically concentrated (in Europeans) and so probably balanced polymorphism
Alkaptonuria	Recessive	Normal	Example of almost harmless inherited disorder of metabolism
Homocystinuria	Autosomal recessive	Variable	Illustration (in some cases) of synthesis of apoprotein with reduced affinity for cofactor
Cystinuria	Autosomal recessive	Normal	An example of a defect in membrane transport
McArdle's syndrome	Autosomal recessive	Several decades	Illustration of heterogeneity of enzymes from different tissues
Pompe's disease	Autosomal recessive	Not beyond infancy	An example of a lysosomal disease
Lactose intolerance	Autosomal dominant	Normal	Example of ethnically-concentrated deficiency
Familial hypercholesterolaemia	Autosomal dominant	Tends to be restricted	An example of a defect in receptor binding
Lesch-Nyhan syndrome	X-linked recessive	Early adulthood	Illustrates damage due to failure of a feedback control
Sickle cell* anaemia	Autosomal recessive	Early adulthood	Mutation leading to amino acid substitution showing critical nature of amino acid sequence
Thalasaemias	Various	20 y (with transfusion therapy)	Illustration of a defect in the rate of protein synthesis

*(see 13.3)

13.14 Acid-base balance

Acid-base balance is that aspect of homeostasis which ensures a stable hydrogen ion environment for the body fluids. Since changes in hydrogen ion concentration (pH) change the ionization states of proteins, and therefore their biological properties, it is evident that such homeostasis is necessary. In the human, life cannot be sustained if the pH of the blood remains below 7.00 or above 7.8. Although acid-base balance is thus an important topic, it is hated by students presumably because of its pronounced physico-chemical foundation. In reality this is not extensive and the topic can be approached in terms of only three fundamental relationships.

1 *The bicarbonate equilibrium*

$$CO_2 + H_2O \rightleftharpoons H_2CO_3 \rightleftharpoons H^+ + HCO_3^-$$ ....(1)

The reversible nature of this relationship implies that an excess of H^+ drives the equilibrium to the left, in the direction of CO_2 release, while H^+ deficiency drives it to the right, increasing HCO_3^-. In other words changes in H^+ tend to be buffered, and the bicarbonate-carbonic acid-CO_2 equilibrium is a volatile buffer, indeed the most important buffer system in the plasma (it is much more significant than the phosphate buffer system with the equilibrium $H_2PO_4^- \rightleftharpoons HPO_4^{2-} + H^+$).

In kidney and erythrocytes there is a zinc-containing enzyme, carbonic anhydrase, which catalyses the combination of water to give carbonic acid and vice versa. It will be recalled that enzymes do not affect the final equilibrium of a reaction, only the speed at which that equilibrium is obtained. Carbonic anhydrase then has an accelerating effect on the attainment of the equilibrium of equation 1. It does not affect the extent of the equilibrium in either direction.

2 *The Henderson-Hasselbalch equation, which is*

$$pH = pK' + \log \frac{base}{acid}$$

In the case of the bicarbonate buffer system, H_2CO_3 is the acid and HCO_3^- the base, so that the relationship is

$$pH = pK' + \log \frac{HCO_3^-}{H_2CO_3}$$

H_2CO_3 is proportional to dissolved CO_2, which is proportional to the partial pressure of CO_2 (this is Henry's law) so that for H_2CO_3, αpCO_2 can be substituted, where α is the solubility coefficient of CO_2 in plasma. Thus

$$pH = pK' + \log \frac{HCO_3^-}{\alpha pCO_2}$$ ....(2)

Under normal circumstances the log term is 20, and since the pK' is 6.1, the pH is 7.4. What the relationship (2) 'says', is that when HCO_3^- rises the pH rises (since pK' is constant); when pCO_2 rises, the pH falls.

3   The haemoglobin dissociation

$$HHB + O_2 \rightleftharpoons H^+ + HbO_2 \qquad\qquad ....(3)$$

To be strictly accurate, only 0.7 mol H^+ dissociates from one haemoglobin monomer when it takes up one mole oxygen, but the above serves as an approximation. In any case, oxygenated haemoglobin (HbO_2) is more dissociated with respect to H^+ than HHb, that is it is a stronger acid. This phenomenon, known as the Bohr effect (Fig. 13.3.4) results in an evolution of oxygen from HbO_2 when acid is introduced (equilibrium driven to the left in 3) whereas high oxygen tensions release H^+ (equilibrium driven to the right). In effect it is another buffer system, comparable in importance to the carbonic acid/bicarbonate one. The plasma proteins themselves act as buffers, thus:

$$H\,Prot \rightleftharpoons H^+ + Prot^-$$

but constitute less than 10% of the total buffering capacity of the serum. In the capillaries, when oxygenated haemoglobin (HbO_2) encounters H_2CO_3, diffusing from the tissues, the equilibrium in equation 3 is displaced to the left and oxygen is released to diffuse into the tissues. There is a rise in bicarbonate, but some of it diffuses out of the erythrocyte, and to preserve electrical neutrality, chloride diffuses in. This is a simple statement of the *chloride shift*. If this did not happen, that is if bicarbonate alone diffused out, an electric charge would build up due to the imbalance of anions across the erythrocyte membrane.

In the lungs, where the oxygen tension is high, equilibrium 3 is displaced to the right and the H^+ ions released drive equilibrium 1 to the left. Thus CO_2 is released and HCO_3^- falls. HCO_3^- diffuses in from the plasma and Cl^- diffuses out. This is the converse aspect of the chloride shift. Total body hydrogen ion homeostasis is achieved by these mechanisms in the blood and lungs, supplemented by others in the kidney, although the buffers are the first line of defence.

The kidney in acid-base balance

There are three distinguishable mechanisms for acid excretion by the kidney.
1   Secretion of H^+. The tubules actively secrete hydrogen ions, resorbing Na^+ in equivalent amounts.
2   Excretion of the HPO_4^- ion, formed in the equilibrium

$$H^+ + HPO_4^{2-} \rightleftharpoons H_2PO_4^-$$

3   Glutamine hydrolysis. The urea cycle in the liver takes up NH_4^+ to produce ammonia but for each NH_4^+ taken up, one H^+ is generated (10.2). Instead NH_4^+ can be taken up by glutamic acid in the liver, to produce glutamine (Fig. 13.14.1). The glutamine has access to the circulation, and

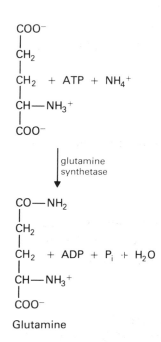

Fig. 13.14.1

239

Fig. 13.14.2

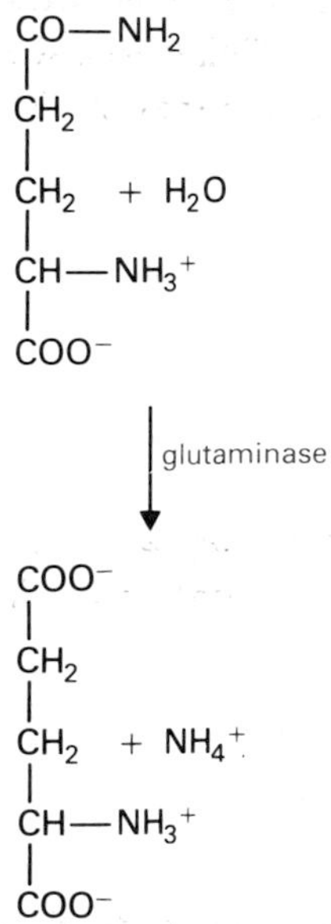

$$
\begin{array}{l}
CO-NH_2 \\
| \\
CH_2 \\
| \\
CH_2 \quad + \ H_2O \\
| \\
CH-NH_3{}^+ \\
| \\
COO^-
\end{array}
$$

↓ glutaminase

$$
\begin{array}{l}
COO^- \\
| \\
CH_2 \\
| \\
CH_2 \quad + \ NH_4{}^+ \\
| \\
CH-NH_3{}^+ \\
| \\
COO^-
\end{array}
$$

Glutamate

can be hydrolysed by the kidney to glutamate and $NH_4{}^+$ (Fig. 13.14.2). The $NH_4{}^+$ is excreted, in exchange for sodium ions. Thus, this renal mechanism avoids the production of H^+ in the liver and the subsequent acidification of the circulation.

The blood and kidney defences operate when there is a tendency for the body fluids to fall in pH (acidosis) and to rise in pH (alkalosis). (The latter term should in reality be called 'basosis' since the alkalis are more specifically defined as the hydrates of the alkali metal oxides, NaOH, KOH, $Ba(OH)_2$ and LiOH. 'Base' is the more comprehensive term, as in 'acid-base balance'. HCO_3^- should not be termed an alkali.) For practical purposes both alkalosis and acidosis are divided into the 'respiratory' and the 'metabolic' types, the former being due to breathing disorders of one type or another. It is convenient to examine these and see straight away how the compensatory mechanisms operate.

1 *Acidosis (metabolic)* This can be caused by:

  (i) starvation, in the course of which the acidic ketone bodies are produced in large quantities;

  (ii) diabetes, which is also associated with excessive ketone body production;

(iii) lactic acidosis, as a result of heavy exercise when lactic acid is produced by the muscles, or as a result of anoxia, when it cannot be converted to pyruvate;

(iv) loss of bicarbonate, in vomiting and diarrhoea;

  (v) ingestion of acids like salicylic acid and of NH_4Cl, which acts as an acid in the course of urea formation in the liver

$$2NH_4Cl + O_2 \rightarrow 2HN.CO.NH_2 + H_2O + 2H^+ + 2Cl^-$$

(vi) kidney failure, when the tubules fail to secrete H^+.

In general terms, the introduction of an acid stronger than carbonic acid displaces the bicarbonate equilibrium (reaction 1, above) to the left. HCO_3^- falls, with a rise in H_2CO_3. Blood pH falls, according to the Henderson-Hasselbalch equation (2) above.

Compensation occurs by the lungs hyperventilating to eliminate CO_2. The fall in plasma HCO_3^-, meanwhile, causes erythrocyte HCO_3^- to diffuse into the plasma and this is replaced by Cl^- in the *chloride shift*. The kidney aids compensation (except of course in cases of kidney failure) by secretion of H^+ and by glutamine hydrolysis).

2 *Acidosis (respiratory)* can be caused by:

  (i) airway obstruction;

  (ii) respiratory muscle weakness;

(iii) central nervous system diseases;

(iv) heart failure.

It is compensated by hyperventilation, by the buffer systems of the blood involving the chloride shift, and by the renal mechanisms.

3 *Alkalosis (metabolic)* can be caused by:

  (i) retention of abnormal amounts of alkali, or loss of acid, as in vomiting the HCl-rich contents of the gastric secretion;

 (ii) administration of sodium lactate or other acid salts. *In vivo* these behave as strong *bases*, in that they take up H^+ on oxidation. This can be seen by considering the stoichiometry of total oxidation of lactate, for example.

$$\begin{array}{c} CH_3 \\ | \\ CHOH + H^+ + 3O_2 \rightarrow 3CO_2 + 3H_2O \\ | \\ COO^- \end{array}$$

Lactate

(iii) bicarbonate administration.

Compensation is initiated by hypoventilation, thus increasing pCO_2 and driving equilibrium 1 to the right. The pH must fall, according to the Henderson-Hasselbalch equation.

4 *Alkalosis (respiratory)* can be caused by:

 (i) head injuries;

 (ii) hysteria;

(iii) drugs causing hyperventilation.

It is clear that this is caused by lowering of alveolar pCO_2 in hyperventilation. It is compensated by hypoventilation, or by increasing CO_2 in the breathed air (for example by putting the mouth in a paper bag).

There are mixed types, of course, and uncompensated types, which are taken to be beyond the scope of our present treatment.

13.15   Immunochemistry

The immune response is the ability of an individual to recognize foreign (non-self) material. In general terms immune activity is manifested through the lymphocytes, of which there are two types, the T-cells (associated with the thymus) and the B-cells (of the bone marrow). The T-cells are responsible for cellular immunity (allograft rejection, anti-tumour activity) and the B-cells for humoral activity (elaboration of the circulating antibodies). The circulating antibodies are otherwise known as the *immunoglobulins* and they have the capacity to combine specifically with their *antigenic determinants*, that is the chemical structures eliciting the immune response. For present purposes 'immunochemistry' is synonymous with 'the biosynthesis and structure of the immunoglobulins'.

The immunoglobulins Five groups of immunoglobulins have been recognized, of which quantitatively the most important is IgG (80% of the total circulating antibodies under normal circumstances). The others are IgA, IgM, IgE and IgD (Table 13.15.1) (you can use some nonsense word like GAMED to remember them) and they appear to differ both in site and function. The immunoglobulins may exist in polymeric forms but the monomeric state has four polypeptide chains: two H or 'heavy' and two L or 'light'. They are linked together by disulphide bonds and there are also

Table 13.15.1 Classes of antibody molecules

Class	Light chains	Heavy chains	Function(s)
IgG	κ or λ	γ	Principal serum antibody; activation of complement*
IgA	κ or λ	α	Antibody of the exeternal secretions — colostrum, intestinal mucosa, saliva
IgM	κ or λ	μ	Cell surface receptor initiating lymphocyte proliferation
IgD	κ or λ	δ	Cell surface receptor on immature B-lymphocytes
IgE	κ or λ	ε	Uncertain, but the reaginic antibody; antigen-IgE reaction triggers histamine release, with urticaria/asthma Perhaps anti-parasitic antibody

Note that no separate nomenclature of the heavy chains is involved; they merely correspond, as Greek letters, to the class of immunoglobulin (gamma for IgG, alpha for IgA, and so on).
*Complement is a cascade system of enzymes necessary for immune haemolysis and bacteriolysis

intrachain disulphide bonds (Fig. 13.15.1). If all the IgG molecules in the blood of an individual are isolated and examined then it turns out that the C-terminal regions of both the light and heavy chains are mostly fixed in their amino acid composition, whereas the sequences closer to the N-terminals are variable. Further, *hypervariable regions* occur in both heavy and light chains. These are segments which display far more variability from molecule to molecule than do other segments near the N-terminals. It seems reasonable to connect these hypervariable regions with the ability of immunoglobulins to combine with a diversity of antigens.

Antigen-antibody interaction In general, antigens:
1   are foreign to the host;
2   are of high molecular weight, usually proteins or polysaccharides (antibodies may be elaborated against nucleic acids but only if they are presented to the immune system as nucleoproteins);
3   have some degree of structural complexity, for example homopolymers are less antigenic than heteropolymers.
The central problem is to account for the specificity of antigen-antibody interaction when there are potentially millions of both naturally-occurring and synthetic antigens. It appears that to initiate antibody synthesis, the antigen is bound to the surface of the B-lymphocyte. Only the antigenic determinant, sometimes a small part of the total antigen, need be bound. Much knowledge of the nature of antigenic determinants in general has come from the use of *haptens*, substances which are not in themselves antigenic but which elicit an immune response when in combination with a large molecule, typically a protein. The immunoglobulin synthesis is directed against the structure of the hapten such that in effect it becomes the antigenic determinant for the protein. Varying the structure of the hapten bound to the same macromolecule has shown that rather small variations in

Fig. 13.15.1 Structure of IgG.
In general it is Y-shaped, with the two short arms having pivotal flexibility. The shape is the result of the alignment, and binding by a disulphide bond, of two heavy and two light chains. Each chain has a constant region and a variable region, the former mediating effector status (that is, determining whether the antibody is IgG or some other type) and the latter mediating antigen recognition. Each chain has regions called domains which are analogous in structure (and therefore function) to domains on neighbouring chains or on separate molecules altogether. One of the domains contains hypervariable regions which appear to constitute the antigen binding sites.

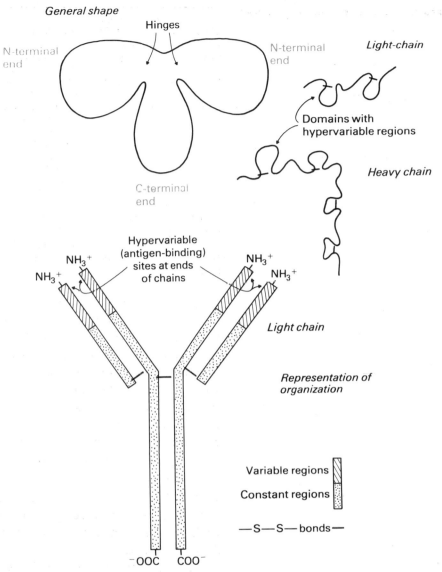

shape may be discriminated against by the immunoglobulins. For example in some organisms antibodies developed against glucose as hapten will not react with galactose, although of course the two sugars differ only in the conformation of a single carbon. Specificity is not however absolute in the sense that a specific immunoglobulin will always combine with a unique substance and only that unique substance. Moreover each antigen does not necessarily elaborate a specific unique antibody.

The basis of antibody diversity and specificity There is no choice but to assume that different antibodies have different amino acid sequences and therefore different conformations at their antigen-combining sites and, further, that these correspond to the hypervariable regions. Any diversity of amino acid sequence must reside in the genome, that is in DNA sequences. The question is, are the DNA sequences for the hypervariable regions generated on the appearance of the antigen by some type of mutation or are they preformed in such a way that they are expressed when necessary, namely on appearance of the antigen? The latter turns out to be the correct version, for the germ line (embryonic) lymphocytes contain hundreds of genes for the variable regions of both the light and heavy chains (V-genes) and several for the constant segments (C-genes) of the various immunoglobulin classes. There are also joining genes (J-genes). Recombinations are thought to be capable of yielding over 10^7 different antibody molecules.

Fig. 13.15.2 Summary of the immune response.

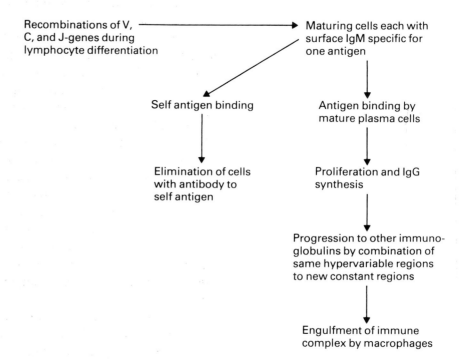

As each embryonic lymphocyte matures it first synthesizes IgM and this is maintained on the cell surface. Each cell produces only one unique IgM, so that there are over 10^7 different lymphocytes (in this sense) in the circulation. If at this stage this immature lymphocyte encounters self antigen, binding of it to the IgM causes the cell to die. Thus antibodies to self are eliminated in embryonic life. The mature cell, normally known as the plasma cell, binding antigen at the surface, is however stimulated to proliferate, forming a clone producing IgG specific for the initially bound antigen. The plasma cell proceeds to make others of the immunoglobulin classes, all of these being identical in the light chains and in the variable regions of the heavy chains. However, the constant regions progress through the structures typical of IgA, IgE, etc. In other words these regions are 'constant' within a given class but vary from class to class.

The combination of antigen with antibody (the *immune complex*) should render the former less noxious and capable of being engulfed by macrophages.

In summary, the sequence of events can be represented as in Fig. 13.15.2.

13.16   Lipoproteins

The internal milieu of all organisms is aqueous, although many species are adapted to the use of lipid deposits as energy stores. Examples are human subcutaneous fat, the camel's hump, and the perigonadal deposits of rodents. However if such water-excluding masses of triacylglycerol appear in the wrong places, for instance in the liver (*fatty liver*) or in the small vessels (*fat embolism*), there can be serious consequences. There is one type of cell assembly which is essentially non-aqueous, namely the membrane in its various aspects. Cholesterol is transported in the blood to at least some of the tissues in order to provide for the membranes, but derangements in cholesterol transport can cause deposition of it under the skin (*xanthomas*) and in the intima of the arteries (*atheromas*). In conclusion, the transport of fatty substances round the aqueous system of the body appears to be precarious. But there are three mechanisms to effect it safely.

1   *Chylomicrons* This term applies to the micelles of newly-ingested fat, to which is added protein and cholesterol by the small intestine. The micelles enter the lymph and are discharged by the thoracic duct into the venous system. After a fatty meal there is observable *hyperchylomicronaemia*, a particular form of *lipaemia*. The chylomicrons have a small content of protein and so may be classified as lipoproteins, but they do not migrate in an electric field.

2   *The serum albumin-free fatty acid complex* Serum albumin binds a number of poorly-soluble substances, of which bilirubin and fatty acids are examples. Such fatty acids have been mobilized from adipose tissue and are on their way to the liver and muscles for oxidation. They have a very short half-life; in other words they are added to and removed from the serum very rapidly, so they have long been taken to represent a labile, metabolically active pool of lipid.

3   *Lipoproteins* The lipoproteins are classified in two ways: by sedimentation behaviour, and by electrophoresis. The lipoproteins are less dense than serum, except for a small part of the 'high density' fraction, and when serum is subjected to high speed centrifugation they float to the surface at different rates. The chylomicrons float without centrifugation and may form a 'cream layer' in hyperchylomicronaemic serum when it is merely left to stand in a test tube.

On the basis of flotation characteristics (flotation being sedimentation upwards) there are three classes other than the chylomicrons: very low density (VLDL), low density (LDL) and high density (HDL) lipoproteins.

On electrophoresis these correspond to pre-β, β- and α-globulins respectively (Fig. 13.16.1). The lipoproteins all contain phospholipid, cholesterol, and triglyceride, but for practical purposes we can regard chylomicrons and VLDL principally as carriers of triglycerides, and LDL and HDL principally as carriers of cholesterol. As an *aide-memoire* write down the lipoproteins in order of increasing density. Write down pre-β, β- and α-. Write down triglyceride and cholesterol:

Increasing density	CHM	No migration	Triglyceride
	VLDL	Pre-β	Triglyceride
	LDL	β-	Cholesterol
	HDL	α-	Cholesterol

Lipoprotein metabolism It was previously thought that the lipoproteins are passive transporters of lipid. Now it is known that they participate in a complex series of lipid fluxes. The protein parts of the lipoproteins, the *apolipoproteins*, take part in the control of these fluxes (Fig. 13.16.1). Triglycerides synthesized in the intestine and the liver are transported in the blood as chylomicrons and VLDL. Some exogenous and endogenous cholesterol is also incorporated. They are taken up by the adipose tissue and skeletal muscle, where their triglycerides are hydrolysed by the enzyme lipoprotein lipase of the capillary endothelium. This enzyme is activated by apolipoprotein CII, a constituent of the chylomicrons themselves. The apolipoproteins, it is thought, can be transferred from one lipoprotein class to another, with the possible exception of apolipoprotein B, which is characteristic of VLDL and LDL. After VLDL and chylomicron triglycerides are hydrolysed in peripheral tissues, their 'ghosts', which are enriched with respect to cholesterol, take on the electrophoretic mobility and composition of LDL. LDL is thus not in itself synthesized *de novo*. HDL lipoproteins, on the other hand, are thought to be synthesized by the liver (Fig. 13.16.1).

Present opinion holds that both LDL and HDL have regulatory roles as regards the level of cholesterol and its esters in the blood. This is important with respect to the pathogenesis of ischaemic heart disease: there is a strong association between hypercholesterolaemia and the formation of the fatty streaks or plaques (atheromas) which block the coronary arteries in heart attacks. LDL is bound by surface receptors on most cells of the body, after which it is internalized and hydrolysed. The cholesterol ester released is hydrolysed and free cholesterol inhibits the enzyme β-hydroxy-β-methyl glutaryl CoA reductase. This is the rate-limiting enzyme of cholesterol synthesis (Fig. 11.7.1). Thus, if LDL can be bound by cells, there is both a removal of cholesterol from the blood and a brake on its intracellular synthesis. In the condition termed *familial hypercholesterolaemia* there is an inherited deficiency of the receptors and patients suffering from it have enormously elevated blood cholesterol and a high probability of early death due to ischaemic heart disease.

HDL contains apolipoprotein AI, which activates the enzyme lecithin cholesterol acyl transferase (LCAT). The enzyme is present in the blood;

Fig. 13.16.1 Lipoprotein metabolism. The green dotted line represents the routes for cholesterol: synthesis in the liver; uptake into VLDL; concentration in LDL after the VLDL loses the bulk of its triacylglycerol; esterification by LCAT; and uptake by HDL, which either passes it back again to LDL or is taken up itself by the liver. In so far as the liver converts it to bile salts, this is an excretory pathway. Evidently there are a series of cycles which ensure that sufficient cholesterol is delivered to tissues but which can allow excretion to prevent over-accumulation.

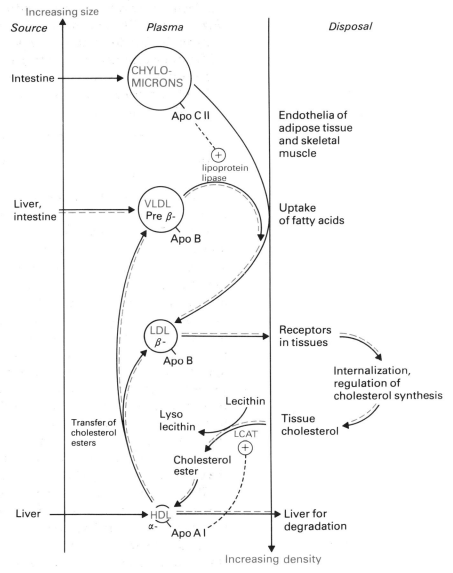

lecithin is a constituent of the HDL itself. The cholesterol is thought to be derived from cell membranes and therefore HDL may play a role in removing excess cholesterol from cell surfaces and transporting it to the liver as cholesterol ester in the interior of the lipoprotein molecule. In the liver HDL is degraded and the cholesterol ester converted to an excretory form, namely bile acids.

In summary the two 'lightest' lipoproteins, chylomicrons and VLDL, are vehicles for triglyceride transport. Their product LDL, along with HDL synthesized in the liver, regulate cholesterol transport and deposition.

Hyperlipoproteinaemias can be 'primary' or 'secondary'. The primary ones, by definition, are inherited. There are five or six recognized types. In general they carry a risk of heart disease if LDL is raised and they can be treated by a low fat diet and/or drugs which clear cholesterol from the body. The secondaries mimic the primary ones but are due to such multivariate pathologies as alcoholism and diabetes.

Alipoproteinaemias are rare, but have given useful information on lipoprotein function. In *hypoalphalipoproteinaemia* (preferred to 'a-alphalipoproteinaemia'), also known as Tangier disease, there is an inherited lack of HDL and the subjects generally die due to massive atheromatous lesions in the coronary arteries. In *abetalipoproteinaemia* there is not a similar risk. Thus the idea has arisen that LDL cholesterol is involved in the pathogenesis of heart disease, but HDL cholesterol, representing a form on its way to the liver for excretion, is associated with protection.

13.17   Photosynthesis

Plant biochemistry has been called the Cinderella among biochemistry topics, since although the importance of plants in the biosphere is immense, it is liable to be neglected in most general courses, and photosynthesis is likely to be the only aspect of plant biochemistry taught. Even then some economies exist, since the study of the subject uncovers analogies to biochemical systems already learned.

Photosynthesis is a self-descriptive term—synthesis using light. In green plants and some bacteria, the light provides the energy for the synthesis of carbohydrate. The total reaction can be represented:

$$6CO_2 + 6H_2O \rightarrow C_6H_{12}O_6 + 6O_2$$

The organisms which conduct photosynthesis, therefore, make do with very simple starting materials; they are *autotrophs*, defined as organisms for which CO_2 is the sole or major source of carbon. (A *photoautotroph* is an autotroph which uses light energy for the assimilation of this simple form of carbon. A *chemautotroph* uses energy derived from the oxidation of inorganic compounds like iron or sulphur. A *heterotroph*, like man, is an organism that depends on complex external organic molecules, namely foodstuffs.) Photosynthesis may be considered as three separate processes, namely absorption of light energy, electron/proton transfers, and carbohydrate synthesis. In experimental systems, carbohydrate synthesis can take place, for a time, in the absence of light. Although this may not be significant *in vivo*, it forms an alternative division of the topic, for purposes of discussion, into the 'light' and 'dark' reactions (Fig. 13.17.1).

Absorption of light energy

Green plants contain the pigment *chlorophyll*, located in the subcellular organelles called *chloroplasts*. There is more than one type of chlorophyll,

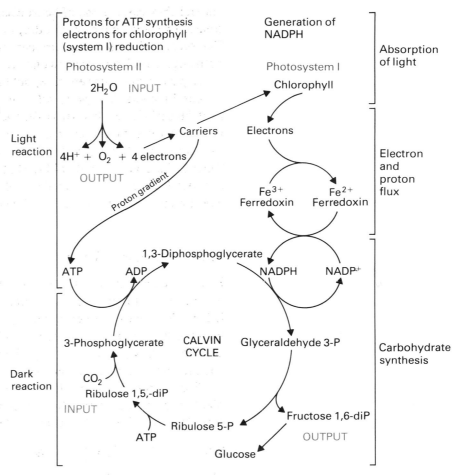

Fig. 13.17.1 Photosynthesis.
It is an observation that the process can be divided into the light and dark reactions. Conceptually, it can be divided into: 1, the absorption of light; 2, electron and proton fluxes; and 3, carbohydrate synthesis.

The stoichiometric summary at the bottom indicates that the incorporation of $6CO_2$ allows the synthesis of one hexose, as would be expected.

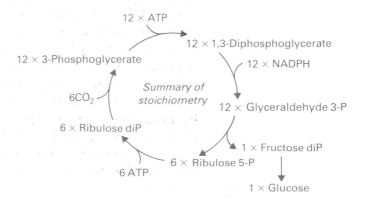

but they are all porphyrins containing magnesium rather than iron. Moreover, the porphyrin is somewhat different from the protoporphyrin IX of haem. Building on the basic porphyrin structure, it can conveniently be

Fig. 13.17.2 Chlorophyll.

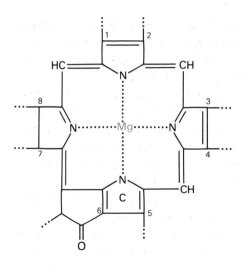

Fig. 13.17.3 **Fig. 13.17.3** Structure of the chloroplast. There are 20–30 of these per cell. Most of the components necessary for light absorption are in the thylakoid membrane. Most of the enzymes necessary for the Calvin cycle are in the stroma.

reproduced as in Fig. 13.17.2. There are substituents at 1–8 which may vary to some extent. An additional peculiarity is the hexanone ring fused to pyrrole ring C and the presence of only 10 double bonds (rather than the 11 in haem).

The chlorophyll is incorporated into the chloroplasts, the subcellular organelles which have some analogies to mitochondria (a double membrane and the presence of DNA). The membrane containing the chlorophyll is the *thylakoid membrane* (Fig. 13.17.3).

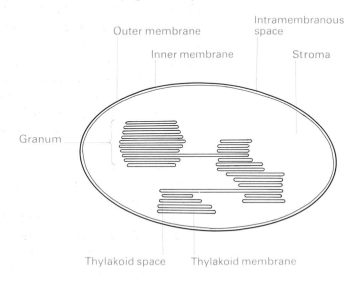

Light energy is absorbed by many chlorophyll molecules in the chloroplasts and transmitted to a 'reaction centre' chlorophyll. This is sufficiently excited either to split water (photosystem II) or to transfer an electron to the iron-containing protein ferredoxin, reducing the iron in the protein (photosystem I). (In fact, there are many photosystems I and II.)

Electron and proton transfer Photosystem I reduces the ferric ion in ferredoxin to the ferrous form. This reacts (although not directly) with $NADP^+$ to yield NADPH. The chlorophyll of photosystem I must then be reduced again, and electrons flow from photosystem II through several carriers. Photosystem II contains manganese and splits water to give protons, electrons, and oxygen.

$$2H_2O \rightarrow O_2 + 4H^+ + 4\epsilon$$

Electrons flow across the thylakoid membrane to photosystem I. If the reactions so far described are written out in the proportions which react,

I $\begin{cases} 4\epsilon \text{ (from chlorophyll)} + 4Fe^{3+}\text{—ferredoxin} \rightarrow 4Fe^{2+} \\ 4Fe^{2+}\text{—ferredoxin} + 2NADP^+ + 2H^+ \rightarrow 4Fe^{3+}\text{—ferredoxin} + 2NADPH \end{cases}$

II $\quad 2H_2O \rightarrow O_2 + 4H^+ + 4\epsilon$

and added, then the resulting reaction is

$$2H_2O + 2NADP^+ \rightarrow O_2 + 2NADPH + 2H^+$$

Thus electrons have been transferred to $NADP^+$ from water, leaving excess protons. It is the proton gradient across the thylakoid membrane which drives the phosphorylation of ADP to ATP. This is thought to be entirely analogous to oxidative phosphorylation in mitochondria (7.1).

Carbohydrate synthesis involves incorporation of CO_2 in a cycle known, after its discoverer, as the Calvin cycle. In the 'light' reactions described above, ATP and NADPH are made available. There is no simple reaction of these starting materials to provide a phosphorylated 1-C precursor, but rather a cycle involving catalytic quantities of carrier molecules, which is familiar enough from, say, the citric acid cycle (7.2). Moreover, there is not a one-to-one stoichiometric transformation of intermediates, but this concept is familiar enough from consideration of the hexose monophosphate shunt (8.4). CO_2 reacts with ribulose diphosphate, and there is reduction and phosphorylation (Fig. 13.17.4). There is then interconversion to produce a six-carbon sugar and regenerate the substrate for CO_2 fixation, ribulose 1,5-diphosphate. These reactions involve transketolases, aldolases, and phosphatases. The stoichiometry is such that twelve molecules of glyceraldehyde 3-phosphate are involved in the provision of one six-carbon (fructose 6-phosphate) and six five-carbons (ribulose 5-phosphate). The latter are phosphorylated to ribulose 1,5-diphosphate (so ATP is needed at two points in the cycle) and then CO_2 is fixed. So if one CO_2 is fixed for each turn, evidently six turns are required to make one fructose unit. With the sugar phosphates acting as carriers, and the consumed NADPH and ATP being regenerated by the light reactions, then the total reaction is as given at the head of the section. It is the exact reverse of the reaction for the aerobic breakdown of glucose to carbon dioxide and water in animals, and is therefore the chemical expression of the complementariness of the two forms of life.

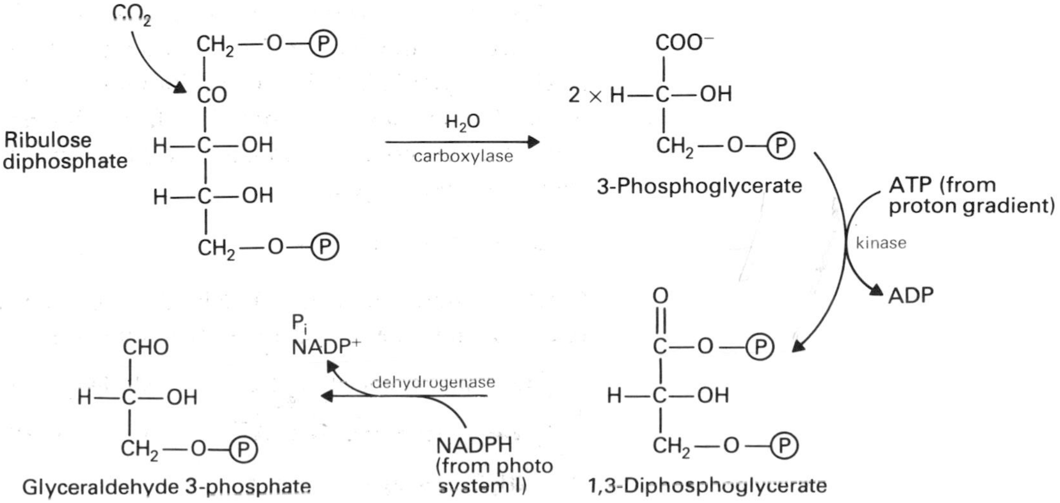

Fig. 13.17.4 Some reactions of the Calvin cycle.

13.18 Vitamin D, milk, rickets, and sunlight

There is an interesting relationship between a factor in the diet promoting bone formation (but which is not essential in sunny climes), the parlous nature of calcium absorption in humans, the failure of adults in some countries to digest milk, and the extent of skin pigmentation. The dietary factor is vitamin D, or rather 'the vitamins D' for there are at least two. They are found in fish oils such as cod liver oil, and to some extent in eggs and butter (margarine is artificially fortified with vitamin D) but other natural sources are few. Vitamin D_3 (cholecalciferol) and D_2 (ergocalciferol) are synthesized in the Malphigian layer of epidermis, when ultraviolet light of wavelength 280–295 nm impinges upon it. The precursors are 7-dehydrocholesterol and ergosterol respectively (Fig. 13.18.1). These natural precursors are not in themselves biologically active and in the absence of sunlight the vitamins themselves must be present in the diet to the extent of about 3 μg/d. In a two-step mechanism involving both the liver and the kidney, cholecalciferol yields 1,25-dihydroxycholecalciferol, an effector substance in some ways reminiscent of a steroid hormone. Ergocalciferol is similarly hydroxylated in liver and kidney. Probably in bone, and certainly in intestinal mucosa, the 1,25-dihydroxycholecalciferol stimulates synthesis of mRNA specific for a calcium-binding protein. In intestine it presumably aids calcium absorption and in bone it concentrates calcium for effective mineralization. *Rickets* is the condition, in children, which results from a deficiency of calcium, due to either a poor diet or lack of sunlight. Subjects are characteristically bow-legged with pelvic deformities. (Not all cases of rickets are curable by administration of cholecalciferol and therefore are not caused by avitaminosis D; in hereditary, vitamin D-resistant rickets the kidney is unable adequately to conserve phosphate. This results in a hypophosphataemia, and since phosphate is consequently resorbed from the bones, there is also obligatory calcium loss with subsequent bone deformations.)

Melanin, the pigment of the skin which makes it appear brown, black, or red, is also synthesized in response to ultraviolet light of about 290–320 nm, but by the skin melanocytes. The precursor is tyrosine and the initial pathway is the same as that for adrenaline synthesis. In many human populations melanin synthesis is independent of sunlight since there is a genetic capacity to maintain a permanently pigmented skin. Heavily pigmented skins are by and large found in warm sunny climates; the dark surfaces would, however, be expected to absorb more heat than lighter ones, so that if it is desirable in the tropics to limit heat absorption, a dark skin would appear to have negative survival value. However it appears that absorption of UV light by melanin inhibits cholecalciferol synthesis. Vitamins D are toxic in excess; therefore it may be that melanin prevents synthesis of toxic quantities of the substance in the tropics.

It is presently supposed that early man evolved in the sunny tropics and

Fig. 13.18.1 Multiorgan processing of vitamin D.
Note the isomerization of the 3β-hydroxyl to 3α-hydroxyl during conversion of 7-dehydrocholesterol to cholecalciferol by ultraviolet light. Ergosterol is processed by an analogous set of reactions. The effector substance is evidently 1,25-dihydroxycholecalciferol. After stimulation of mRNA synthesis this is metabolized by side-chain scission and excreted in the bile and faeces.

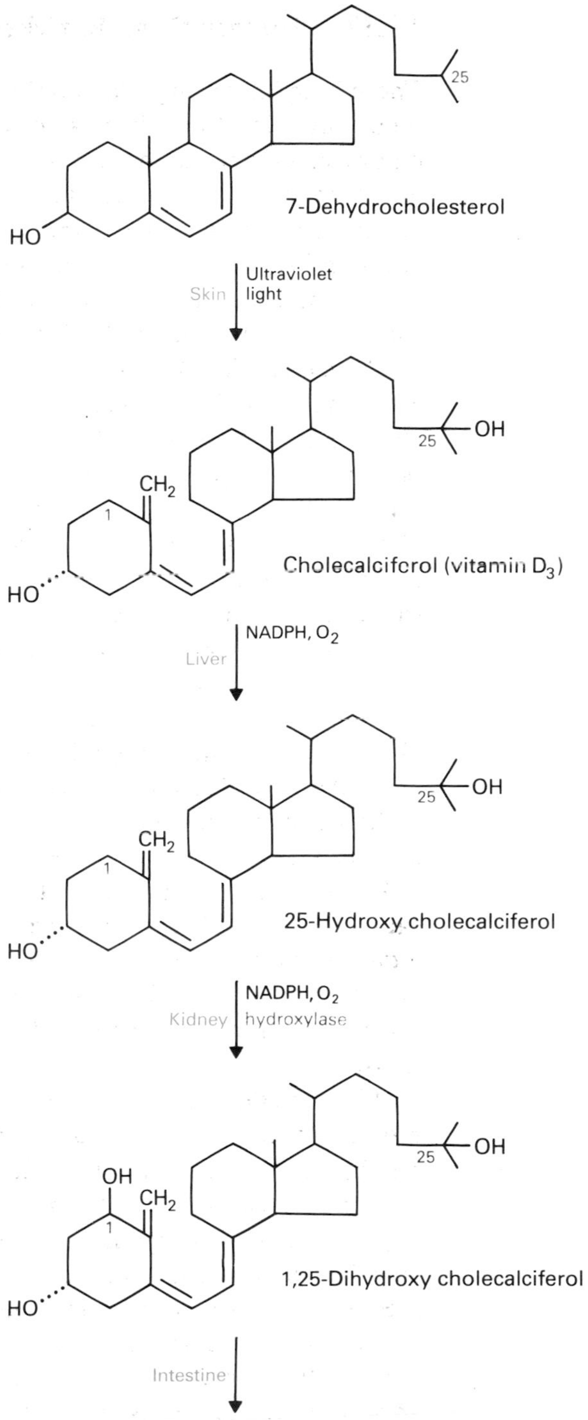

7-Dehydrocholesterol

Skin | Ultraviolet light

Cholecalciferol (vitamin D$_3$)

Liver | NADPH, O$_2$

25-Hydroxy cholecalciferol

Kidney | NADPH, O$_2$ hydroxylase

1,25-Dihydroxy cholecalciferol

Intestine

Binding by receptor
mRNA synthesis
Synthesis of calcium-binding protein
Enhanced calcium absorption

had a dark skin, preventing synthesis of toxic amounts of vitamin D. Subsequent migration northwards to temperate climes presumably resulted in the reverse situation—the risk of too little synthesis of the vitamin. (In modern times the same risk is run by Asians migrating to dark Northern European cities.) It is thought that there was selection for lighter skins which allowed the scanty sunlight to synthesize enough of it from its precursor steroids. So the northern races became 'white', but maintained the safeguard of tanning or freckling when exposed to strong sunlight.

Vitamin D activity serves to enhance calcium utilization, but for this to happen, obviously there must be sufficient calcium in the diet. A major dietary source of calcium is milk, which also contains the disaccharide lactose. This may be hydrolysed in the small intestine

$$Lactose \longrightarrow galactose + glucose$$

Lactase, the enzyme involved, is synthesized by all individuals during the perinatal period but many races, especially in the tropics and sub-tropics, lose this capacity as adults. (This is termed 'lactose intolerance'.) Now it appears that lactose hydrolysis in some way aids calcium absorption, so that lactase-synthesizing adults, who predominate in cold northern lands, are in this respect at an advantage which to some extent offsets the lack of sunlight. Anything which can enhance calcium absorption, even marginally, is valuable, for it is thought that even under optimal conditions, as much as 90% of the calcium entering the gut may be lost in the faeces.

In addition, there are other risks to calcium absorption (Table 13.18.1). Most people throughout the world gain a large proportion of their calcium from cereals such as wheat, which is of course baked into bread of various types. Cereals and therefore breads contain a compound called phytic acid which binds calcium and renders it unavailable for absorption in the intestine. It is present mostly in the outer layers of the cereals and so it is more concentrated in those breads made with unextracted (whole wheat) flour, like brown loaves and chappatis.

So we are in a position now to work out the worst case scenario. This is a woman who does not receive much direct sunlight, has a dark skin, consumes no milk because she is intolerant, and has a diet high in unextracted cereals which are rich in phytic acid. Her diet may also be deficient in the calcium-rich foods such as cheese and fish. Probably during repeated pregnancies she is drained of calcium due to the demands of the fetal skeleton and her own milk.

Conversely, you are at risk of absorbing too much calcium and having it precipitate in awkward places such as the kidney (nephrocalcinosis) if you have a white skin and expose yourself to much sunlight, eat cheese and fish rather than brown bread, and have chosen parents who transmitted the adult lactase gene to you (Fig. 13.18.2).

Table 13.18.1 Factors influencing calcium absorption

Enhancing	Inhibiting
Sunlight	Lack of sunlight
Vitamins D	Rachitic diets
Phytase*	Phytic acid
Lactose	Lactose intolerance
Acid	Alkali

*Enzyme hydrolysing phytic acid, also found in cereals.

Fig. 13.18.2 Climates, skins, and vitamin D.

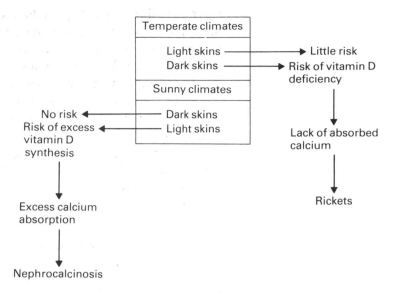

13.19 Receptors

A receptor is a structure inside or on the surface of a cell, having the capacity to bind specifically either a single compound or a class of compounds. As a prime example, receptors for certain hormones are situated on the surfaces of the cells of the target organs for these hormones. Insulin, ACTH, adrenaline, and glucagon come into this category. Receptors for other hormones, for example oestrogens, androgens, corticoids, and thyroid hormones, are internal. But using one cell type or another, receptors have been demonstrated for such diverse entities as antigens, parasites, viruses, neurotransmitters, growth factors, and drugs. The substance bound is called the *ligand*. If the ligand produces some response, it is called (this being pharmacological parlance) an *agonist*; a substance which prevents the response is an *antagonist*. It is apparent that general discussion of agonists binding to receptors and causing a biological response might be rather wide-ranging. Thus, the ones you are most likely to be asked about are summarized below. There is one general point to be made about receptors—each forms part of an amplification system whereby a single agonist molecule can ultimately result in the synthesis of a large number of product molecules. This is generally a protein or an enzyme, and the amplification operates by some sort of cascade mechanism (12.2).

The insulin receptor

The insulin receptor has been intensively investigated, but notions of its structure are not well advanced. It is a glycoprotein of four subunits, disulphide linked. Two of the subunits are 'heavy', two 'light', so that there is a superficial analogy to immunoglobulin structure (13.15).

Whatever its structure, the receptor can distinguish insulin from all other substances. This is not to say that it binds a single unique substance. 'Insulin' refers to proteins of a family whose composition varies somewhat from species to species. Receptors for insulin generally bind insulin from other species; they also bind insulin-like substances, such as nerve growth factor and relaxin.

Surprisingly, if you examine insulin receptors in a given tissue, their number is not static. Moreover their insulin affinity can increase or decrease. Diminished affinity makes apparent the phenomenon of 'insulin resistance'. Obese subjects are known to have an abnormal glucose tolerance, that is they do not clear a dose of absorbed glucose as fast as non-obese subjects and this is now known to be due to a diminished affinity of their receptors for insulin. The molecular basis is unknown. But by what means could this reaction of surface receptors with insulin, whatever its mechanism, influence events within the cell? Again, this is not definitely known, but insulin seems to affect phosphorylation–dephosphorylation cycles in regulatory enzymes. The receptor–insulin complex may have a catalytic role in itself or be an effector within the cascade. In general, the effects of insulin are antagonistic to those of glucagon and adrenaline. Since these two hormones ultimately cause phosphorylation of some key enzymes (phosphorylase, hormone-sensitive lipase), rendering them more active, it might be expected that insulin has a dephosphorylating effect. This is true in some cases, but phosphorylation is observed in others. Thus in adipose tissue, acetyl CoA carboxylase is phosphorylated and activated; since this is the rate-limiting step in palmitate synthesis, an observed effect of insulin in adipose tissue is lipogenesis. Significantly, a role for cAMP has not been found in mechanisms of phosphorylation of enzymes stimulated by insulin.

Catecholamine receptors

The cell surfaces of many tissues have receptors for the neurotransmitters noradrenaline and dopamine, as well as adrenaline. Adrenaline and nor-adrenaline are also synthesized by the chromaffin cells of the adrenal medulla. The receptors for adrenaline have been differentiated into two types, termed α- and β-, on the basis of their capacity for binding antagonists to the catecholamines. However there is some agreement that the β-receptors function by stimulating cAMP formation upon adrenaline binding. Thus the chain of events is as shown by Fig. 13.19.1. In this way adrenaline activates phosphorylase and lipase so that its observed effects are glyco-genolysis and lipolysis.

Acetylcholine receptors

These receptors take historical precedence. Early, they were postulated to bind to the acetylcholine synthesized at the nerve endings and released in

Fig. 13.19.1

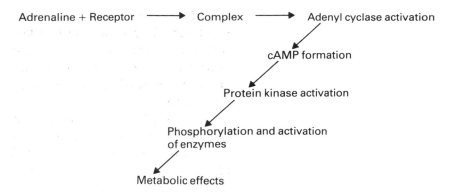

Adrenaline + Receptor ⟶ Complex ⟶ Adenyl cyclase activation

cAMP formation

Protein kinase activation

Phosphorylation and activation of enzymes

Metabolic effects

response to the depolarization waves conducted along the axon. They have been found to be of two types. One, associated with the neuromuscular junction, is called the nicotinic receptor since nicotine produces the same response as acetylcholine. The second is the muscarinic receptor since muscarine, but not nicotine, mimics acetylcholine; it is associated with the contraction of smooth muscle. The structure of the nicotinic receptor has been partially characterized—in electric rays and eels it is in great concentration in the electric organs and can be extracted with detergents. It turns out to be a pentameric protein. Otherwise the mode of action is as obscure as that of the insulin receptor. In some way reaction with acetylcholine causes an increase in the membrane permeability for sodium and potassium and so membrane depolarization is propagated from one neurone to another.

Steroid receptors

Steroid receptors are cytoplasmic, not on the cell surface. Steroids appear to enter cells freely; their binding to receptor causes some change in the receptor whereby it can enter the nucleus. By some means the receptor-hormone complex causes the nucleus DNA to produce more (or less) mRNA. In the case of oestradiol, which is a protein anabolic hormone, it is known that binding causes a conformational change in the receptor so that it binds some other compound. This triple complex has a high affinity for DNA and thus influences transcription. The observed effect is stimulation of uterine growth, although all the intermediate details remain obscure.

Low density lipoprotein (LDL) receptors

These are surface receptors on most cells. LDL is a cholesterol-rich fraction of the lipoproteins and has a specific apolipoprotein which binds to the receptors. Once bound, the LDL is internalized and its components hydrolysed. The cholesterol released is thought to inhibit intracellular cholesterol synthesis. In the inherited condition known as familial hypercholesterolaemia the receptors are lacking so that the LDL cholesterol is not taken up by the tissues. Not only is the intracellular regulation of cholesterol synthesis impaired, the circulating LDL cholesterol is not adequately discharged from the blood. Patients with the condition tend to die young with massive atheromas causing heart disease.

Summary

1 Receptors are ubiquitous and in respect of hormone action at least they are part of an amplification system for a stimulus.
2 They are not static entities but vary in number and affinity for ligands.
3 Any deficiency in receptor numbers or affinity is potentially a cause of disease.

13.20 Vitamins

'Vitamins' constitute a diffuse subject, but fortunately much knowledge of them is picked up in other areas; indeed, in biochemistry you can seldom get away entirely from vitamins, when you consider the role of thiamine in oxidative decarboxylation (9.2), of folates and cobalamins in one-carbon transfers (11.13), of biotin in carboxylations (11.14), of riboflavin and nicotinamide in oxidation reactions (7.1), of calciferols in calcium metabolism (13.18)—to name but a few. Here we consider vitamins under the following heads—chemical nature, dietary requirements, sources, physiological roles, and deficiency syndromes. What generalizations are possible?

Table 13.20.1 A vitamin summary. All the items below 'C' are known as B vitamins.

Vitamin	Allowance	Good sources	Poor sources	Deficiency syndromes	Toxicity	Metabolic role(s)
A	1 mg	Fish oils, green veg.	Vegetable oils	Nyctalopia, xerophthalmia	Yes	Some basic cellular role, visual cycle
D	5 mg	Fish oils, eggs	Most foods	Rickets, osteomalacia	Yes	Metabolites induce calcium-binding prote
E	10 mg	Vegetable oils	Animal fats	None in man	No	Antioxidant, free radical scavenger
K	100 mg	Green veg., mouldy foods	Animal foods	Haemorrhagic disease of newborn	No	Prothrombin synthesis
C	60 mg	Green veg., fruit	Animal foods, potatoes rice	Scurvy, associated anaemia	No	Essentially unknown
Thiamine	2 mg	Whole grains, eggs	White bread, polished rice	Beri beri	No	Oxidative decarboxylations (TPF
Riboflavin	2 mg	Liver, milk, eggs, green veg.	Rice, fruits, potatoes	Dermatitis, glossitis	No	Redox reactions (as FAD and FMN)
Niacin	20 mg	Wheat, fish, beef, pork	Potatoes, green veg., fruit, eggs, milk, cheese	Pellagra	No	Redox reactions (as NAD and NADP)
B_6	2 mg	Most foods		None in man	No	Transamination reacti
Biotin	0	Most foods	Meat, cereals	None in man	No	Carboxylations
Panthothenate	0	Most foods		None in man	No	Part of Coenzyme A
B_{12}	3 µg	Liver, meat, eggs, milk	All plant foods	Pernicious anaemia, megaloblastic anaemia	No	One-carbon transfers
Folacin	0.4 mg	Liver, kidneys, green veg.	Chicken, pork, milk	Megaloblastic anaemia	No	One-carbon transfers

Fig. 13.20.1 Structures of the fat-soluble vitamins.

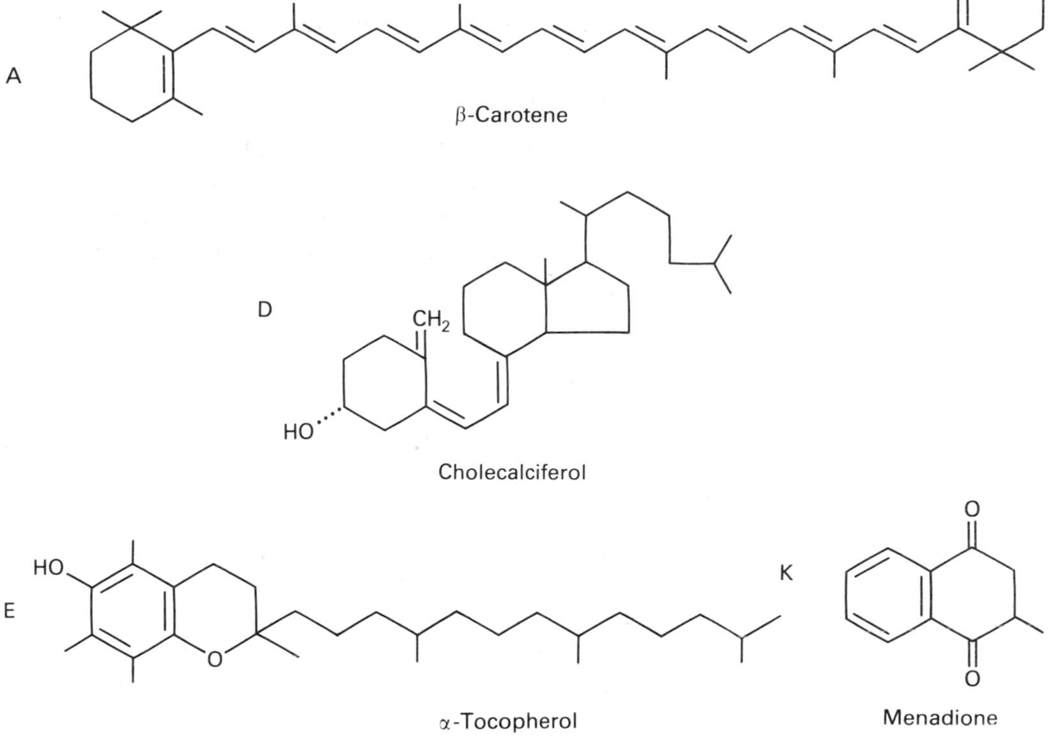
A
β-Carotene
D
CH2
HO
Cholecalciferol
E
HO
O
α-Tocopherol
K
O
O
Menadione

1 *Chemical nature* There is a long-standing distinction between the fat-soluble and the water-soluble vitamins. This dates back to their discovery, when it was thought that there was only one fat-soluble principle and one water-soluble one. Further fractionation soon revealed further complexities. But the distinction is still a valid one, since the fat-soluble vitamins tend to occur together, dissolved in the fatty portions of foodstuffs. Moreover dietary fat is necessary as a vehicle for their absorption. In fat malabsorption syndromes, avitaminoses due to deficiency of all the fat-soluble vitamins are liable to occur. Moreover in the chemical sense the fat-soluble ones have a certain similarity, at least in respect of their isoprenoid side chains. Vitamins A, E, and K all have such side chains (Fig. 13.20.1) but even the vitamins D or calciferols are synthesized *de novo* from isoprene units. However, the water-soluble vitamins are extremely diverse in their chemical nature.

As regards chemical stability, the vitamin which suffers most degradation in cooking procedures is ascorbic acid. It can almost completely disappear during prolonged stewing or boiling. Vitamin A is labile in an oxidizing environment. Riboflavin is photolabile. Folate and thiamine are thermolabile in neutral or alkaline solutions. Vitamin B_{12} is labile in acids and alkalis. These points are summed up in Table 13.20.1.

2 *Requirements* Some vitamins are needed in quite large amounts (as expressed by the recommended daily allowance or RDA, which is itself

Notes

Synthesized by intestinal bacteria

Least stable vitamin—destroyed by cooking

Partial needs met by tryptophan metabolism

Synthesized by intestinal bacteria

Vegetarians at risk; synthesized only by soil bacteria

variously defined) but others are needed in comparatively minute quantities. It would not be productive to memorize these; what is easy to remember are the orders of magnitude (Table 13.20.2).

3 *Sources* A knowledge of poor and rich sources is useful to biochemists and non-biochemists alike. There is no need to differentiate between individual dishes or species (for example, between shrimps and lobsters) but be aware of the value of the broad classes of foodstuffs, which are as follows (some notes on protein and minerals being included).

(1) *Milk* A distinction must be made between human and cow's milk and reference is usually only made to its role in the nutrition of young children, to whom it is the sole, or major, food. Cow's milk is poor in vitamin C; human milk is often poor in thiamine, depending upon the thiamine nutrition of the mother. Both human and cow's milk are poor in iron. Of course both are prime sources of high quality protein.

(2) *Eggs* They are rich in protein and iron. They are a good source of vitamin A and B-vitamins but have no vitamin C.

(3) *Cereals* These are rich in protein, calcium and iron, although the two minerals may be poorly absorbed. They are generally adequate sources of the B-vitamins, but poor in vitamins A and C. Against that, the B-vitamins, calcium, iron, and protein predominate in the outer layers so that low extraction flour (most of outer layers removed) and polished rice are proportionately poorer in these nutrients.

(4) *Green leafy vegetables* These are rich in vitamins A, E, most B, and C and have moderate amounts of protein, calcium, and iron. Again, little of the two minerals may be absorbed under normal circumstances.

(5) *Fruits* Vitamin C is the only well-established nutritional benefit of most fruits.

(6) *Meats* Of course meats are primarily a source of protein, iron, and phosphates. They are generally rich in the B-vitamins, but poor in A and C.

(7) *Seafood* Again seafoods are generally a source of protein, and to some extent calcium.

(8) *Fish oils* These are the well-known rich sources of vitamins A and D, as in cod liver oil.

(9) *Gut bacteria* There is some synthesis by gut bacteria of absorbable biotin and vitamins K.

4 *Physiological roles* These are summarized in Table 13.20.1. In general, the roles of the B-vitamins are fairly well-defined in man. Each of the others has one or more fairly well-defined role, not in itself sufficient to explain fully the range of biological activity in each case.

5 *Deficiency syndromes* (*avitaminoses*) These are summarized in Table 13.20.1. In general the outward manifestations of the avitaminoses are not directly explainable in terms of known roles of the respective vitamins in metabolism. Two vitamins are known to be toxic in excess. These are D (overdosage with cod liver oil, but see section 13.18) and A (overdosage with cod liver oil or ingestion of polar bear liver, a hazard of Arctic explorers).

Table 13.20.2 Vitamin requirements

Required in large amounts (10–30 mg/d)	Vitamin C
	Vitamin E
	Nicotinamide (niacins)
Required in moderate amounts (1–10 mg/d)	Thiamine
	Riboflavin
	Vitamin B_6 (pyridoxines)
Required in small amounts (< 1 mg/d)	Vitamin A
	Vitamin B_{12}
	Folates (folacins)
Not required in the diet at all	Vitamin D (in sunny climates)
	Vitamin K (synthesized by gut bacteria)
	Biotin (synthesized by gut bacteria)

These figures refer to adult men. Women need less in proportion to a lower body weight. Pregnant and lactating women as well as growing children need proportionately more. Requirements are thought to diminish in old age.

In man there are no known syndromes due to nutritional deficiencies of biotin or pantothenate. Some avitaminoses are apparent in children but not in adults, for example vitamin E deficiency can occur in premature infants treated with oxygen, and vitamin K deficiency (haemorrhages) can occur in neonates with sterile guts, not yet containing bacteria to synthesize the vitamin. Vitamin D deficiency manifested as rickets tends to be more severe in children, since their bone metabolism is so active. Pyridoxine deficiency has occasionally been observed in infants on feeds accidentally deficient in it.

Note on megavitamin ingestion

A common self-dosage of vitamin C by people with common cold symptoms is 1 g/d. This could hardly be obtained from natural sources; you would have to eat about 7 lb of oranges to obtain 1 g. The prefix 'mega' implies that ingestion is of this order of magnitude, that is it could not be obtained from natural food, or only with grave difficulty. It has been established that if vitamin C is taken in megadoses for treatment of common colds, it reduces their duration by an average of 0.1 d. This is not clinically significant (if statistically significant) and the mechanism remains unknown. However there is some scientific respectability for megavitamin dosage, resting on studies of methionine degradation. Methionine is broken down as follows:

$$\text{Methionine} \xrightarrow[1]{} \text{homocysteine} \xrightarrow[2]{} \text{cystathionine} \xrightarrow[3]{} \alpha\text{-ketobutyrate}$$

Reactions 2 (catalysed by cystathionine synthetase) and 3 (catalysed by cystathioninase) both require pyridoxal phosphate (vitamin B_6). There are two conditions, both associated with mental retardation, that result from deficiency of the two enzymes. Named after the accumulation of the relevant metabolites in the urine, they are termed homocystinuria and cystathioninuria respectively. It was found that the conditions in some instances could be significantly relieved by the administration of massive doses of pyridoxal phosphate. The rationale is that although these syndromes may appear to be due to a lack of cystathionine synthetase and cystathioninase, the enzyme proteins are not lacking, although they are inactive or catalytically inefficient. One aspect of catalytic inefficiency may be lack of affinity for B_6 coenzyme, and so massive administration of the coenzyme allows enough binding for significant activity.

Other B-vitamins have been administered for mental disorders, with equivocal results. Vitamin E has been given in large amounts to prevent ischaemic heart disease, but this has fallen into disrepute. It appears to be successful, however, in relieving intermittent claudication (pain and lameness due to impaired circulation in the legs). The physiological basis for this is unknown.

13.21   Detoxication and drug metabolism

The body has to cope with the ingress of a large number of foreign and/or toxic substances. In terms of particle or molecular size the largest of these tend to provoke an immune reaction (13.15); the smaller molecules are disposed of by hepatic mechanisms, which serve to render them more diffusible, excretable, and (usually) less toxic. We can classify the substances subject to detoxication-type reactions into a number of categories, not necessarily exclusive of one another, and these are listed in the vertical column in Table 13.21.1. The first two are straightforward enough, but the inclusion of 'hormones' may need some comment. If an effector is elaborated in response to a stimulus, then when the stimulus has passed the effector should be inactivated to allow a return to the starting or basal state. For example, insulin is the hormone responding to ingestion of food, ensuring that absorbed glucose is converted to glycogen and fat and that amino acids are utilized for protein synthesis. In the fasting state insulin should be inactivated to prevent possible catastrophic decrease in blood glucose.

Under the heading 'metabolic end-products' are placed substances which, when produced by the body, should be eliminated before they rise to noxious levels—an obvious example is ammonia, which is converted to the less toxic but equally diffusible urea. In the same category bilirubin, toxic to the brains of young children, is rendered more diffusible by conjugation to glucuronic acid. To these can be added hydrogen peroxide, arising during flavoprotein-catalysed reactions such as

$$\text{Xanthine} + O_2 \xrightarrow[\text{flavoprotein hydroxylase}]{\text{Iron-sulphur}} \text{urate} + H_2O_2$$

Moreover free radicals such as the superoxide ion O_2^- are thought to

Table 13.21.1 Overview of detoxication reactions

	Oxidation	Reduction	Hydrolysis	Conjugation
Drugs				
Aspirin				
Chloral				
Poisons				
Picrate				
Alcohol				
Benzoate				
Phenol				
Hormones				
Steroids				
Insulin				
End-products				
Peroxides				
Bilirubin				
Superoxide*				

*The reaction of superoxide ion

$$2O_2^{\cdot -} + 2H^+ \xrightarrow{\text{superoxide dismutase}} H_2O_2 + O_2$$

involves a dismutation, that is simultaneous oxidation and reduction of the substrate. It is thus not so easily categorized. The superoxide dismutase acts as a 'free radical scavenger'.

result from the spontaneous oxidation of the iron atom in a large number of ubiquitous cellular proteins such as haemoglobin and cytochromes, as:

$$\text{Protein-Fe}^{2+} + O_2 \rightarrow \text{protein-Fe}^{3+} + O_2^{\cdot -}$$

A ready source of rather toxic materials is the bacterial population of the large intestine, for many of its metabolic products are absorbed to some extent. Even teetotallers gain (if that is the correct word) perhaps one to ten grams of alcohol each day by this means.

There are large numbers of *xenobiotics* (that is compounds foreign to a living organism) such as colourings and flavourings in food, contaminants like insecticides and herbicides and congeners in alcoholic drinks which might or might not be classified as poisons. Many drugs are steroids and subject to the same type of reactions as the natural hormonal steroids. So the categories in the table are very broad ones.

The horizontal headings give four equally broad categories of detoxication reactions. The liver is considered the main locus for them, although it is certainly not the only one. Oxidation and glucuronidation are thought to take place mostly in the microsomes (the products of centrifugation deriving from the smooth endoplasmic reticulum).

Representative reactions are as follows.

1 Oxidation

(1) Alcohol is oxidized in the liver by the enzyme alcohol dehydrogenase.

Fig. 13.21.1

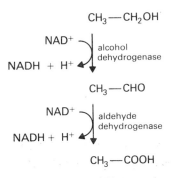

The product, acetaldehyde, is probably even more toxic and is also subject to dehydrogenation. The final product is acetate, which is largely converted to fat (Fig. 13.21.1).

(2) Steroids are hydroxylated (often prior to conjugation with sulphate) in reactions involving microsomal enzymes and microsomal cytochrome P-450. Cytochrome P-450 is so named because it is a cytochrome similar to those in the mitochondrial respiratory chain and has an absorption maximum at 450 nm wavelength following reduction and exposure to carbon monoxide. It is a coenzyme in many types of oxidation reaction, but best known in connection with hydroxylations using molecular oxygen.

$$RH + O_2 + NADPH + H^+ \rightarrow ROH + H_2O + NADP^+$$

The enzymes involved are termed *mixed function oxidases* since two substrates, RH and NADPH, are simultaneously oxidized. RH is typically a steroid, but many classes of substances may be oxidized in this reaction, including drugs and aromatic hydrocarbons.

The process involves at least four steps:

(i) the substrate combines with the oxidized form of P-450 (that is with the iron or the porphyrin group as Fe^{3+});

(ii) the complex is reduced by one electron from NADPH (iron as Fe^{2+});

(iii) the reduced complex reacts with O_2 to form an oxygenated complex which accepts the second electron from NADPH;

(iv) this decomposes to yield the oxidized steroid and the regenerated oxidized P-450.

(3) The hypnotic chloral is oxidized to the alcohol trichloroethanol

$$CCl_3-CHO \rightarrow CCl_3-CH_2OH$$

2   Reduction

(1) Hydrogen peroxide is removed in a reaction involving the selenium-containing enzyme glutathione peroxidase. Glutathione is a sulphydryl-bearing tripeptide represented as GSH

$$2GSH + H_2O_2 \xrightarrow{\text{glutathione peroxidase}} GSSG + 2H_2O$$

Hydrogen peroxide is also decomposed by the iron-porphyrim enzyme, catalase:

$$2H_2O_2 \xrightarrow{\text{catalase}} 2H_2O + O_2$$

The two enzymes possibly complement each other in some way.

(2) Picrate, a health hazard of those working in the explosives industry, is reduced to picramic acid (Fig. 13.21.2).

(3) Insulin contains disulphide cross-linkages, so that it can be inactivated by their reduction to —SH. This is catalysed by the enzyme glutathione insulin transhydrogenase. Reduced glutathione G—SH, is simultaneously oxidized to G—S—S—G.

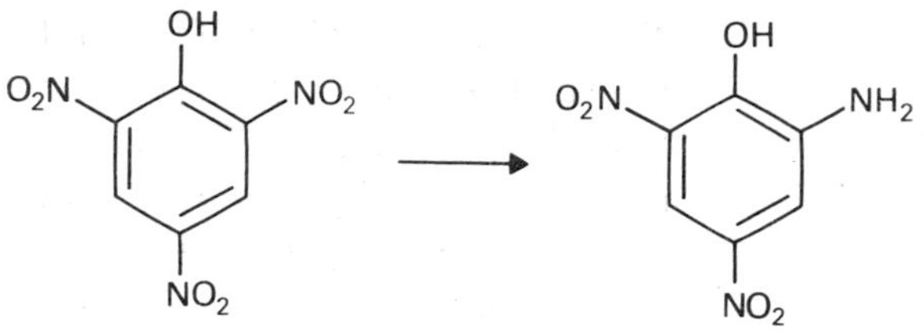

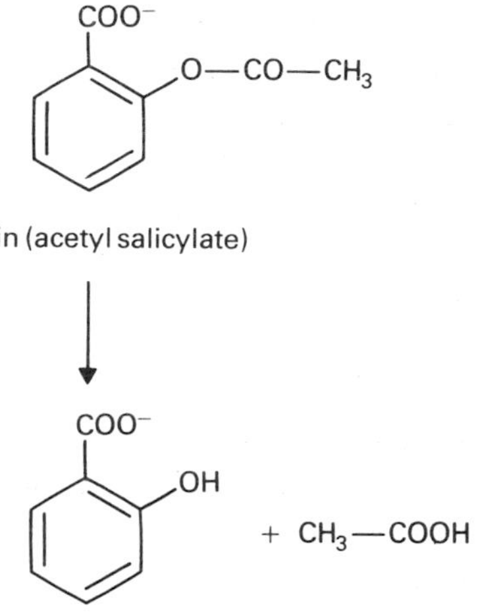

Aspirin (acetyl salicylate)

$+ CH_3—COOH$

Fig. 13.21.3

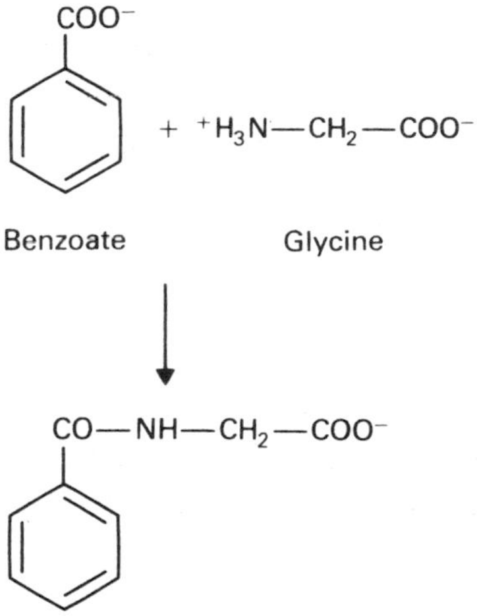

Benzoate   Glycine

Hippuric acid

Fig. 13.21.4

Phenol

Fig. 13.21.5

Fig. 13.21.2

Picrate   Picramic acid

3 Hydrolysis

Aspirin, or acetyl salicylate, is hydrolysed by an esterase (Fig. 13.21.3). There do not appear to be many useful examples of hydrolysis in detoxication, but presumably protein hormones are ultimately hydrolysed to amino acids. Insulin is hydrolysed after reduction, for example.

4 Conjugation

(1) Benzoate, used as a food preservative, is conjugated with glycine to yield hippuric acid (Fig. 13.21.4). Historically this is a famous reaction and was formerly used as a liver function test. Benzoate was given by mouth, the urine collected for a specific time, and the hippurate crystallized from it. Poor liver function resulted in a low yield.

(2) Bilirubin is conjugated with glucuronic acid

$$\text{Bilirubin} + \text{2 UDP glucuronate} \xrightarrow[\text{transferase}]{\text{glucuronyl}} \text{bilirubin diglucuronide}$$

(3) Phenols are often sulphated (Fig. 13.21.5).

Other conjugation reactions may involve methylation, acetylation, or coupling to glutathione.

Repeated administration of a drug may increase the activity in the liver of the enzyme or enzymes known to metabolize it. This is particularly true of the microsomal cytochrome P-450 enzymes. Repeated dosing with phenobarbital enhances not only cytochrome P-450 enzymes but also glucuronyl transferase. Evidently there is something in the way of generalized microsomal stimulation.

Cytochrome P-450 linked enzymes can sometimes potentiate toxicity rather than reduce it. Thus benzpyrene (found in tobacco smoke) is converted to a highly mutagenic and carcinogenic epoxide (Fig. 13.21.6).

The question arises as to whether the liver has ready, or inducible, a whole range of enzymes specific for the vast number of synthetic xenobiotics which potentially could be introduced into the body. This is somewhat reminiscent of the dilemma over whether the body has, in the immune system, antibodies specific for every possible antigen (13.15). The immune system *does* maintain lymphocytes capable of synthesizing every necessary antibody, but there are no specific enzymes for all the xenobiotics. Rather, the detoxicating enzymes are of low specificity. And although many of the products of the reactions are more diffusible and less toxic, this is by no

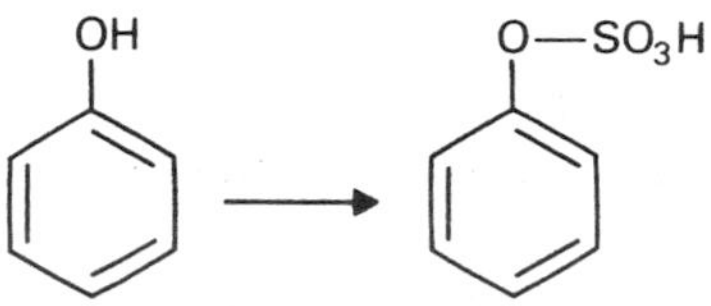

Fig. 13.21.6

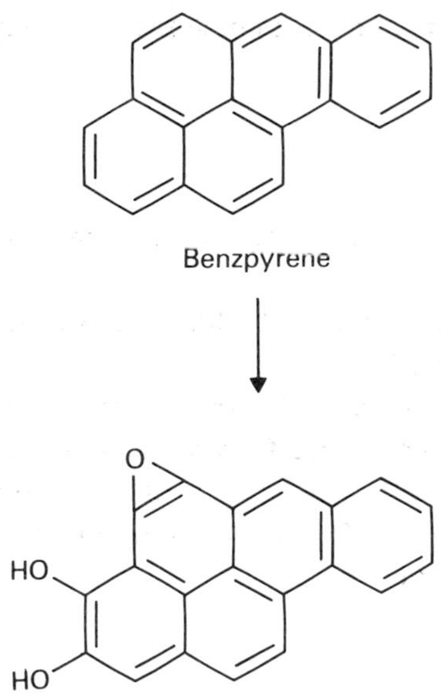

Benzpyrene

↓

Benzpyrene
7,8-dihydrodiol-9,10-epoxide

means always the case (1.1 above). There may therefore be more subsequent reactions, perhaps involving conjugation, before a substance is finally excreted. Thus, apart from Table 13.21.1, we can make the following generalizations about detoxication:

(1) the enzymes involved are of low specificity;

(2) the products of the first reactions may require further processing before they are excreted;

(3) enzyme activity is not static but can be enhanced by prolonged administration of certain xenobiotics;

(4) toxicity can sometimes be enhanced, rather than reduced, by the system.

13.22   Genetic engineering

Insertion of the DNA from the cells of one species into another, such that the recipient replicates the inserted DNA and therefore expresses its genetic information, is the principle of genetic engineering. Alternative terms are 'genetic manipulation' and 'recombinant DNA technology'. The social and economic relevance of this area has been well publicized and some of its details are rapidly filtering down to elementary biochemistry courses.

DNA transfer from the cells of one species to another was first discovered in connection with acquired drug resistance; for example, *Salmonella* types were able to acquire resistance to antibiotics from *Escherichia coli*. The resistance factors were transferred in small extrachromosomal DNA molecules known as *plasmids* which can replicate in the host cell without lowering its viability. The 'engineering' process involves the synthesis of DNA with the desired information in a form (recombinant DNA) in which it can be inserted into host cells, and the 'tools' consist of enzymes. It is interesting to note that in much of the literature on the subject, the strictly correct convention of noting that such and such an enzyme catalyses a reaction is dropped. So efficacious, so purposeful seem the enzymes, that they are said to 'join', 'split', 'anneal', and so on. Indeed, the discovery of enzymes was essential to the whole body of expertise; they can be listed as those which catalyse synthesis of DNA, and those which catalyse its hydrolysis.

Synthesis of DNA

There are four enzymes in this category:

1   DNA polymerase, which catalyses polymerization of deoxyribonucleotide phosphates, with single-stranded DNA as template—the result is a new complementary strand (cDNA) base-paired to the template;

2   reverse transcriptase (RNA-dependent DNA polymerase) which catalyses synthesis of a double-stranded DNA molecule containing the information present in the RNA genome;

3 terminal transferase, which catalyses the extension of DNA at its 3' terminus by addition of deoxyribonucleotides;

4 DNA ligase, which catalyses the covalent bonding of fragments of newly-synthesized DNA.

Scission of DNA

There are only two enzymes in this category, but the second is a large class with similar function.

1 S_1 nuclease catalyses the hydrolysis of phosphodiester linkages in single-stranded, but not double-stranded, DNA and RNA.

2 Restriction endonucleases, of which about 200 are known, cleave double-stranded DNA at sites specific for each one. They are mainly from bacteria and it is thought that their natural function is to inactivate viruses. Thus they recognize specific sequences in DNA some 4–6 nucleotides in length. EcoRl endonuclease, for example, recognizes

```
          ↓
5' --G--A--A--T--T--C-- 3'
3' --C--T--T--A--A--G-- 5'
                   ↑
```

and catalyses hydrolysis at the points shown (↓). Note that the two sequences are the same when read from the 3'-end in each case or from the 5'-end in each case; the description *palindromic* is applied. Restriction endonucleases appear to recognize only such sites and evidently after cleavage of the RNA there are two tails which remain complementary to each other.

```
-----G     A--A--A--T--T--C--

--C--T--T--T--A--A     G------
```

A principal aim of recombinant DNA technology is to achieve bacterial synthesis of large amounts of non-bacterial protein which is in short supply for one reason or another. Insulin is the obvious example. Porcine and bovine insulins are not only derived from finite biological sources, but they tend to be immunogenic and the pure human protein would be much preferable for clinical use. Other examples are interferon (in fact a group of proteins synthesized by mammalian cells when infected by viruses) and growth hormone. Social concern arose from the fear that insertion of DNA into bacteria might produce harmful organisms or that there would be accidental multiplication (cloning) of genes for cancer or for toxins. These fears have not been realized, and insulin at least is now about to become available by means of DNA recombinant technology.

Production of insulin by *E. coli*

Several approaches may be employed, but let us consider the one illustrated in Fig. 13.22.1. In this the starting material is mRNA prepared from the

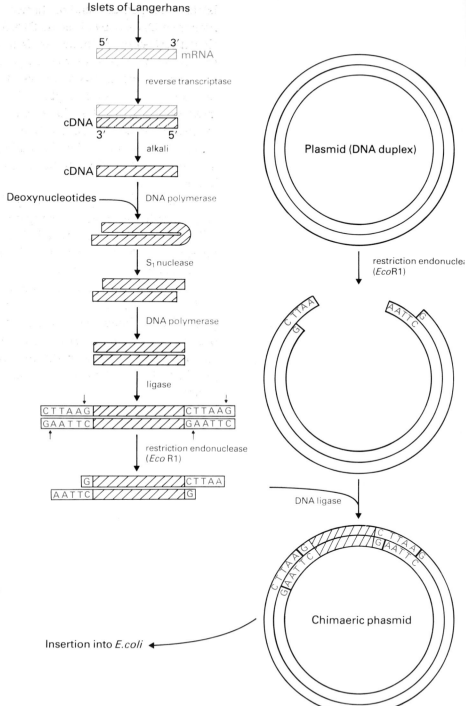

Fig. 13.22.1 Incorporation of foreign DNA into a plasmid.

mRNA prepared from human pancreatic islets is mixed with deoxyribonucleotides and reverse transcriptase. A double-stranded hybrid is obtained, containing complementary DNA (cDNA). The RNA strand is digested with alkali. DNA polymerase, in the presence of more deoxyribonucleotides, is used to produce a double-stranded DNA which is, however, continuous at one end. S_1 nuclease removes this loop and then DNA polymerase is again used to produce a double strand, with no loose ends. To this is joined synthetic polydeoxyribo-nucleotide sequences which are going to be susceptible to hydrolysis by the restriction endonuclease EcoRl, using the enzyme DNA ligase. This gives a duplex but with TTAA 'tails'.

Meanwhile the plasmid is opened with EcoRl, and because of its specificity for the palindromic sequences the openings have 'tails' which are exactly complementary to the tails on the DNA synthesized. It is then a matter of recombination or 'annealing'. The chimaeric plasmid must be inserted into E. coli.

Islets of Langerhans, and the battery of enzymes described above is used to prepare a DNA copy of the mRNA for insertion into a plasmid. The role of EcoRl restriction endonuclease is crucial and a study of the diagram shows that it is the ability of this enzyme to catalyse scission of palindromic sequences which allows synthesis of an insert with tails complementary to the tails on the modified plasmid. After the *chimaeric plasmid* has been inserted into *E. coli* it is still necessary to employ special techniques to allow expression of the information on the chimaeric plasmid. Moreover, insulin is synthesized as proinsulin (in reality, as 'preproinsulin') (13.12) and the unwanted peptides have to be removed by proteolysis before the product is biologically active.

An alternative approach is to synthesize nucleotide sequences coding for preproinsulin, insulin, or the separate A and B chains. To these are tagged tails complementary to the tails formed on plasmids after restriction endonuclease attack. Annealing is performed and the chimaeric plasmid inserted into the bacterium as before.

Yct again, the insulin gene may be connected to a gene which is 'natural' to *E. coli*, such as that for the enzyme β-galactosidase. The chimaeric plasmid containing the sequences for the tandem β-galactosidase-insulin protein is inserted, and expression of the insert can be induced by growing *E. coli* on lactose rather than glucose medium. Naturally at the end there is the problem of separating the functional insulin from the unwanted β-galactosidase.

14 Some Lists

14.0 Roundup of principles

Respectable post-renaissance science is supposed to proceed by the deductive method: facts/observations are gathered, classified, and listed; then principles/hypotheses/laws emerge from consideration of facts. The principles are the more advanced stage of the process; facts are easier, they need no logical correlation. Most of basic biochemistry patently consists of collections of facts; this cannot be helped in the early stages as they are supposed to be recalled for examinations. But if you can also recall principles, so much the better. If you can *only* recall principles, I fear you cannot pass examinations; they might gain you enough credit to have another try. The principles are limited, as below, to nine. (This total has an untidy appearance due to a prejudice, in lists, for the number of one's fingers.)

1   The chemical substances making up living organisms are all of the same type. This applies not only to elements (C, N, O, H, Cl, S, of the second and third periods of the periodic table) but also to compounds. That is, all living cells comprise proteins, carbohydrates, DNA, RNA, and lipids, as well as a relatively standard range of low molecular weight substances. Glucose and fatty acids, for example, are found in all cells. Moreover, in living things all catalysts are proteins; most membranes are largely composed of phospholipid; all genetic information is incorporated in nucleic acids. There is a basic structural unity.

2   These compounds which comprise living cells are constantly being broken down and resynthesized, that is they are in a state of turnover or flux, even though total amounts or concentrations remain relatively constant (dynamic equilibrium). It is not completely clear why this should be so. All or most tissues have some capacity for renewal, and the constant 'labilization' of tissue components obviously makes possible their replacement by newly-synthesized and perhaps more perfect molecules. There is an energy cost in the degradation-regeneration cycle but it is presumably worth sustaining in terms of survival.

3   The functional macromolecules of life have specific topography, that is each has some specific surface shape further specialized to bind to other macromolecules or low molecular weight substances. Thus enzymes, receptors, antibodies, and polynucleotides all function by specific surface-to-surface interaction.

4   The information for macromolecular structure resides in the base sequences of DNA and RNA, and the code contained in these sequences is universal throughout life.

5   A gene codes for a single polypeptide chain. (There are exceptions, such as the DNA of ϕ174 phage. In this virus the codes overlap, which can be seen as an economy measure with some disadvantages.)

6   Reactions can only proceed spontaneously if they result in a loss of free energy. There are mechanisms to avoid this constraint, such as the coupling of exergonic and endergonic reactions, and carboxylation reactions (11.15) to provide intermediates with high group transfer potentials.

7   All organisms are able to conduct both catabolism and anabolism, though not necessarily the complete catabolism and anabolism of a specific compound. Where catabolism and anabolism both occur, they occur by different pathways to facilitate control.

8   Control of metabolism is above all exercised through control of enzyme activity (although there are a number of ways, in turn, in which this can be done—the pattern of enzyme control is diverse and can be achieved through feedback loops, cascades or compartmentation mechanisms, as discussed in 12.2).

9   Economies appear to be practised universally. That is, cycles and pathways are arranged to yield optimal amounts of ATP; put another way, we cannot at the moment conceive of better, alternative pathways, say by direct oxidation rather than dehydrogenation, which would give more ATP. Substances are not, in the same cell at the same time, synthesized and degraded cyclically, unless possibly in a futile cycle (12.3) to generate heat. Where one energy-yielding substrate, say glucose, is plentiful the synthesis of another, say fatty acids, is suppressed. Moreover low molecular weight substances are often utilized for many purposes, for example glycine is a constituent of proteins, and a precursor of purines, porphyrins, bile acids and creatine; acetyl CoA yields steroids, fatty acids, and ketone bodies.

The preceding list is a universal one. What principles apply solely to mammals, or solely to man? Well, of course the real generalizations are above. But mammals are a large group and their metabolism is evidently quite different from that of bacteria. Most biochemical research has employed rodents and rabbits, with extrapolation to man. Ruminants, though, do not absorb glucose from the alimentary canal—no glucose drinks for them to give energy—they ferment carbohydrates to fatty acids, which are absorbed. Hibernating bears do not excrete urea, they split it and combine its nitrogen with glycerol to replenish amino acids; this is not 'respectable' in terms of the rat-rabbit-man axis. So if a list is to be made, it should be for a single species: there follows a short one for man.

1   Fat cannot be converted to carbohydrate, still less to protein.

2   Carbohydrate is convertible to fat.

3   On a weight basis, the brain consumes the most glucose and oxygen.

4   Blood glucose for brain function is maintained by the liver from glycogenolysis or gluconeogenesis.

5   The feeding-fasting cycle is a shift from glucose to fatty acid oxidation and back again. Amino acid oxidation, depleting tissue protein, is a last resort.

14.1 Checklists

The major pathways and cycles are first set out (Table 14.1.1). These are the basic canon of metabolism. Anybody taking a biochemistry course needs a working knowledge of them.

Next there is a list of the prominent metabolic activities of some organs and tissues in the human subject (Table 14.1.2).

Table 14.1.3 should be checked against Table 14.1.2. It narrows down the activity from organ to the site within the cells of that organ.

Information on specific actions of various hormones is scattered throughout the preceding pages. Table 14.1.4 is a round-up of them. It omits hormones which are not important in basic biochemistry, such as angiotensin and the digestive hormones.

Table 14.1.5 is a checklist of competence in metabolism.

Finally you may have noticed that there is much concern about the mechanism of getting things into and out of mitochondria. This occurs in glycolysis, fatty acid oxidation and synthesis, for example; it is a consequence of the compartmentation of substrates for metabolic control. Table 14.1.6 summarizes the permeability of the mitochondrial inner membrane.

Table 14.1.1 The major pathways and cycles

Ref.	Cycle/pathway	Significance
7.2	Tricarboxylic acid cycle	Energy-yielding oxidation of two-carbon units
8.2	Glycolysis	Anaerobic oxidation of sugars
8.1/11.1	Glycogenolysis/glycogenesis	Cyclic synthesis/breakdown of easily-mobilizable energy source
8.4	Pentose phosphate pathway	Production of NADPH and pentoses
11.13	Protein synthesis	Enzymes, structural components, hormones, antibodies, contractile systems
11.12	DNA replication	Transmission of genetic information
11.10	Nucleotide synthesis	DNA and RNA synthesis
10.1	Deamination/transamination	Utilization and transformation of amino acids
10.2	Urea cycle	Disposal of amino acid nitrogen
11.3	Gluconeogenesis	Replenishment of glucose from amino acids
11.9	Porphyrin synthesis	Formation of haem and cytochromes
10.4	Bilirubin formation	Disposal of erythrocyte haem
10.5	Urate formation	Disposal of nucleotides
9.1	Fatty acid oxidation	Energy from fatty acids
11.6	Ketone body synthesis	Hepatic production of oxidizable substrates for export
11.4	Palmitate synthesis	Principal means of fatty acid synthesis
11.7	Cholesterol synthesis	Synthesis of an essential cell component and hormone precursor

Table 14.1.2 Organ specifity

All	DNA replication
	RNA synthesis
	Protein synthesis
	Tricarboxylic acid cycle
	Glycolysis
Liver	Glycogenolysis/glycogenesis
	Gluconeogenesis
	Fatty acid synthesis, desaturation
	Triglyceride and phospholipid synthesis
	Cholesterol and bile salt synthesis
	Lipoprotein synthesis
	Serum albumin synthesis
	Detoxication (urea, bilirubin, drugs)
	Ketone body synthesis
	Vitamin and iron storage
Skeletal muscle	Glycogenolysis/glycogenesis
	Glycolysis to lactate
	Fatty acid oxidation
	Transamination
	Synthesis of ATP from ADP
Adipose tissue	Assimilation of exogenous lipid
	Glycerol synthesis
	Fatty acid synthesis
Brain	Glucose oxidation
	Ketone body oxidation
	Glutamine synthesis
	Neurotransmitter synthesis
Kidney	Glutamine hydrolysis
	Gluconeogenesis
	Ketone body oxidation
	Secretion of hydrogen ions, creatinine, and other excretory materials
Erythrocytes	Glycolysis
	Hexose monophosphate shunt
	Oxygen binding
	Glutathione synthesis
	(*No* protein synthesis or tricarboxylic acid cycle)

Table 14.1.3 The principal activities of cell organelles

Organelle	Metabolic activities
Nucleus	DNA replication
	RNA synthesis (transcription)
Mitochondrion	Oxidative phosphorylation
	Tricarboxylic acid cycle
	Fatty acid oxidation
	Pyruvate dehydrogenase activity
	Ketone body production
	Urea cycle (in part)
	Transamination
Endoplasmic reticulum	Protein synthesis (translation)
	Fatty acid desaturation
	Drug and steroid oxidation
	Lipoprotein synthesis
Golgi apparatus	Addition of carbohydrate to proteins
	Secretion of proteins
Lysosomes	Storage of hydrolytic enzymes (digestion of foreign inclusions and autolysis)
Cytosol	Glycolysis
	Gluconeogenesis
	Glycogenesis/glycogenolysis
	Pentose phosphate pathway
	Palmitate synthesis
	Urea cycle (in part)
	Transamination
	Cholesterol synthesis
	Porphyrin synthesis (in part)

Table 14.1.4 Endocrine activities

Hormone	Organs/cells	Effects
Insulin	Muscle Liver Adipose tissue	Glucose uptake (therefore hypo-glycaemic) glycogen synthesis, protein synthesis, lipid synthesis (response to the fed state)
Glucagon	Liver	Glucose release (therefore hyperglycaemic)
	Adipose tissue	Fatty acid release (antagonistic to insulin)
Adrenaline	Liver	Glucose release (therefore hyperglycaemic)
	Adipose tissue	Fatty acid release (antagonistic to insulin)
Cortisol	Liver	Gluconeogenesis and therefore hyperglycaemia (antagonistic to insulin)
Aldosterone	Kidney	Salt-retaining
Thyroxine, triiodothryronine	Most cells	Stimulation of energy metabolism, possibly by altering permeability of mitochondrial membrane
Androgens	Muscle Bone	Protein anabolic Closure of epiphyses (maturation)
Oestrogens	Uterus Bone	Protein anabolic Calcium deposition
Progesterone	Uterus	Anabolic
Parathyroid hormone	Kidney	Decreased calcium excretion, increased phosphate excretion
	Bone	Calcium mobilization (therefore hypercalcaemic)
Calcitonin	Kidney	Increased phosphate excretion
	Bone	Inhibition of parathyroid-induced resorption
Somatotrophin (STH)	Most cells	Anabolic, growth stimulating (also hyperglycaemic)
Adrenocorticotrophin (ACTH)	Adrenal cortex	Stimulation
Follicle-stimulating hormone (FSH)	Ovary Testis	Promotion of follicle growth Stimulation of seminiferous tubules
Luteinizing hormone (LH)	Ovary	Development of corpus luteum
Prolactin	Mammary gland	Lactation
Thyroid-stimulating hormone (TSH)	Thyroid	Stimulation

Table 14.1.5 Catabolism and anabolism in the human body

Totally* synthesizable and degradable	Synthesizable only not fully degraded	Degradable only	Neither synthesized nor degraded
Nucleotides	Cholesterol (to bile salts)	Essential amino acids	Vitamins
DNA	Porphyrins (to bilirubin)		
RNA	Purines (to urates)		
Glycogen and glucose			
Fructose, galactose			
Fatty acids			
Phospholipids			
Proteins			
Non-essential amino acids			
Organic acids (e.g. citrate)			

*i.e. from simple two- or three-carbon precursors

Table 14.1.6 The permeability of the mitochondrial inner membrane

Impermeable to:	But permeable to (as carriers):	Ref.:
NADH	Malate, glycerol phosphate	8.2
Acyl CoA	Acyl carnitine	9.1
Acetyl CoA	Citrate	11.4
Oxaloacetate	Pyruvate, aspartate	8.2
Amino acids in general	Glutamate, aspartate, pyruvate	10.1

15   Examinations in Biochemistry

Examinations in biochemistry, as in other subjects, may take a variety of forms, but most of the time students are confronted with some variation of (a) the essay (b) the multiple choice (MCQs) or (c) the oral or *viva voce* ('viva' for short). In the context of dread and horror the viva probably comes out on top; as we shall see however it is in this type that you are likely to get most help and understanding. Let us run over some aspects of (a), (b), and (c) as they apply to biochemistry. Practical tests are omitted from the discussion since the approach to them is very localized and where they are still conducted they are generally a straight test of practical skill, in respect of which a textbook has no place.

(a) Essays

The essay paper is intended to be a test of ability to marshal facts, order and correlate them, and then set them down in some coherent manner. Naturally the questions should be designed to permit this. At the elementary level often they will not do so, because it makes things easier all round to ask for a chunk of the lecture material. If there is any deviation from this straight-forward approach, for example if the relationship between two systems, say carbohydrate and fat metabolism, is asked for, there is a standard complaint of the part of the examiner when he sees the answers. This is that the wording on the question has triggered off a conditioned reflex whereby everything the candidates can associate with both topics is conveyed pell-mell onto the paper with no order or selectivity. The keyword 'relationship' in the question has been ignored, he will complain.

    You may assume if you are in an elementary class that who ever is marking your essay question is doing the same for the whole class; in other words, he is marking more than a few. Ideally he will approach each answer book with a fresh mind and assess it in an unbiased manner against pre-viously-established criteria. But such conditions never obtain. He becomes increasingly bored and irritated. There is no doubt that an answer which adds to his irritation, whatever its factual content, is going to receive fewer marks. Ways of irritating the examiner can be listed as follows, though it is clear that they are all variations on the same theme.

1   The conditioned reflex, as described above. Think for a few moments about the wording of the question. Take time to consider the specifics of the answer.

2   Filling up space without really saying anything. It is no good coming out of the room and saying 'Well, I wrote ten pages', or 'I managed to stretch it

out to a page and a half'. It is what these pages contain that is important, all over the world. No examiner with numerous scripts to correct wants to spend precious time wading through padding. He can recognize it with extreme ease and it will not increase his benevolence.

3 Being a 'smart Alec'. Admonitions to the examiner are never kindly received. However they will certainly call attention to you and more especially to your answer. What might have been perused with minimal attention will instead receive fierce scrutiny. It had better be good, then.

4 Needless repetition. In biochemistry it is often possible to present virtually the whole of an answer in the form of a nicely annotated diagram. It is common however to see answers in which beautiful diagrams are also described step by step in sentences on the next few pages. This form of insurance is unnecessary. Examiners are more than happy to make their assessment on the basis of a diagram. They are easier to scrutinize, familiar in form, and they take less time.

In summary it makes no sense to irritate your examiner or to attempt other than a straight answer to a straight question. There is no prospect of essays being marked by computer; a computer would certainly be less irritable, but also less flexible.

One final point on essay questions. Candidates are usually dismayed by the wide range of knowledge they may have to demonstrate. Examiners on the other hand are constantly surprised at the limitations of the questions available when they get down to setting them. So check the previous papers. There is not much room for the examiners to manœuvre. They tend to be limited to standard pathways and cycles (Table 14.1.1), so know these at the very minimum.

(b) Multiple choice

The fundamental difference between MCQs and essays is supposed to be that the former can be marked objectively, even by computer, whereas the latter are subject to wide variation in their marking due to human error and bias. There is a more interesting difference from the student's point of view. In an essay paper the skill of the examiner goes into the marking procedure, whatever it may be, and it is exercised *in camera*; in the MCQ, the skill is required in setting. This means that in MCQs any deficiencies or anomalies, or even straight errors, are plain for all to see and when they occur students can very well make the most of them. There are several recognized formats for the MCQ.

(i) The true-false type

Subsequent to a common stem, usually five statements are provided. You are required to mark each one as being either true or false, using some specified sign such as T or F. An example might be:

Insulin
- *a* Raises blood glucose (F)
- *b* Accelerates glycolysis (T)
- *c* Accelerates protein syntheses (T)
- *d* Accelerates cholesterol synthesis (T)
- *e* Accelerates gluconeogenesis (F)

In this type any number of the items may be true, and any number false. It is for this reason that it is also called 'indeterminate'. A variation in the instructions is sometimes this: you are to write 'A' if all are false, 'B' if all are true, 'C' if only b, c, and d are true, 'D' if only . . . etc. So the correct answer to the above would be 'C'. In this particular format it is sometimes called 'multiple completion'. Usually one mark is given for a correct response, and one taken off for an incorrect response. Statistically, guessing over a large enough number of items ought to achieve a score of zero.

(ii) 'Real' multiple choice

In this type, the stem is usually followed by five assertions, only one of which is true. Mostly, you are required only to mark the true item as true; you can leave the others (the 'distractors') blank. As an example:

In promoting glucose release from the liver, the predominant hormones are
- *a* Insulin (F)
- *b* Adrenaline (F)
- *c* Glucagon (F)
- *d* Both glucagon and insulin (T)
- *e* Both adrenaline and glucagon (F)

Whereas in type (i) above all the statements are independent, this is not the case in the 'real' MCQ. The predominant hormones could not be glucagon and adrenaline, as well as glucagon and insulin, at one and the same time. Since only one can be correct, you have a 1:4 chance of guessing the correct response (that is marking as 'true' the one which is true). Usually either four marks are given for a correct response and one is taken off for an incorrect response, or one is given for a correct response and one quarter taken off for an incorrect one. An unattempted question gains zero. Again, over a large enough number of questions, guessing should make zero. Unlucky guessers might even get less than zero. What happens if you fail to put anything against the true option but put false against one or two which are indeed false? This is like saying 'I don't know which is correct, but I'm certain that one or two are wrong'. The procedure varies; there should be definite instructions on these complex matters.

(iii) Progressive questions

In this type the later part of a question depends upon the answers to the first part. Thus:

> A patient complains of thirst and hunger and is found to have a fasting blood sugar of 180 mg/100 ml and detectable glucose in the urine. Your provisional diagnosis is
>
> a Insulinoma   (F)
> b Renal glycosuria   (F)
> c Diabetes mellitus (T)
>
> You would attempt confirmatory diagnosis by
>
> a Glucose tolerance test (T)
> b Insulin assay   (F)
> c Kidney function test   (F)

Obviously if you get the first part wrong, you have no hope with the second. The first part is usually marked as for 'real' multiple choice. So $1/(3-1) = 1/2$ would be deducted for a wrong answer. If the first part is correct, the second part is marked the same way. If it is wrong, the second part receives 0.

(iv) Matching block questions

In this type you have to match items on one list with those in another:

> Indicate an important action of each hormone in list A using list B

	A		B		Answers
a	Insulin	1	adrenocorticotrophic	a	.....4.....
b	Adrenaline	2	hyperglycaemic	b	.....2.....
c	ACTH	3	hypertensive	c	.....1.....
d	Cortisol	4	hypoglycaemic	d	.....5.....
e	Vasopressin	5	gluconeogenic	e	.....3.....

This type is quite difficult to set, since the pairs must be absolutely exclusive of each other. It is no good at all if two numbers of A have some sort of association with one in B. The necessity to avoid this creates much difficulty. But if you have this type there is a bonus in that if you make four correct associations the fifth is automatic. A wrong connection would receive a deduction of one mark, so this type has certain similarities to (i).

There are yet other types, such as the 'assertion-reason' and the 'diagram identification' but in biochemistry you are highly likely to come across (i) and (ii) above most of the time. Others may be tried on you; the chances are that you are participating in an experiment and the examiners will revert to (i) and (ii) again as the snags become apparent.

By common consent worldwide, the nature of the questions as well as the marking system should be communicated to the candidates beforehand so that they do not have to waste time on unfamiliar sets of instructions. If the questions come from a computer bank so much the better, they will be quite standardized.

With this theoretical background, we can formulate some rules for an approach to MCQs.

1 Make absolutely sure about the instructions; especially, make absolutely sure whether or not there is a penalty system for incorrect responses. If there is any ambiguity or obscurity, agitate.

2 Decide about guessing. If there is no penalty system, it is worthwhile to guess on every question you are not sure about. Leave none blank. If there are penalties, you have to be more careful, but you can still increase your chances by careful guessing.

First of all, and this only applies to (ii) and (iii) above, if you do not know the correct answer, try to decide which of the distractors are highly improbable. In (ii) above, you need pretty minimal knowledge to know that insulin is hypoglycaemic rather than hyperglycaemic. Insulin is put in to confuse the issue. You can therefore eliminate responses 'a' and 'd' even if you are not sure whether adrenaline alone, glucagon alone, or glucagon and adrenaline together significantly affect the release of glucose from the liver. This narrows the choice down to three and so you have a 1:2 chance of guessing correctly. Remember you only have a quarter mark or so off if you are wrong.

There is a fair probability that there will be a good balance of 'trues' and 'falses' as correct responses in type i. The balance may not be 50-50 but it should be near that. Are most of the responses you are fairly sure about 'trues' or are most of them 'falses'? If most of the ones you are sure of are 'trues', then probably most of the ones you are going to guess are really 'falses'. If you have completed sufficient responses of which you are sure and you add them up and they are sufficient to allow you to pass, leave well alone. If you then go on to guess, you may bring your mark down, even to below the pass level. If you add up the responses you have made and you are fairly sure that these are correct and they do not add up to a pass, you might as well proceed to guess the remainder. You may thereby pass. If you penalize yourself by too many marks deducted, well, you were failing anyway.

3 Look for clues. Really, they should not exist, but they do occur. For example, in the real multiple choice above (ii) the word 'are' has inadvertently been left in the stem, revealing a sense of grammatical propriety. You might then rightly suspect that the correct response is a plural one. If you can eliminate insulin as ridiculous, you settle on 'e' with no trouble. In general, statements which are true tend to be more lengthy than statements which are false. This is because of the need to attain complete clarity and unambiguousness in true statements; qualifications and exclusions abound. False statements are, however, nonsense to start with; short nonsense is just as presentable as lengthy nonsense.

(c) Vivas

In your junior years vivas are probably not given to all students, even in end-of-year examinations. You are called if you are in line for some sort of distinction or if you are borderline. (One or two might be asked to appear even if they are in the medium range, but this will be for some special reason, perhaps to see what sort of showing they can make in a viva in comparison with their effort in the essay or multiple choice. This is however unusual, being justly regarded as needless torture.) The first thing to realize is that unless you have made an unspeakable nuisance of yourself, the group of people facing you round the table really want to pass you. If there is an external examiner present, he especially wants to pass you for a number of reasons: he is probably humane *per se*; also, passing people rather than failing them gives the visit that comforting glow of success; and he wants if possible to please the internal staff too (none of this will hinder his being asked back). The internal staff want to pass you even more. There is the teaching reputation of the department at stake. Some of them, having taught you for a year or so, may even personally like you. Some may even feel insecure and feel that passing you and as many as possible of the others vindicates their performance.

With all this hammering at the cast iron structures of their integrity, the panel will decide on a starting questioner, who may kick off like this:

'What is the difference between the fate of glucose in the muscle and in the adipose tissue?'

Silence.

(He really wants to know if you know that in adipose tissue glucose is by and large broken down to two-carbon units for fatty acid synthesis, whereas in exercising muscle it yields lactate and pyruvate!)

'Well, what happens to glucose in one of these? Let's take muscle first.'

Silence.

'Well, how in general is glucose broken down? What is the pathway in most cells for its "lysis"?'

'Ah, glycolysis.'

(You should know this, really you should. On the analogy of an auctioneer calling for bids, a sort of starting price has been obtained. The examiner now wishes to build on it.)

'So what happens in glycolysis?'

'Glucose is broken down.'

(Patiently, he refrains from pointing out that this has just been established. Since glycolysis is just the Greek for 'breaking down glucose' you are going round in circles.)

'Well, does glycolysis take place in muscle?'

'Yes, I think so.'

'Indeed it does. Well, what does it produce in muscle?'

Silence.

'Have you heard of pyruvate?'

'Oh, yes. Pyruvate, of course.'

'What might it give in adipose tissue, then?'

(It is reasonable to suppose that it is something different. He started off by inviting some contrast. But maybe you still don't know.)

'Have you ever heard of acetyl CoA?'

'Oh, yes! Acetyl CoA.'

'Well, then?'

'Ah, pyruvate in muscle—to give energy—acetyl CoA in adipose tissue.'

'But what happens to acetyl CoA in adipose tissue? What is adipose tissue made up of?'

'Fat.'

'Right, I think we got there at last.'

Nobody is saying that you will receive a very good mark for something like the above—after all, he told you most of it himself. But at least, his goodwill manifest, you have gone along with the process. You achieve some credit for allowing the examiner to be educative in the Socratic sense, to drag out of you what he is prepared to believe is lurking in there somewhere.

There are two other small points about vivas. First of all *do* go over the paper. The examiners will have your answer book on the table in front of them, and when inspiration fails, they may look for errors in it to see if you can now rectify them. Sometimes, when they run out of inspiration completely, they will ask you to name an area which you either have studied thoroughly or have an interest in, so that they may question you on that. Have some topic ready, and make sure you really do know it.

Index

Generally whole phrases are indexed, that is 'glucogenic amino acid' rather that 'amino acids, glucogenic' appears. Also, important substances are indexed as their abbreviations, for example 'NAD$^+$' rather than 'nicotinamide adenine dinucleotide'.

THIS
CARELESS
LIFE

RACHEL McINTYRE

First published in Great Britain in 2017
by Electric Monkey, an imprint of Egmont UK Limited
The Yellow Building, 1 Nicholas Road, London W11 4AN

ISBN 978 1 4052 7368 8

58395/1

A CIP catalogue record for this title is available from the British Library

Typeset by Avon DataSet Ltd, Bidford on Avon, Warwickshire
Printed and bound in Great Britain by CPI Group

For Tim

⊙ 1 LIV

1 July, 10 a.m.
When Olivia Dawson-Hill opened the front door she had no idea what she was letting herself in for.

'I'm Cass,' the woman on the step said. She held out a business card between elegantly manicured fingers. 'And you must be Liv.'

Liv took the card. Read it. Frowned.

Cassandra Verity
Assistant Director
Pretty Vacant Productions

'But I spoke to Tony last night. He said we were all fixed for this afternoon.'

'There's been a change of plan. He's going to call you later.'

The woman smiled.

Liv assessed the visitor with an expert eye: flawless skin, discreet make-up, glossy deep brown hair. White linen dress, understated but expensive. High-end high street? Leather sandals, *definitely* designer, and by them a large black and silver case.

The woman pulled her handbag higher on her shoulder and Liv's mouth, poised to utter, *What change of plan?* dropped open at the sight of the distinctive gold *P* dangling from the strap.

'Oh. My. God. Is that a genuine Pandora?' she said in hushed tones, reaching out a tentative hand. 'I mean, I've been on the waiting list since it opened, but even so, they said at least Christmas. How did you . . .?'

'Beautiful, isn't it? I guess you could say I've got friends in high places.' The woman held the soft tan leather up as though it were some holy artefact. 'Now is it OK if I bring this inside?'

'This' was the large black case. Plastered in wasp-striped stickers that read *Fragile Audio-visual Equipment*. A dizzying flare of excitement shot through Liv.

This was it!

Right now, waiting inside that black vinyl box was The Future.

Tucking the business card in the pocket of her jumpsuit, Liv paused, holding it there for one . . . two . . . in an attempt to disguise the sudden trembling in her fingers.

Deep breath.

Smile.

'Of course. Come in.'

Cass stepped forward, bringing with her a drift of some ultra-subtle fragrance Liv recognised. Like wild herbs and sea air and sunshine mixed up together. *So* familiar. What was it called?

But the answer dangled stubbornly out of reach, vanishing entirely as Liv caught sight of Jez's precision parked BMW, and behind it . . .

Bloody Jez! He *knew* not to leave the gates open. She tutted, flipping the security panel open to key in the code. The gate mechanism groaned; two Rottweilers hurtled across the walled courtyard, clattering their long chains over the cobbles.

Beyond the high wall, glinting tractors and tiny stooping figures dotted the endless fields. Squinting against the sun, she could just make out a horse going round the training ring and a smaller dot that must be Mum. Good job she hadn't seen the open gates. Liv waited until they clanged shut, cutting High Acres off from the surrounding farmland, then raised her voice above the ferocious barking.

'Sorry about that. My dad's super security conscious.'

'It's fine,' Cass said, lifting the case over the threshold. 'Wow. I remember thinking what a lovely space when I saw your application footage, but it's even more impressive in real life. Love, love, *love* the staircase.'

Pride mingled with Liv's fizz of excitement, although if she wanted to be picky, 'lovely' didn't *quite* do the hallway justice. Presiding over the entrance, the show-stopping stairs, designed to her own vague specifications ('wooden steps with glass up the side') had featured in *two* interiors magazines, attracting adjectives such as 'stunning' 'dazzling' and 'sublime'.

'Spectacular,' Cass murmured, brushing her fingertips lightly along the wall.

Exposed brickwork lined the way to the lounge, and the ceiling, stripped back during the renovations, arched high above them in a skeleton of beams. And even though her mum had tried to put her foot down

4

over the inset floor lights Liv flicked on now, Liv got her own way in the end. She always did.

Mind you, every time she pressed the switch, Mum's mocking voice rattled in her head: 'Low-level lighting will guide you to the nearest available exit.'

Hilarious.

Just after the door – the *locked* door – that separated Liv from her parents in the main house, Cass paused to examine a gallery of framed photos: Liv with other pearly-toothed girls, lifting champagne flutes outside a grand marquee. Sleek hair streaming behind them, the same friends captured mid-shriek on a waltzer. Liv, alone, in front of a glowing Ferris wheel. Liv posing, arms in the air, against a vivid blue sky. 'Stunning pics. I take it they're recent?'

Liv nodded. 'That one's my eighteenth in April. That's Greece at Easter. Those ones are from the prom. We had this 1950s theme with a proper fairground. The photographer works for *Vogue*; all the prom pics were amazing.'

Cass nudged the corner of a frame slightly to straighten it. 'Lucky you. Only July and you've already had an unforgettable year.'

Unforgettable? Well, that was one way of looking at

it, but not entirely for reasons Liv cared to remember. She swallowed hard. *Do not think about HIM now.* 'It's been eventful,' she agreed, careful to keep her tone light. 'Would you like a cup of coffee? Green tea?'

The woman shook her head. 'Thanks, but we don't have much time. It's better if we just get straight on with the casting.'

'Yeah, sure. Follow me.'

But Cass had turned back towards the front door. A shaft of sunlight streamed through the stained-glass panels, throwing kaleidoscope patterns on the wooden floor.

'What a truly lovely home you have,' she said.

'The windows came from a church that was being demolished. And the floor.' Liv pointed her toe at the patinated boards, worn to a shine by decades of worshipping feet. 'You can see if you look closely.'

'Amazing,' Cass said, gazing down.

As they reached the end of the passageway, a thumping beat joined the *click-clack* of Cass's sandals. She raised her eyebrows. 'I'm guessing the others have already arrived.'

'Yep. Everyone got here early.'

Liv twisted the circular handle on a carved door which, until the architect snapped it up, had apparently eavesdropped on a lifetime of confessions. She elbowed it open; the hinges gave a groan.

Tacky. That's how Mum described the chandelier dangling from the beams, the aqua geometric print rug, white gloss furniture and huge L-shaped sofa Liv had chosen. Even the open-plan kitchen, with the sleek cupboards and cavernous pastel-pink fridge Liv literally *worshipped*, had only elicited a disparaging, 'Not very practical, is it?'

On the rare occasions Mrs Dawson-Hill ventured through the door from the Land of Boring Bland, she screwed her face up like she'd accidentally stumbled into raw sewage.

Liv could have asked the designer for a Mum-friendly scheme: pine, flowery cushions and easy-to-clean flagstones.

But she didn't.

With the door open, the wall-mounted TV was revealed as the source of the music. More specifically, the video of a suit-wearing male singer flanked by a flock of bikini-clad dancers.

Sprawling on the rug beneath, transfixed by the screen and tapping the remote control to the beat, was Declan Duffy.

'Switch the telly off,' Liv hissed at him.

'What for?'

'Just do it.'

Something of Liv's irritation must have got through because he jerked upright, fumbling for the button. Although the music stopped, the dancers continued gyrating silently.

'I said off, not down.'

Liv snatched up the remote control, killed the screen and stood in front of the TV.

'OK, everyone, listen up. There's been a change of plan. Tony can't make it so he's sent someone else. This is Cass.'

Three expectant faces gazed up and Liv, conscious of the woman's presence, had the strange sensation she was seeing her friends for the first time.

Declan Duffy, stubble scuffing his chin and dark circles, almost bruises, ringing his currently red eyes. On most guys you'd think *knackered*. On Duff they weirdly added another dimension to his apparently irresistible bad-boy appeal. (Irresistible to other girls, that is. Not Liv, who'd known him since primary and therefore knew *exactly* what he was like.)

Perched on the edge of the sofa: Hetty Barraclough,

brown hair tugged back in a ponytail, knuckles whitening around an iPhone. Hesitant smile hovering on her scrubbed face even as she shrank inside her baggy grey sweatshirt.

Liv suppressed a sigh. A *sweatshirt*. What exactly was Hetty thinking? Liv wouldn't use that rag to clean the floor, let alone audition for a TV show in it.

Not that Liv had any intention of cleaning this or any floor.

Finally, Jeremiah Livingston, almost-but-not-quite touching Hetty. Not a crease wrinkling that immaculate shirt, not a smudge on those blinding trainers. Owlish behind Harry Potter glasses, with thick eyebrows that were currently lowered in a frown as he asked, 'Sorry, what do you mean by a change of plan?'

Cass set the heavy-looking case down by the side of the coffee table. Taking the *actual* Pandora off her shoulder, she placed it on top then swept her gaze across the three seated friends.

'Hello, Jeremiah — Jez, isn't it? Nice to meet you. And Hetty too and . . . you must be Declan.'

'Call me Duff.'

He unfurled himself up off the rug in a single fluid movement, stealing the opportunity to not-so-subtly check himself out in the oversized mantel mirror. 'And you are . . .?'

Oh *please*. Liv raised her eyes to the ceiling. Like the dogs going mental over the screech of the gates, Duff's flirt offensive was so depressingly biological. Every interaction he had with a female started with that cool up-down once-over; the instinctive, preening *hey hey hey*.

But the woman fixed her attention not on him but on the retro clock above his head and the white digits that flipped to 10.08 before she replied in a neutral and distinctly non-flirty way.

'Hello, everyone. Great to meet you all in the flesh; I've been hearing so much about you. My name is Cassandra Verity, but please call me Cass. I'll be taking you through the casting for *This Careless Life* today.'

Duff didn't even appear to register the slight; instead he rocked back on his heels, unselfconsciously watching the visitor snap the tabs on the black case, tip the lid and extract several smaller, squishier bags.

That was the thing about Duff. His ego was galactic. Like a constantly inflating ball of vanity expanding beyond the earth, beyond the solar system, it bounced through wormholes, emerging in parallel dimensions where billions of super-cocky identi-Duffs blatantly sized up anything woman-shaped.

As always, Liv felt torn between admiring his self-confidence and massively wanting to give him a slap.

A sharp intake of breath caught her attention. Hetty. Mouth open, about to speak and nervously turning the phone around in her hands.

'*What?*' mouthed Liv.

Hetty positioned her lips into the determined smile Liv recognised from school functions. 'Sorry, Cass. Hello. I, er, thought the audition wasn't till two? Are we still going to have time to rehearse?'

Twist, twist, twist went the phone.

Cass rasped a Velcro strap and straightened up. 'I'm sorry about the short notice. I've had to juggle the timing because I've got a plane to catch this afternoon. I know you've been dealing with Tony, but something urgent cropped up so I'm taking over. But if you really want Tony to fit you in after he's seen the other candidates, I'm sure he would –'

'No,' Liv cut in before Hetty could. 'Today, now, with you is fine. Totally fine. Beyond fine. Awesome.'

Cass grinned. Her teeth really were lovely: very straight, very white against her olive skin. Bending her shiny dark head over the table, she unzipped the various bags and set a selection of silver and black tech on the coffee table. Tripods, cameras, a fuzzy-headed microphone . . .

'Fantastic. Does mean we're on a super-tight schedule though, I'm afraid. We need to set the equipment up asap so we can make a start.'

'Can we help?' Jez asked.

Cass, expertly screwing a stand into one of the cameras, used her head to indicate the white gloss bookcase, crammed as always with an avalanche-in-waiting of fashion magazines.

'Sure. Can you put this . . . there?'

'Allow me,' Duff said, leaping up to take the camera. Jez snorted softly and Liv bit back a tut. *So* predictable.

'Thanks. And these two . . .' The woman pointed first at the worktop of the (shiny, untouched) kitchen and then at the corner desk stacked with Liv's (shiny, untouched) revision books. 'There and there should do it. And camera number four. Let's see . . .' She scanned the room. 'There.'

With a flash of irritation, Liv scooped the scattering of jewel-bright lip glosses, nail polishes and random earrings along the sideboard and into a drawer. What a total mess. She'd have to talk to Dad about . . . was it Martina or Marina? She couldn't remember. Whatever the new cleaner's name was, she was a joke.

Yesterday, Liv had caught her vacuuming *around* a pair of cerise Victoria's Secret knickers in her bedroom. And when Liv suggested that maybe she could, you know, pick them up first? Martina/Marina had stared blankly at her with eyes ringed in glittery blue liner, *then carried on.*

Absolute joke.

'Lens caps off and angled at the sofa, please,' Cass continued. 'Have a wiggle, check the tripod's steady. Press the green button on the top. And don't worry about not rehearsing, Hetty; spontaneity gives the best results. Makes the whole process more . . . honest.'

She placed her bag on the counter that divided the kitchen from the living space. The gold *P* chinked against the granite as Cass rifled through, pulling out first a mini laptop case and then a black gadget studded with tiny buttons. From her position by the door, Liv had a perfect view of the computer screen Cass was now adjusting to suit the light filtering in through the blinds.

'Nearly there, guys. Duff, may I borrow you to check the angles?'

'Sure, what do I need to do?'

'Look pretty,' Cass said, flicking her gaze between the sofa and the screen.

Liv's eyes automatically followed, tracking across her friends. Jez's watchful, relaxed expression giving nothing away, hands clasped in his lap.

Then Duff. Every inch of that six-foot-plus gym-toned, buffed and waxed body currently squeezed between Hetty and Jez radiated natural-born poser. Liv would not be *at all* surprised to learn that Mrs Duffy had an ultrasound image somewhere of foetal-stage Duff pouting like a pro.

Unlike acute photo-phobe Hetty who, even when she smiled, gave the impression she would rather be facing a machine gun than a lens. At the prom she had avoided the crush of the booth, ducked to the back for the group shots. And now, hugging a velvet cushion to her chest, she was playing the part of person most unlikely to audition for a TV show to perfection. Liv damped down a surge of exasperation. Hetty had *promised* she would go along with it.

'Hets?' she said in an undertone, miming dropping the cushion.

But if Cass noticed Hetty's nerves, she didn't seem

bothered. One final tap on the keyboard and four LEDs blinked on; a red eye staring from each of the cameras placed around the room. Four Duffs materialised on the quartered screen, each fiddling with his phone from a slightly different angle.

'Duff, can you say something so I can test the sound levels?' Cass said, setting a microphone on the table.

'Anything in particular you'd like me to say?'

A line spiked into jagged peaks in the top left of her screen.

'That'll do, thanks.'

She moved the microphone a few centimetres closer to the sofa and wagged a finger at Duff's phone.

'Sorry, no devices. They interfere with the equipment. Disabled Wi-Fi or switched off, please.'

Duff swiped his thumb over the screen and placed it on the table. 'No worries.'

Liv quickly slipped hers out of its case and touched the little aeroplane icon. *Wi-Fi Off.* When she looked back up, Jez's domed forehead had loomed into shot. The light caught his glasses and the lenses flashed opaque white.

Lines zigzagged in the corner of the screen as he asked, 'Do we need to sign anything? Contract? Disclaimer?'

Cass clicked her fingers and pointed her index finger at him. 'Excellent question. Yes, I have paperwork. There's *always* paperwork.'

She reached inside the Pandora, drawing out a document wallet containing a sheaf of A4 sheets attached to thin plastic clipboards, each with a printed label.

'Here we are. Jeremiah . . . Henrietta . . . Olivia . . . Declan. These are the agreements you submitted with your initial applications. Legal tells me I have to read through the section marked "Audition" to make sure you know what you're getting into.'

'Sounds ominous,' Liv said on her way to sit down. Then laughed to show she didn't mean it.

She curled up next to Hetty, tucking her feet under her. Hetty grinned and flexed her eyebrows in a *this-is-it!* kind of way, which immediately made Liv feel like an evil bitch. But honestly, that sweatshirt! Liv had a photo of Hetty aged *fourteen* wearing the exact same garment on a school skiing trip. Which, in clothing years, made it older than the actual Pyrenees.

'It's just to make sure we're all completely clear,' Cass said.

Draping one arm along the leathery back of the sofa, Duff fanned himself with his copy. 'No need.

We trust you a hundred per cent, Cassandra.'

A lion on the scent of a zebra. That's what he reminded Liv of. A lion too full of itself to realise this particular zebra was way, way out of its league.

'That's good to know but have a quick look to re-familiarise yourself,' said Cass, her brown eyes trained on the laptop.

Jez curled his shoulder away from Duff's loosely dangling hand and studied the contract, one side, two sides before taking Hetty's copy. He flapped his fingers, offering to do the same for Liv but she shook her head and pulled the contract to her chest.

Cass looked up expectantly. 'Got a pen? Great. Any questions, fire away, otherwise please tick the boxes after each point.'

She began to read in a can't-put-your-finger-on-it accent. Posh, yes, but with a hint of something under it. Something not entirely English?

'I understand I am auditioning to take part in a vehicle for Pretty Vacant Productions with the working title: *This Careless Life*.

'By agreeing to participate in the audition process, I confirm I am eighteen years of age.'

Tick.

'I understand this audition will never be broadcast or made publicly available.'

Tick.

'I give Pretty Vacant Productions permission to continue to obtain and research my online presence including accessing my social-media profiles.'

Tick.

'I understand that participation could have a profound impact on my life and those around me.'

Tick.

Pause.

Bookcase. Kitchen. Sideboard. Desk.

'One . . . two . . . three . . . four.'

Cass jabbed her pen at each camera in turn. 'Audiences expect full disclosure, guys. In this job, festering skeletons have a nasty habit of tumbling out, sooner or later. And it will be much better for you if you're transparent from the start. So think carefully now – is there anything you need to tell me?'

Duff spread his arms wide. 'What you see is what you get with me. No skeletons, no secrets. Guaranteed.'

Cass laughed softly.

'Everyone's got secrets.' Her deep brown eyes met Liv's. '*Everyone.*'

Then she raised one perfectly shaped eyebrow a tiny fraction and panic jolted through Liv.

Wait. Did she know about —?

No.

No way.

It was impossible. Apart from herself, only two other people knew: Hetty (who would never blab) and *him*, and he'd been careful to the point of paranoia. No names on texts, separate SIM cards, no likes on each other's posts . . . literally nothing to give away that they were anything more than vague friends of friends.

A rustling interrupted her thoughts. Cass had turned to the next page in the contract, but her gaze still rested on Liv.

Liv's heartbeat speeded up. *Oh God. Please don't.*

But then Cass's thoughtful face broke into a warm smile. 'No one wants to unload their burning confessions? OK, so if we're all happy with the Big Brother stuff, let's move on.'

As Cass carried on reading, Liv's nerves gradually unknotted. No. Absolutely no way Cass knew about *him*. Part of the test, wasn't it? A lucky guess to put her on edge. Like those TV psychics who cast a million generalisations in the air and wait for the audience to bite.

Liv's mind floated out of the room, up the stairs and into her dressing room to pull open a few drawers, rifle along the hangers . . . Oh dear. A terminal crisis loomed on the clothes front that a trip into Manchester couldn't cure. Two, maybe three hardcore shopping days in London with Dad's credit card would do the trick. Should she invite Hetty along? Persuade her to slip into something a little less comfortable for once? Maybe even brave a manicure?

Liv inspected her nails throughout the section on 'disclaimers and exclusions'. Rubbed at a faint scuff mark on the side of her sandal during 'investigative access'. Listened to Duff crick his neck from side to side while Cass droned on about 'post-interview courses of action'. Whatever that meant. Duff yawned and tapped his finger against the plastic clipboard. The rhythm wormed its way into Liv's brain: *get on with it . . . get on with it . . . get on with* —

'This is the last one, I promise,' Cass said, reeling Liv out of her trance. Caught mid-yawn, she coughed unconvincingly and pasted on her hanging-on-every-syllable expression, a look she'd recently perfected thanks to Hetty's adoring-boyfriend-monologues.

'I understand if I choose to withdraw from today's

audition process, the entire team will be disqualified.'

'All in or all out,' Jez murmured.

'Exactly. So . . .'

Cass mimed ticking the box.

Next to Liv, Hetty stopped chewing the end of her pen and tilted her head to one side. 'What if we do get chosen and then one of us wants to drop out later on?'

What?

Snapped out of her relaxed state, Liv cast her friend a *shut up* glare and quickly clarified. 'I think what Hetty means is if someone is ill or has an accident.'

'Still, it's a valid question.' Cass held up a finger. 'Bear with me one second . . .'

There was a pause while Cass rummaged through her beautiful Pandora. Liv nudged Hetty, urgently mouthing, '*Don't say drop out.*'

Before Hetty could respond, Cass withdrew something with a flourish.

'Aha. I knew it was in there somewhere. Hope you don't mind, but I like to make some notes by hand. Helps me keep things straight in my mind.'

It was a notebook bound in battered red leather, somewhere between A4 and A5 in size and held together with a fraying ribbon. Cass pulled one end and, as the bow unravelled, the thick book crackled open.

Quirky. The yellowed pages reminded Liv of those junior-school projects where you recreated the Magna Carta by wiping ye olde wet teabag across ye olde history homework. Unexpected that this fashion-savvy woman with her sleek tech and leather Pandora would even possess such an ancient piece of tat. What next, a quill?

'Kind of old school, I realise,' Cass continued, 'but I'd be lost without it; I've had it for years. Centuries even. Anyway, where were we? Oh yes. Dropping out.'

She cracked the notebook's spine, smoothing the pages flat with her long tanned fingers as though she were ironing it.

'Once this process is underway, the last thing anyone needs is to chop and change. We need everyone fully on board, otherwise it just won't work. Honestly, there'd be no shame in calling it quits at this stage, Hetty.'

Oh God. Liv could have sworn her heart actually stopped beating. *Please, please, please don't bail. I need this so bad.*

'We're all committed, one million per cent, honestly. I swear. She just –'

Liv hadn't got to the end of the sentence before the sofa creaked at Jez's sudden movement.

Bracing his hands on his knees, he said, 'Liv, Hetty is entirely capable of speaking for herself, if only you'd give her the opportunity to –'

'Thanks, guys, but you don't need to worry about me. I'm not going to quit,' Hetty interrupted quietly, but with an unmistakeable undercurrent of determination.

'Fantastic. So glad to hear it,' Cass said, drumming on the notepad, *rat-a-tat-tat*, with a glance up at the clock. 'Two seconds more while I jot this down.'

Liv exchanged an anxious glance with Hetty who gave an almost imperceptible shrug in return. *Well, I can ask, can't I?* it seemed to say.

What was Cass writing? From across the room, it was impossible to decipher the tiny black squiggles crawling across the page. Was she 'jotting down' *Hetty is a flake?*

Whatever it was, Het's question had tapped right into the nagging *she can't hack this* that had been buzzing round Liv's skull for the last three weeks like a wasp. Much as Liv loved her friend, she had to admit that when she first read the Pretty Vacant pop-up, Hetty's name hadn't exactly leapt to mind. To be brutally honest, it hadn't even crawled.

It had been the day after her History A-level exam (her *last ever* exam). A day that dawned on a clear sky of limitless future possibilities . . . for everyone else anyway. For Liv, it reverberated with the slam of a thousand doors.

She'd tried.

And as Results Day would undoubtedly prove, she'd quite spectacularly failed.

Liv woke to the sound of hailstones hitting her bedroom skylight like ping-pong balls shot from a celestial cannon. Bleary-eyed, she stared up at the grey-black clouds pressing themselves against the glass, billowing portents of doom.

Summer holidays. Ha ha *ha*.

With a sigh, she fired up her Mac, rolling her finger over the touchpad to open the footage she'd been editing and re-editing until the early hours. Crucial stuff, this: her final video before the National Beauty Blogger Awards.

And it was absolutely vital she nailed the pitch. Last year, she'd lost out on Most Inspirational Newcomer to that suck-up extraordinaire, *Sonya Sunshine*. A thought that almost twelve months later still made her want to puke. Or punch something. Preferably Sonya.

The Cinderella Project was Sonya's baby: sourcing prom dresses to donate to teenagers living in poverty. Or as Liv preferred to call it, a pathetically transparent attempt to win votes.

Seriously, how dumb would you have to be not to get that?

As dumb as a National Beauty Blogger Awards judge apparently.

Well, this year Liv had been nominated again, and this year she was coming home with it. No way could she politely clap *again* while Sonya delivered a vomit-inducing speech accessorised with fake tan, fake nails and even faker tears.

She'd rather *die*.

But with her crossed fingers about to press *upload*, the screen flashed and a pop-up appeared.

CASTING CALL
**If you're over 18 and you've just left school,
Pretty Vacant Productions could be about to offer YOU
the opportunity of a lifetime!
Visit www.pvp.tv.org to find out more!**

Liv's heart pounded as she clicked on the link.

Well, hello!

Your A levels are history, you've kicked the dust of the schoolyard from your shoes and left your cares at the gates. Results Day, uni places ... they're just specks on your horizon. We're talking freedom, baby. The freedom to live your life without a care. Freedom you'll never experience again. This summer is YOUR summer.

The lowdown: we need FOUR friends to be the stars of *This Careless Life*, Pretty Vacant Productions' new six-week special. PERSONALITY plus, that's top of our wishlist.

Before you apply, remember we have some PRETTY strict guidelines. So if every individual can answer a massive *Yeah, baby!* to the following questions, then PLEASE drop a 60-second video HERE explaining WHY we should choose YOU.

Are you 18?

Have you just finished A levels?

Are you in the UK between July 1st and August 31st?

And above all: will the nation's 16- to 25-year-olds LOVE you or love to HATE you?

So if you'd LOVE to share your outrageous, uber-exciting or totally ridiculously INCREDIBLE post-school/pre-uni summer ... get in touch now!

'Yes, I'm 18! Yes, A Levels! Yes, I am in the UK!' Liv murmured to the screen. 'And love me, they will all love me.'

And just like that a whole new door, one she hadn't known even existed, had flown open.

Choosing three friends? That was a no-brainer. Or at least it would have been, except with the exams done, Freyja, Scarlett and Touko had departed St Benedict's and jetted home to Brazil, Russia and Japan. Meaning Liv had to fish from a much smaller pond solely stocked with weekly boarders, and local ones at that.

Liv frowned, tapping her forefinger against her lips. Only two days until the closing date for entries. Who was still around? Who would inspire both love and hate?

Jeremiah Livingston? Head boy and all-round good guy. He'd be up for anything to promote his charity stuff for Connecting Together. And Declan Duffy, naturally. With his plus-size ego a-gogo, he'd do anything to promote himself. That left one more . . .

Letting the nation analyse six weeks of your life?

Really, Liv only mentioned it because she thought Hetty might be able to suggest someone. So when – unbelievably – Hets had shrugged the suggestion, 'How about me?' Liv's jaw dropped so fast it bruised her kneecaps.

'You serious?' she said, failing to hide her shock.

'I'll have to check with Duncan,' Hetty replied, 'but I'm sure he'll say it's a brilliant idea.'

What's it got to do with him?

Liv hid the retort behind a beaming smile.

In gaining a boyfriend, Hetty had apparently excised her decision-making lobe. Ever since she hooked up with him, all you got from her was *Let me check with Duncan. I'll have to ask Duncan.*

Still, His Royal Duncan-ness graciously granted his permission, and with Duff, Jez and Hetty on board, Team Liv was ready to roll.

Amazing how the pieces just fell into place. Perhaps if Liv were the kind of girl who believed in horoscopes and all that destiny crap, she might have called it fate.

But waiting to find out if they'd got through? Well, that was every childhood Christmas Eve squished into one tortured ball of anticipation, frustration and panic. *Come on. Come ON.*

Until finally, on a day she'd planned to spend wrapped in a cocoon of duvet and misery trying to forget she'd ever even met *him* . . . the phone rang and on the other end was Tony from Pretty Vacant Productions saying congratulations, they'd reached the final three and were they all still up for it?

Liv had immediately flung back the duvet, palmed a night's worth of tears from her cheeks and assured him that yes, she wanted this *so* bad. In fact, she literally could *not* want this more. Yes, they were free for the casting on July 1st. And yes, her friends would be literally *ecstatic* to be through to the next stage.

When the phone call ended, she flopped back down on the bed, gripping the phone like a lucky charm while Tony's question bounced round her head.

Was she up for it?

Liv was up for many things.

She was up for not obsessing over her A-level results.

She was up for making her parents realise uni was not the only option.

And she was *definitely* up for being a TV star.

But on the day of Tony's phone call, she was mainly up for never, ever shedding a single tear over a man for as long as she lived.

Things were about to change. She knew it.

'*Your whole life is about to change.*'

She felt the whisper of breath; sensed a light grip on her shoulder; a body close to hers. A drift of Eau de Expensive Whatever-it-was enveloped her.

Except . . .

Liv blinked.

How did —?

Cass hadn't moved. Definitely not. There she was, immaculate in white, still leaning over the worktop with the now deflated Pandora sagging by her elbow. Notebook completely open; mouth completely shut.

Liv wrapped her arms around herself, rubbing at the sudden ripple of goosebumps.

What the —?

Her mind frantically scrabbled to latch on to any explanation other than *I am cracking up*, but then Cass put her pen down and looked directly at her.

'Exciting, isn't it? Knowing today could be life-changing.'

Liv nodded. Stunned.

'Yes.' Jez put on his mature I-am-leader voice. 'And on behalf of all of us, I'd like to say thank you for this incredible opportunity.'

'My pleasure,' Cass said. 'I've got a great feeling about you guys.' The dark curtain of her hair swayed as she turned her head towards the wall clock. 'Let's push on. Now I'm going to focus on one person at a time; show your application video followed by a group conversation.'

She stepped forward and with a collective wriggle they all straightened against the button-backed sofa. Attentive. Ready. Even Duff slouched marginally less.

From where Liv sat, still dazed, some proportional illusion made it seem as though Cass were towering over them. When she spoke, far from a whisper, her voice filled the room.

'With that in mind, what I need you to understand is this process is tougher than you might think. There'll be questions you may not want to answer and answers you certainly won't like. By the end, you'll feel drained, emotionally and mentally. Maybe you'll decide you're not up to it. Maybe you'll decide you're happy the way you are now. But if you go ahead, you'll be rewarded with an opportunity to change your lives forever.'

Your whole life is about to change.

The mysterious whisper echoed in Liv's mind. Too much caffeine, too little sleep. That was it. And stress.

Exam stress, prom stress, HIM stress. Who wouldn't lose the plot with all that going on?

Her ears were playing tricks on her, that was all. Liv shook her head slightly as though fending off a mosquito and focused her attention on Cass, who was holding a printout of the contract up in front of her chest.

'Pep talk over. Now it's time for you all to sign on the dotted line, which you will find in this box . . .' She pointed a beautifully polished fingernail at the space in the bottom left corner. 'Here.'

Liv complied, capping her pen almost before Cass finished her sentence. She kept her hands occupied by unfastening and re-knotting the belt on her silk jumpsuit, tying it in a floppy bow on her left hip.

Next to her, Duff scrawled his signature, the pen scratching as it looped extravagantly across the bottom of the page, refusing to be constrained by the box. Hetty's ponytail bobbed as she wrote tidy, blocky letters in the allotted space.

What was Jez playing at?

With a dismayed glance, Liv noticed him hesitate, the groove in his forehead deepening while his hand hovered.

Get on with it . . .

'Jez?' Cass prompted gently.

He frowned. 'I need clarification on a few points before I can commit.'

Liv's heart sank. Oh God, what now? Jez Livingston. Undoubtedly the loveliest of guys in a multitude of ways, but very, very occasionally (like now) so pedantic he made her want to claw his smug face off.

'Fire away,' Cass said, glancing down at her book.

'I understand the audition won't be made public, but how much control do we have over what happens next?'

For a start, that wasn't even Jez's *own* voice. Too deep. He'd borrowed it off his dad, trying his courtroom act on for size.

'Complete control of what you say and do,' Cass answered.

'But do we have the right to veto?' Judge Jez continued, pompously. 'If the footage is edited to present us . . . unsympathetically?'

Liv's toes clenched in her sandals.

Shut up!

Cass raised her palms in a conciliatory gesture.

'Look, I promise no one is going to judge or humiliate you. My sole remit is to get you to show yourselves and what you're about within these four rather lovely walls.

After that it's your call: you can take this opportunity forward or you can pretend I was never here. No one will be forcing you to do anything.'

Jez pursed his lips and continued channelling his inner Yoda.

Most monumental arse I am.

Liv twisted her hair into a ponytail then let it fan over one shoulder. Duff dragged his palms down his face with an exaggerated sigh, *drum-drum-drumming* his heels on the floor.

And then – finally! – Jez signed his name.

'You'd better get used to autographs.' Cass said, plucking the form from his hands.

The tension popped like a balloon. Liv's laughter bordered on hysteria.

'Anything else before we start?'

'Yes,' Liv said, delving down the side of the sofa. She brandished her selfie stick. 'One tiny thing. Please could we take a team selfie?'

'If you make it very quick,' Cass replied.

Jez shook his head and gave a disapproving tut.

'Honestly, you're addicted. Olivia Dawson-Hill and The Stick of Narcissus. Cass, I apologise, I realise you don't have time for this.'

Cheek! Liv was about to tell Jez where he could shove both the stick *and* his comments when Cass burst in to a melodic peal of laughter.

'Stick of Narcissus. Very good. I love that. Very appropriate.'

She was still grinning broadly when Liv took the picture. Four friends with their glamorous visitor beaming in the middle. Cass turned to *click-clack* back to the laptop and Liv quickly uploaded the image.

#topsecret #fingerscrossed #ohmygodohmygod #lifechanging

'Phone off now, please,' Cass said, glancing at the notebook. 'And we really need to make a start. You're first, Liv.'

Liv's heart gave a tight squeeze. *Show time!*

'One sec.'

She smoothed the ends of her shiny dark hair down en route to the mirror. Grimacing at her reflection, she ran her tongue over her teeth and a practised finger over both immaculately threaded eyebrows.

'Ready?'

Liv nodded, cleared her throat. 'I'm ready.'

⊙ 2 LIV

1 July, 10.28 a.m.

'Eeek. Cringe.' Liv peeped at the TV through splayed fingers. 'What an absolute hound.'

The image showed her frozen in the act of kissing a red-soled shoe; the exposed brick of the coach-house wall visible behind her.

'Don't be stupid,' Hetty said, reaching over to pull Liv's hands away from her face. 'You look beautiful, just like always.'

Truthfully? Hetty was right. OK, Dad's nose loomed large (so unfair he refused to let her have surgery), but the rest of it . . . skin, hair, eyes. Perfect.

'This is the video from the start, guys,' came Cass's voice from the corner. 'I've got high hopes for you, Liv.'

She pressed a button and the caption *Olivia Dawson-Hill Pretty Vacant Productions #1* appeared on the screen. Liv swished her silky hair to the other side and leaned forward, propelled by anticipation. The atmosphere reminded her of the opening night of a play or a film premiere, that same tension crackling like static in the air. The expectant expressions.

Then they were all looking at her face on the TV *(hair, good; make-up, great)*. No trace of her dad's flat, northern vowels in the voice that filtered through the speakers.

Olivia Dawson-Hill Pretty Vacant Productions #1

Hi, everyone [*waves*]. My name is Olivia Dawson-Hill, but my friends call me Liv. This is my application for *This Careless Life*.

OK, so I'll start with my favourite topic: me! I've just left school, finished my A levels [*pulls face*] and waiting for the results in August [*closes eyes, sighs noisily*]. Don't ask. But the things you really need to know about me are: I literally live for fashion and beauty. I swear, without shopping, I would die. Look. [*Camera turns to film room and image jerks through into another.*]

This is my bedroom [*camera pans over white walls, oak beams*] and this [*flings door open, automatic spotlights shine on rails of clothing*] isn't all of it either; I've got another wardrobe in my old room over in my parents' side of the house.

And of course ta-daaaa! Shoes ... [*Camera zooms up and down on tiered storage. Liv's hand appears to take a pair, lifts the still-attached price tag*] £200. [*Laughs.*] Not even been worn. I can't help it; I'm a shoe-aholic.

[*Buries her face in the shoes, inhales deeply*] Aaah, happiness is ... the smell of new shoes. And new lipsticks, new clothes, new handbags ... [*laughs*]

[*Camera POV returns to Liv.*] So why should you pick me? That's easy! You want personality plus and I can give you that. I'm great on camera. I've got my own beauty vlog ... maybe you've heard of it? *Miss Olivia Loves?* I'm on target for a million subscribers across my channels by the end of this month. Crazy, isn't it? All the major brands are contacting me, sending me samples. It literally is a full-time commitment; I've worked so hard. And I got nominated at the Beauty Blogger Awards last year for Most Inspirational Newcomer. I didn't win [*laughs*] but the party was In. Cred. Ible. One of the best nights of my life. I'll remember it as long as I live.

What else? [*Looks at ceiling.*] Oh yeah. Money. We're, like, seriously minted. My dad has an agricultural business and my mum breeds racehorses of the Grand-National-winning kind. We've got this place [*pans camera round*]. Well, this is all mine. Mum and Dad have got the big house next door. Then we've got one in the South of France, where I'll be going in a few weeks, unless you pick me for the show [*laughs*].

So [*camera back on Liv, counts off points on fingers*] I look good, I sound good and my life is outrageous so please please pretty please, Pretty Vacant, put me and my team through to the next round. You won't regret it. Mwah! [*Blows kiss at screen.*] Sorry if that was more than sixty seconds! [*Leaning forward.*] Mwah!

Cass pressed pause, leaving Liv's cherry-red lips suspended mid-pucker on the screen.

Just out of shot were the lamps fitted with diffusing bulbs she'd bought from a theatrical accessories website. Insanely expensive, but judging by the results, worth it. The slight golden tone cast a filter-free I-just-got-back-from-St-Tropez glow on her skin and popped her green eyes; added a glossy sheen to her long dark hair.

Totally worth it.

The air fluttered around her: Duff had risen to his knees and was flapping both hands in mock-hysteria.

There were times when Duff made Liv laugh until her lungs ached and tears poured down her cheeks. Now was not one of them.

'Fashion is my life,' he squeaked breathlessly.

'Shut up.'

'Shoes! Handbags! Make-up!'

'Shut *up*!'

Liv rammed a velvet cushion in his face, not quite muffling his falsetto cries.

'Cut me and I bleed lip gloss!' Fending her off he emerged red-faced, choking on stray feathers and his own hilarious words. He wrestled the cushion from her, threw it along the sofa, narrowly missing Jez, and strolled up to the mirror.

'Cass, don't be fooled by this vapid display. Liv's not dumb. And she's not that shallow.' He dragged his fingers through his mussed-up hair and continued thoughtfully. 'Well, obviously she *is* shallow, but she's not *that* shallow, if you get my drift.'

Liv blew out her cheeks. 'Cass, everyone, please ignore Idiot Boy,' she said, more harshly than she'd intended.

'Hey, I'm defending you here,' Duff protested, tucking in then re-tucking his shirt.

'Don't worry,' Jez said, handing her the cushion. 'You can get your own back when it's his turn.'

'She loves me really. She just can't admit it. Can you, Liv?'

'Yeah, sure I do. Totally,' she deadpanned in return, crossing her legs.

Duff slumped back on the sofa, putting an arm around her. 'I've known Liv since she was four years old, Cass. And I can tell you she's got hidden depths. Very secret depths.'

What?

She recoiled, pushing him away. 'What *are* you on about?'

'Take no notice of him,' Hetty said quietly, placing a gentle hand on her knee. 'He's winding you up.'

'You know what I mean. Hidden depths.' He straightened himself, tipping her an overblown wink. 'Or hidden shallows at least.'

Liv paused a beat to steady her voice. 'I have no idea what you're talking about.'

Cass quirked her lip, placing her pen on top of her notebook, and Liv cringed. What she must think of them, this bunch of so-called young adults bickering like kids in a playground? But Cass remained apparently unfazed by the commotion. Positively enthused in fact.

'Great banter. Sparky. Gives a flavour of your personality, Liv. Passion, conviction . . . qualities I love to see. If you can channel that, I think we'll have a fantastic result today.'

Really?

'Thank you.' Liv masked her surprise behind a neutral expression. Tucking a strand of hair behind her ear she wedged the cushion back on the sofa, hoping her posture suggested a professional attitude. And a more camera-loving nose.

'Right then, the next thing is the individual interview, so you guys . . .' Cass's eyes swept across the sofa. 'Can you sit on the floor or find a chair out of shot, please? Liv, can you sit closer to the middle?'

Shoulders back and down. Don't fidget. Liv shifted along the sofa and placed her not-at-all-shaky hands under her thighs while the others relocated. Duff back to sprawl across the rug; Hetty and Jez neatly propped either side of the floor lamp, a pair of human bookends. When Hetty crossed her legs at the ankles, Jez did the same.

'You're certainly entertaining to watch,' Cass murmured, adjusting the laptop. 'Perfect. OK. Liv, remember to keep your head up so the cameras have you the whole time. Ready?'

Deep breath. Swish back hair. Nod.

'Great. So my first question is: there are hundreds of TV shows available, why would watching your life grab the audience?'

Yes!

Over the last three weeks, Liv had absorbed Louis the acting coach's advice: *pitch, pace, intonation, gesture, eye contact*, preparing and polishing every pause, every giggle, every expression for maximum impact. And yes, Dad would probably have a fit if he saw the final bill, but the sessions had been worth every penny. Because along with his years of TV audition experience, Louis had brought a list of possible questions.

And *Why you?* had been right at the top.

Liv's shoulders softened and she set her mouth to automatic, starting the well-rehearsed monologue to rave about the outrageous nights with friends, the exclusive clubs, crazy shopping sprees, glamorous beauty launches, amazing holidays . . .

And yet . . . she couldn't shake off that wink. *Secret depths*, and now, propped up on his elbows on the rug, why the hell was Duff still smirking?

What did he know?

Halfway through gushing over her *insane* clothes allowance ('I'm such a Daddy's girl, he can't say no! It must be more than some people earn in a year!') she faltered before tailing off into an apology.

'Don't worry, you're doing great,' Cass said, pressing

the tips of her fingers together. 'Let's try a different tack. Your beauty channel . . .'

'*Miss Olivia Loves*.'

'Fab name,' Cass commented, jotting on her notepad. '*Miss. Olivia. Loves.* What are you most proud of with that?'

Liv faked a confident smile and twisted her torso slightly to avoid catching Duff's eye again.

'I'm so grateful to the subscribers. Even when it was just me waffling on about a few products I'd bought in Selfridges, the feedback was phenomenal. And then, when brands started putting me on their mailing lists, it snowballed. The audience figures kept rising and so did the positive comments. Then, after the blog was featured in *Grazia* . . .'

Louis' advice echoed: *Don't forget to pause when you name-drop the mag.*

Right on cue, Cass looked up. '*Grazia*? Impressive.'

Liv grinned. 'Yeah, it went mad after that. I got nominated for the National Beauty Blogger Awards; went to the ceremony in London nearly a year ago. It was surreal, you know? Actresses, models . . . all famous people. And I got to meet loads of other bloggers and the goodie bag was to die for. Chanel. Dior. Tom Ford . . .'

'It certainly sounds like a night to remember.'

'It was.'

But thinking about it now, the standout memory wasn't rubbing shoulders with celebrities, or the food, or the pulse-racing thrill of being part of a major industry event. No, it was the moment when she'd looked up to see HIM leaning against the bar, watching her with those blue, blue eyes.

Instant heat had rushed up in a wave from her feet. She'd forced herself to count the foil stars sprinkled on the table. *Don't look.*

When he strolled over and placed a bottle of champagne in front of her, her heart almost tore through her dress. She watched beads of condensation slide down the sides of the glass.

Up on the podium, the award for most life-enhancing lip product was accepted by a company representative whose lively speech contained enough witty one-liners to ramp the atmosphere from cheerful to hysterical.

But Liv barely noticed. Pulling up a spare chair to the table, he'd wedged himself closely between Liv and her neighbour, his thigh deliberately pressed against hers. He still didn't speak. She could hardly breathe.

To loud cheers and applause, the speaker descended

the steps. Some sombre music started and behind the equally sombre presenter, a slideshow lit the screen. The audience hushed and Liv stared straight ahead, barely registering the miserable grey girls who, frankly, had no business being the focus of a beauty industry award.

He was running his fingertip up and down her arm, and even though she continued staring at the montage of misery, her subconscious had hopped in a cab and raced through the busy streets back to the hotel. Only unlike when she'd checked in two hours earlier, she wasn't alone.

She imagined her staccato heels ringing through the marble lobby. She imagined the door to her sumptuous room soundlessly closing behind them. Onstage, the presenter droned on about someone fleeing from somewhere; a single suitcase; fitting into a new country, a new school; struggling to pay for the essentials. Absent parents.

Blah-blah-blah.

Well, tonight there would be no parents getting in Liv's way either.

Tonight there would be no clingy girlfriend expecting him.

Tonight there would be just the two of them. Alone.

She shivered.

Would this gloom-fest never end? Droning on in that uber-worthy tone about the Cinderella Project, the presenter tapped the screen and the same girls appeared, smiling in a selection of tacky prom dresses. The audience cheered. Liv didn't.

Her mind was busily compiling a dramatic montage of its own, complete with bridesmaids' dresses abandoned, cakes left unbaked, cancelled venues, guests apologised to, honeymoon plans forever shelved . . .

And right at the heart of it, him breaking the news to his fiancée: *I'm sorry. It's over. I've fallen in love with a girl named Liv.*

He poured another glass of champagne and leaned so close his breath tickled her neck. 'I shouldn't be here, you know. I really shouldn't be here.'

By the time Sonya Sunshine climbed the steps to a standing ovation, Liv had forgotten about the award.

'And are you single at the moment?'

Lost in the memory of that night, it took a second for Liv to take in Cass's question.

The woman smiled, adding, 'Only you haven't mentioned a boyfriend or a girlfriend or anyone special.'

Liv clasped her hands between her knees. 'No, I'm

not with anyone right now.' *Pause.* 'I mean, there was someone, but we broke up.'

'Someone?' Cass prompted.

On the rug, Duff kicked his legs out straight and flung himself on to his back with a theatrical groan.

'Seriously, Cass, *please* do not bring up the whole Olivia's mystery man thing. We've had a whole year of it and I can't take any more.'

Hetty prodded him. 'Not now, Duff.'

Sparks of panic darted through Liv. *Bring it back to the blog.* She opened her mouth, but the words stalled in her throat.

Duff had shot upright and fixed his attention on Cass.

'Actually, you know what? This would be the perfect time for her to finally tell us who he is. Get her skeletons out.' He mimed air quotes. 'Do the "Big Reveal".'

'Except there isn't anything to "Reveal",' Liv mimed back and pressed her lips in a tight line.

Anyone, any *normal* person, would quit it. Not Duff though.

Instead, he turned down the corners of his mouth. 'I am trying to help. He put you through hell for months, Olivia. Whoever he is — was — I'd love to know. I'd like to, you know.' He punched his open palm. 'Have a chat with him.'

'No!'

'He's right,' Cass said. 'It's far better to get your secrets out in the open now.'

Liv's mind snagged on the s-word. *No. No no no no no*.

'Liv?' Cass's deep brown eyes were boring holes into Liv's head. Her throat, suddenly as dry as sawdust, made a clicking sound and she swallowed.

'Did I mention my blog has a quite a number of male followers?' she said, turning desperately to Cass. 'I've been totally blown away.'

The woman gestured for Liv to look straight ahead. 'A secret boyfriend. Now that's interesting. Tell me more.'

'We should never have got together.'The words took her by surprise, blurting out before she even realised she'd opened her mouth. What was going on? She never talked about this now, not even with Hetty. Full stop, period, on pain of death never.

'Let me guess,' Duff said, jumping to his feet. He ticked a list off on his fingers. 'Your golf coach? Tennis coach? Personal trainer? The guy who cleans your swimming pool? Or maybe it's not a man at all . . .'

He waggled his eyebrows. Liv stayed silent and he made a noise like a talent-show buzzer. '*Nuh-uuuh*. No? OK then, it's –'

'No one,' insisted Liv, too high and panicky even to her own ears. Cass was still staring at her. Oh God. She felt the words bubbling up. She felt the blood drain from her cheeks and battled the overwhelming compulsion to blurt it out.

'It was –'

Ping.

All eyes immediately swivelled in the direction of the sound.

'Hetty Barraclough,' Duff said, putting his hands on his hips. 'Was that your mobile telephone?'

Like the flick of a switch, the urge to confess was gone. Liv felt herself go limp.

'Sorry,' Hetty said, her finger swiping across the screen. 'I only got it yesterday. I haven't worked out how to use it properly yet.'

'*Hetty!*' Liv said, fake-scandalised. 'Cass said no devices.'

'It's Duncan,' Hetty said as though that explained everything.

'So what?' Jez said.

'No problem,' Cass said, smiling. 'Duncan's your boyfriend, isn't he? I think you mentioned him once or twice in your application.'

No man of mystery jibes for Hetty. And knowing her obsession with Duncan, 'once or twice' was probably a polite understatement.

Hetty nodded, eyes glued to the illuminated display. 'He wants to know what time you think we'll be finished.'

Liv's irritation grew less fake. 'Hetty! This is our audition.'

But Cass didn't seem too offended. 'If everything goes according to plan, you should all be on your way around half past one.'

One final flourish of taps and Hetty put the phone on the coffee table. Jez put his glasses back on.

'All done? Let's try something else,' Cass said, turning her attention back to Liv. 'How about if one day, *bang*,' she snapped her fingers, 'everything vanished. The houses, the shopping, the money . . . everything. What would you do?'

Liv's heartbeat steadied. No more talk of secrets. OK, she hadn't rehearsed this exact one, but Louis had warned her they'd throw a few curveballs her way.

'That's a tough one! Erm, I suppose we'd cope. People do. I mean, we've got friends and family. But I really can't see it happening. My dad's far too careful for that.'

'And has your family always been well-off?'

'As far as I can remember,' Liv said.

Cass unbent a folded corner and ran her finger down the page. 'Interesting. My notes say Frank Dawson . . . left school at fifteen. Let's see . . . went from running the family fruit and veg stall to owning the UK's largest agribusiness. Is that right?'

Liv held in a sigh. They'd got the riches, why did people insist on bringing up the rags? Seriously. Why was that humble beginnings stuff always such a big deal?

She forced her smile a little wider. 'Yes, I'm very proud of him. But my mum's family, the Hills. All this . . .' She waved her arm towards the window. 'The farm, the stables, the land . . . it's been in Mum's family for years.'

'Interesting.' Cass pursed her lips and scribbled something down. Turning her attention away from Liv, she said, 'Your turn, guys. You know Liv better than anyone, right? How do you think she'd change without money?'

The other three looked at each other.

'I'm sure she'd be OK,' Hetty said slowly. 'You know, it'd be different from what she's used to, but she'd adapt. Get a job; she's got a lot of talents. I'm sure she'd be fine.'

Duff rubbed his eyes with the heel of his hand. 'I can't really see that. Sorry, but I think you'd fall apart in the real world.'

'For your information, I do live in the real world.'

Duff laughed, 'I love this girl! You think this –' he waved vaguely round the room 'is the real world?'

'And how are we defining the "real world"?' Cass said.

Jez put his dad's voice on. 'Debt. Wondering how you'll pay the bills or buy food. Lack of opportunities. Poor housing. Being stuck in a cycle of –'

Duff clicked his tongue in disagreement. 'Nah. If you're poor, you go get a job and work at making money, same as everyone else. You can't expect it on a plate. That's the real world. Life's what you make it, right?'

'No!' Jez's head snapped up. 'That's a complete myth. Poverty is –'

Cass held up a hand for silence, releasing a fresh wave of familiar summer perfume.

'Guys, this is fascinating stuff, but time is not on our

side. We can debate it later. Liv, remember to keep in camera shot, please.'

Realising she had slumped deep in the sofa, Liv pulled her shoulders back and lifted her chin.

'As I said at the start, you need to be prepared for some difficult questions.' Cass put her hand to her chest. 'Devil's advocate here. You don't need me to tell you there are plenty of small-minded people ready to spill their vitriol across social media. Imagine I'm the nastier side of the internet calling you out for being selfish or a spoilt rich kid, what would you say?'

Liv chewed her lip. Of course, *Miss Olivia Loves* attracted random nasties; that went with the territory. But her subscribers usually slayed the trolls before she even read their comments. And, not that she'd admit this to anyone, least of all Cass, she actually enjoyed watching these strangers froth at the mouth on her behalf.

'I'd say it's not true,' she replied carefully. 'I mean, I know I'm lucky, but it doesn't mean I don't care about other people. I'd tell them all the things I've done for charity, like donating stuff. Clothes and things.'

'Speaking for myself,' Jez chimed in. 'I am committed to using my good fortune to help others and I'm sure Liv feels the same.'

Duff rolled his eyes so hard they were in danger of getting stuck round the back of his head and Liv suppressed a smile.

'That's right,' Hetty added. 'And we did fundraising at the prom, didn't we? That was your idea.'

'Yeah.' Liv warmed to the theme now. 'And I gave a genuine Hermès handbag for the raffle at Jez's Valentine Auction.'

Ouch. She plastered on a happy smile and rode the pang of regret that struck whenever she thought about that beautiful bag. Jez had whipped her into a state without giving her the opportunity to reconsider. And now her Hermès bag was gone forever, a victim of emotional blackmail.

'And the fashion show last summer,' Jez was saying in Cass's direction. 'I set up a charitable foundation with my parents, Connecting Together. We empower young people who find themselves in challenging circumstances, providing financial assistance as well as —'

'I remember from your application,' Cass cut in. She'd taken her phone out of the bag and was swiping at the screen. 'Most of the applicants only talked about raising their own profile, but yours really stood out.

It's so . . . refreshing. One of the reasons we wanted to see more of you.'

Jez shrugged, aiming for nonchalance, but Liv could practically *see* his head expanding.

'If we get accusations about not caring about people less fortunate, then we have evidence to prove that's not the case.'

'Well, I'm very pleased to hear that,' Cass said, 'and I'm sure it'll be really interesting for the audience, challenging their perceptions of privileged young people. But like I said, you need to be prepared for some difficult moments. And on that note, Liv, I've got a photo I'd like to show you.'

Photo?

The panic returned. Bigger, harder and sharpened into actual terror.

Blood whooshed in Liv's ears. How could Cass have the photo?

'Photo?' she said, only the slightest catch betraying her emotions.

She reached out a shaking hand, but Cass pulled the phone out of reach, angling the screen in her eyeline.

Almost sick with dread, Liv obeyed.

At first she couldn't quite tell what it was, but that

really didn't matter because it *wasn't* the photo she'd taken almost a year ago. The photo he'd persuaded her to send him. The photo he'd promised he'd deleted. The photo he'd used to buy her silence.

She breathed out slowly, dizzy with relief.

Liv couldn't quite fathom what she actually *was* seeing. The dim image showed a hump of something fabric. Bedclothes?

'There's a girl, lying down . . .' she said.

'And?'

'She's got her back to the camera, so I can't see much. She's got long dark hair. She's wearing a pale dressing gown with stars on it. Looks skinny, really skinny. Erm . . . room's quite dark. Messy. It's really dark.'

Cass flicked her thumb and index finger at the screen. 'Let's zoom in.'

'Um,' Liv said. Her eyes narrowed, trying to focus, then widened. 'Oh.'

'What is it?' said Hetty.

'The girl, I think . . .'

'What?' Jez said.

Liv peered at the screen. The clothes piled on a chair. The red lights of a radio alarm clock. The hump on the bed. The dark shape of the head with its long, long tail of hair. Something tugged at the back of her mind and then was gone.

'Do we know her?' said Duff.

Liv shook her head. 'It's not that. Just . . . Cass, is she *dead*?'

⏵ 3 LIV

1 July, 10.53 a.m.

'Dead?' Duff was getting to his feet. 'Let me see.'

But Cass had already blanked the screen and returned to the laptop. 'What makes you say that?'

Obviously, Liv had never seen an actual dead body, but still there was something wrong, something *lifeless* about the way the girl lay there. Not a definite Grim-Reaper-by-the-bed thing, just . . . a something.

Liv picked up a cushion, the velvet soft against her arms. She rubbed at the pile with her fingertips.

'Not sure, the way she was lying maybe. And there were empty packets next to her. Paracetamol, stuff like that. I think.'

Duff lifted his hand up to his mouth, speaking from behind it in a pantomime whisper. 'It's fake.

Something to see how you react to the unexpected.'

He folded his arms, looking at Cass.

'Who is she? The girl in the photo?'

And then the weirdest thing happened. The skin on the back of Liv's neck tightened. Either the room had suddenly got very cold or her own temperature had dropped. And she remembered her gran's saying when she shivered, that somebody had just walked over her grave. The hairs on Liv's arms stood up.

And because she was actually staring at Cass, she definitely *knew* it could not be her because she was answering Duff, saying, 'I'm going to come on to that in a minute.'

But a soft puff of breath whispered against her ear.

Don't you remember her?

'Are you OK?' Cass asked from across the room.

No, I'm hearing voices, Liv wanted to say. *Either this is a masterclass in freaky mind-games or I am in the early stages of a mental breakdown.*

But instead she visualised her therapist. Heard the soothing Welsh lilt that always made her think of sinking into a warm bath. *Imagine a string, like the wick of a candle, emerging from the top of your skull. Now imagine a hand gently pulling the string, lifting your head.*

Gently, gently. And straighten. And elongate. And —

'Liv?' Hetty said softly.

Conscious of everyone watching her, Liv took a deep breath. 'Sorry. Just the photo, it was shock. Who is she? Was she?'

'I know it's a distressing image. The girl in the photo was called Evangelina Bakharina. She came to the UK to work.'

Duff snorted. 'Hey, new slogan for your mum, Hetty. *They come over here, steal our jobs, kill themselves in our bedsits.*'

'Not appropriate,' Jez said.

Duff rolled over on the rug and winked at Hetty. 'You know that's a joke, right?'

'Except "jokes" are traditionally humorous,' Jez said.

Duff made a *pah* noise, sat up and circled his bent knees with one arm. 'Whatever, bro. Your researchers must have told you about the Britannia Party, Cass? Hetty's mum's the leader. If you can't trace your family back to the Vikings . . .' He drew a finger across his throat. 'Game over.'

Jez rolled his eyes. 'The Vikings weren't even British. *Bro.*'

'That's not true. About my mum, I mean. She's not

61

anti anybody,' Hetty said in Cass's direction. 'Jez is right about the Vikings though. They were from –'

'Yes, I do know about your mum. It's all in here.' Cass patted the notebook, then flicked through it. 'I'm sure we'll get on to that later, but at the moment I want to focus on Evangelina. Who, coincidentally, shares a birthday with Liv. Same date, same year.'

Liv gave an uncertain smile.

What did this have to do with anything?

'So Evangelina was happy at first, even though the work was hard and she put in long hours –'

'Ah, the sex trade,' Duff contributed knowingly. 'You should see them in Amsterdam. We went for my brother's stag weekend. There's hundreds of them. You knock on the window and they just pull a curtain across and –'

'Evangelina wasn't involved in anything like that,' Cass interrupted. 'She was working legally. The pay was the problem. You see, she'd come to the UK on the promise of decent wages, but that never materialised.'

'That's a shame,' Liv said automatically.

Was this girl one of her followers or something?

'How much was this girl getting paid?' Jez asked.

Cass pulled a face. 'A little under ten pounds a week.'

Duff paused from tracing his finger round the geometric patterns in the rug and looked up. 'Ten quid a week?'

'Terrible,' Hetty said. 'That's slave wages. And just so you know, Duff, that's an issue my mum actually campaigns about.'

Liv adopted a suitably sympathetic expression, then cast a sideways glance at the Pandora bag. *Beautiful.* Come to think of it, it was actually pretty similar to the Hermès Jez conned out of her. Did anyone bid on it at the auction? There was a slim chance he still had it somewhere. When Cass left, she'd ask.

'Isn't there a minimum wage?' Hetty said. 'Why didn't she go to the police?'

'Or take herself off to Amsterdam. At least she'd get paid properly.'

'Not funny,' Jez said.

'Hey.' Duff held his hands out, palms up. 'Don't blame me; blame Hetty's mum. She's the one making life hard for them.'

'That is so not true,' Hetty said hotly.

'Give it a rest, will you, Duff?' Liv said.

And if Jez said he didn't have it, well. *Friends in high*

places — wasn't that what Cass had said? If today went OK, maybe she would put a word in, ask them to bump Liv up the list.

She felt her cheek muscles give a bored twitch. *Where exactly is this going?*

Cass clapped her hands together. 'I'm sure you're wondering where this is going.'

Liv's expression immediately switched to *Me? No, of course not.*

'Well, Evangelina went to speak to someone to sort the situation out. Not just for herself, but for the others too. She put her coat on and headed across the muddy fields where she'd been working fifty hours a week towards the farm owner's beautiful house.'

Fields? A warning light flashed around the edges of Liv's memory.

'It wasn't dark, but it was raining,' Cass continued. 'Torrential summer rain. Evangelina slipped on her way over. Got covered in mud.'

Unease prickled up Liv's spine. A hazy memory was taking shape. *Oh God.*

'She worked on a farm?' Jez said.

Cass nodded. 'That's right. Until a year ago anyway. Do you remember now, Liv?'

Her heart thudded; she heard herself say, 'Sorry. I don't.'

The TV screen flickered to life: *One of the best nights of my life. I'll remember it as long as I live.*

Cass placed the remote control on top of her notebook. Carefully straightened it without a word.

'Are you saying she worked *here*?' Hetty said.

'Your dad pays his workers ten pounds a week?' Jez said simultaneously.

Duff laughed. 'No wonder you're rolling in it.'

'No!' Liv said loudly. 'There's no way Dad would let that happen. He's really careful about stuff like that.'

'But you do know something about it, don't you, Liv?' Cass insisted. 'Because Evangelina came to you for help.'

Oh.

The memory solidified. The doorbell. Muddy puddles in the courtyard. A stranger's shadowy face.

'So you did know her,' Duff said, pulling himself to a sitting position.

'Liv?' Hetty said, furrowing her forehead. 'Is that true?'

'It wasn't like that!' she protested, her voice thin and whiny.

'Then what was it like?' Cass said. 'Tell us.'

⏪

⊙ 4 LIV

Twelve months earlier

Sorry, have to cancel. Youngest fell out of a tree. In A&E.
Maddy x

Liv stared at the text message in disbelief.

Fell out of a tree?

Her first instinct was to ring and ask, no, *demand* Maddy honour her commitment. But what was the point? She'd only do a rush job to spite her.

She had to take a deep breath before she could trust herself to reply.

That's terrible. Lx

And it was terrible. Truly, cataclysmically, disaster-level terrible. Because how the hell was Liv supposed to find a replacement make-up artist at such short notice?

With a growl of frustration, she threw first her

phone and then herself on the unmade bed. Three hours until the London train and she was nowhere near ready.

She rolled over, careful not to disturb the suit bag. Slid out the dress with a shivery wave of anticipation. The sea-green beaded vision was so pretty she could cry. In the end, she'd lost count of how many dresses she'd tried on before she found this one. *The* one.

She let herself picture his face when he saw her wearing it. His eyes. His smile.

Of course, he hadn't promised he'd be there. And if his fiancée decided to stick her talons in, he wouldn't be able to get away. But he said he'd try and that was enough.

Turning her attention to the array of bottles and jars on her dressing table, she padded across the carpet. One flick of a switch and the bulbs around the mirror instantly lit her face. She peered in close, angling her cheek this way and that. Tilted her chin up and down. Not like she'd never done her own make-up before, was it?

'Mirror, mirror on the wall,' she murmured, watching her reflection select a sable brush. 'We totally got this.'

Forty, fifty minutes of deep concentrated blending, smudging, patting passed before her dad's voice drifted in from the landing.

'Are you in your room, love?'

'Come in.'

She spritzed the setting mist in the air and closed her eyes, feeling the fine spray settle on her skin.

He poked his head round the door. 'You look beautiful.'

'Really?' She gestured down at her velour tracksuit.

'Really,' Dad replied firmly.

And that was what was so great about Dad. He always knew exactly the right thing to say. Mum would have frowned and made some helpful-not-helpful comment like, 'Have you finished your homework?' or 'Shouldn't you be revising?' or 'What is this thing you're going to again? Beauty what?'

'I'm thinking ahead, about the traffic to the station,' Dad continued. 'We should probably get going soon, just to be on the safe side.'

'I'll be ready,' Liv said. Was the lipstick too matte? She picked up a tube of clear gloss.

'Now are you absolutely sure you'll be all right with the hotel and everything?'

'It's all sorted,' she said uncapping the tube.

Dad rubbed his chin. 'You know, maybe I should come with you. Or your mother could.'

Liv's hand froze halfway to her lips. 'Honestly, there's no need. I'll be fine.'

'I'm sure we'd be able to get another room,' he said, walking over to the window.

She watched her reflection dab on the gloss, panic in her eyes. 'Dad, seriously. I can take care of myself.'

After what felt like hours, he sighed. 'All right, I'll stay put. But ring me when you get there. And don't even think about the Tube anywhere, you get cabs. Proper black cabs, not one of those off the street.'

He pulled the curtain to one side. The summer sky was dark and dull, the first fat raindrops spattering against the glass.

'Rain,' he said. 'I'm going to the office, but if you buzz across when you're ready I'll come back. Save you getting wet.'

'I won't be long.' She reached up to give him a glossy kiss on the cheek. His hair had been white ever since Liv could remember, but there was less of it now. Wisps like cotton wool floated around his ears. When he bent to return the kiss, he put a hand to his hip, wincing.

Liv felt a sharp stab of guilt. 'Are you sure you're OK driving to the station? I could get a taxi.'

He shook his head. 'I'm fine, love. Walked up to the

top field today, forgot how far it was, that's all. Anyway, you'll never get a taxi to come out in time.'

True enough.

Liv had inherited her thick dark hair and long legs from her mum and her green eyes from her dad, but somehow their devotion to the land had bypassed her entirely. God, she hated living in the middle of nowhere. Life here was just so . . . primitive.

As soon as he'd gone, Liv shucked off her onesie, hooked her foot around it and kicked it in the direction of the laundry basket, where it lay in a heap, like fleecy road kill.

Slipping the dress on, being careful not to smudge her face or mess up her hair, excitement bubbled through her. She stepped in front of the triptych of mirrors in the dressing room, hardly daring to ask herself the Big Question.

Will he want me?

Liv set the filter to vintage, enveloping herself in a flattering golden haze and sucked in her cheeks for a selfie. Not that she needed to: half an hour with the contouring palette could carve cheekbones on a football.

#beyourownmakeupartist #ownyourlook

She clicked to upload to *Miss Olivia Loves* and, sure

enough, within nanoseconds the *ping ping ping* of likes began.

Hopefully Maddy would see it. Potential blog post there: outdo the make-up artist.

The phone vibrated in her hand, giving her a start. When she opened on the message, she bit her lip to suppress a grin. He'd seen it!

You look gorgeous. Champagne for two. Tonight. xxx

She sat down heavily on the bed and hugged her phone to her chest for a second, and thought back to the day it began.

May half-term. Saturday. Liv had been at the Country Club with Hetty, idly scrolling through her newsfeed. *North-west England is hotter than Spain, hotter than Australia, hotter than the sun*. Every second post was a throbbing thermometer screaming *Hottest May on record!*

On the lounger, Hetty combed her fingers through her limp hair.

'I'm melting,' she said as she shuffled her legs up into the shade.

Liv peered over the top of her sunglasses.

'I don't get why you're wearing *leggings*. It's twenty-seven degrees.'

'Just am.' Hetty shrugged and blew damp strands

of hair away from her pink cheeks.

Liv didn't press the point. Just getting Hetty out of the house since her dad died was a major triumph. She hoicked the straps of her bikini top back on her shoulders and sat up, perching her sunglasses on top of her head.

'I could do with a drink. Shall I go to the bar?'

'Yeah, please,' replied Hetty, fanning herself with her hat. 'Coke, diet. Loads of ice. Get them to put it on Mum's tab.'

Liv pulled the white sundress over her head; slid her feet into sparkly flip-flops and picked up her phone.

'Back in a minute.'

The Country Club was totally heaving. Every lounger occupied, every inch of the pool crammed. Walking the length of poolside, she was aware of appraising eyes. Stomach in, shoulders back. She kept her head up and focused on the striped awning of the bar terrace.

And the man sitting under it.

She stopped abruptly, one flip-flop caught between two slats of decking. Pain spiked from her toe to the top of her skull and her eyes blurred with tears. Gritting her teeth, she discreetly flexed her foot. But he was too busy talking to the waitress to notice, thank God.

Like half the girls she knew, she'd had a crush on

him, like, *forever*. And she'd heard the rumours he'd come back and now, here he was.

Skirting the edge of the pool she ducked in through the side door. This wasn't a boy. Boys she could handle. *This* one, slouched on a wooden chair, stirring sugar into a coffee, was a man. No, not even just a man. He was a *myth*.

His phone shrilled as she stood at the empty bar, listening to the low murmur of conversation from his direction. Getting the drinks took an age. Yet another waitress struggling to master basic English. First there was no lemon and she had to trot off to the kitchen, then the ice bucket needed filling. Liv breathed in, leaning oh so casually on the bar, twiddling a straw and playing with her mobile. And during that whole time her skin tingled, hyper aware he was watching her.

'Just stick it on Mrs Barraclough's bill,' she said. The girl stared uncomprehendingly. Liv sighed impatiently through pursed lips.

'B-A-R-R-A—'

'Let me.' A hand appeared, holding a ten-pound note. 'Keep the change.'

Her elbow caught the glasses, rattling the ice as she turned, meeting blue eyes, bluer than seemed possible. A flicker lit them, like turning up a gas flame. A flicker that hinted at not altogether innocent thoughts.

'Hello, Liv. Long time no see. You're looking stunning, if I may say so.'

His flattery warmed her skin like the sun.

To any listener it would have seemed the most casual of conversations. Yes, school was going fine. Yes, one more year. No, she couldn't believe how hot it was. Yes, his girlfriend was around somewhere, probably gone for a dip in the pool. Yes, she might be interested in some part-time promotional work.

Yes, of course he could have her number.

Then, after he'd added it, he touched a finger first to his own lips, then to hers, widening his deep blue eyes. *Don't breathe a word*, they said.

As she rounded the corner, about to step back into the bright sunshine, she knew that if she turned around she would catch him watching her.

She turned.

He was.

When she handed a rather warm Coke to a rather annoyed Hetty, something inside her had melted along with the ice.

'What are you so happy about?'

Hetty had emerged from under the shade of

her hat and was observing Liv curiously.

'Nothing,' she lied, throwing her phone and sunglasses on the lounger. 'I'm off for a swim.'

The cold shock of the pool didn't halt the secret shivers of excitement. Neither did watching his fiancée (aka 'the BBC's fastest-rising star') execute a faultless dive. Liv pulled herself up on to the tiled edge, dangling her legs in the water and feeling the reflected sunlight play over her skin.

Her lips still tingled from his touch.

After that, nothing. No call, no message, no fancy-meeting-you-here chance encounters at the club (even though she practically grafted herself to the poolside for the whole of half-term). Nothing for nearly two weeks.

She'd just – *just* – given up when Number Withheld flashed up on the screen.

'Hello, Liv.'

The familiar voice turned her bones to liquid. She sat down on the end of her bed with a thump.

'Friday night, eight o'clock. You free?'

She wasn't; she had arranged to go the cinema with Hetty. But she heard herself answering, 'Sure.'

'Great,' he said. 'And, Liv, let's make this is our little secret, yeah?'

1 July, 11.03

'You were at home.' Cass's voice pinged Liv back to the present. 'The day of the awards ceremony in London. Can you tell us what happened, please?'

Liv paused.

She knew what she *wanted* to say: *How is this relevant and actually how do you even know because I'd completely forgotten about it until right now?* Or maybe: *What the hell has any of this got to do with the audition?*

'Look at the camera,' Cass reminded her.

Obediently lifting her head, Liv kept her eye on the red light straight ahead and slipped smoothly into her monologue.

'OK,' she began brightly. 'Being nominated for an award for my blog was such an incredible honour that . . .'

Cass cut in with a sympathetic smile. 'Maybe we could just fast-forward to the doorbell?'

For God's sake! Liv stretched her smile for the camera.

'Sure! Let me see . . . OK, so it was my old room, at Mum and Dad's next door. I looked down into the courtyard and someone was standing there.'

'Someone you recognised?'

She shrugged, fiddled with the belt on her jumpsuit. 'I don't have anything to do with the field staff. Dad deals with all that and he's got managers and things.'

'And your dad wasn't home?'

Liv shook her head. 'And Mum was up at the stables or the paddock or somewhere. I tried to ignore the bell, but the girl kept ringing and the dogs were going mental so I went downstairs to get rid of her.'

Details came back now. Flinging the door open without even a *What do you want?* Let alone a *Can I help you?* The reek of earth and rotting straw invading the jasmine-fragranced hall. Actually standing *inside* the porch – not outside on the step – was a mud-splattered figure in a vivid red parka. A young woman with a hood veiling her hesitant smile. Raindrops glistened on the fake fur trim.

'Hello. Please may I speak with Mr Dawson?'

She spoke precisely. Not swallowing the final vowel the way locals did, but pronouncing each separate syllable carefully. *Door. Sun.*

Liv wrinkled her nose and lifted her foot. The buckle on her metallic strappy sandal had rubbed an unsightly red mark on her ankle bone.

'He's not here. Come back tomorrow.'

But instead of leaving, the cheeky bitch had placed one mud-caked boot on the threshold. Small crumbs of soil dropped perilously near Liv's pedicured feet.

'Please, it is urgent.'

Sitting under the gaze of the cameras, Liv relived her revulsion. The girl leaning in to grip the door frame. Those fingernails embedded with dirt; the smell of cold mud emanating from her cheap coat; the rain pouring behind her and the frenzied riot of barking.

Anger had boiled through her. What if Dad appeared now? What if she missed the train?

Liv had tugged on the door handle. 'I told you, come back tomorrow.'

'I can't go to the office. This is . . . private.'

The hood tipped slightly. Liv glimpsed sharp cheekbones framed by long dark hair. Striking eyes, as green and slanted as a cat's.

'Please,' the girl said with such desperation that Liv loosened her grip on the door handle. Maybe if Dad promised to deal with it when he got back from the station . . .

She was about to pick up the house phone when a spray of rain gusted into the hall and Liv recoiled as though the girl had thrown acid on her. She stared in horror at the scattering of damp spots staining her dress.

'Look what you've done! Do you have any idea how much this cost?' She shook the silk material.

'No, of course you don't. God, it's *ruined*.'

'I am sorry. Wait —' The girl ferreted round in her pocket and drew out a disgusting rag of tissue and *actually went to touch Liv with it.*

Liv exploded. 'Get off me! What the hell are you playing at?'

She jerked the door, hard, slamming it against the girl's foot. Liv saw pain flash across her face despite the thick rubber boots, but she didn't leave. Instead she wedged her elbow against the frame.

'I am sorry to disturb you, but please tell your father I urgently need to speak to him about Pavel . . .'

'Pavel?' Cass interrupted sharply.

Liv frowned. 'I'm pretty sure that's what she said.'

'OK. And then . . .?'

'It was getting late,' Liv said, hating the whine creeping into her voice. 'I'd been looking forward to this night for months. If my dad got involved in some big farmer drama I knew we wouldn't make the train and I'd miss the whole thing.'

'Do you think your dad would have *wanted* to help her?'

Liv nodded, stroking her thumb down the velvet cushion. Of course he would. For all the tough businessman act, he had the softest heart imaginable.

'Did you give him the chance?'

There was a pause.

'For the camera, please,' Cass said. She'd picked up her pen and was making notes again. Notes that Liv knew wouldn't be complimentary.

You've blown it, you've blown it, her mind chanted.

'I told her to come back the next day; it would have been fine,' she said.

'But she didn't come back and it wasn't fine.' Cass pointed her pen at the mobile phone. At the photo of the girl who may or may not have been dead. 'You're slamming the door on her, you're about to go out. What happens next?'

Liv fidgeted with her belt, kept her eyes down. 'I told her if she didn't leave, I'd set the dogs on her.'

Hetty exhaled softly, but didn't speak. Liv focused on the wall and in particular a long stringy cobweb that revived her annoyance with useless Martina/Marina.

Maria?

'Look, I was running late. The farm workers aren't allowed up to the house. She shouldn't have been there in the first place.'

'And did your dad ever sort it out?' Jez interrupted. 'Or does he still pay them ten quid a week?'

Liv rolled her eyes so hard they actually hurt. 'Of

course not. It must have been a . . . misunderstanding.'

'Did you set the dogs on her?' Hetty said.

'How can you even ask me that? I just said it to make her go away.' She folded her arms, defensive now. 'Don't look at me like that. I only did what the rest of you would.'

Duff shrugged. 'Maybe I'd have listened to what she wanted.'

Liv coughed out a laugh. 'Oh, right. Because you're such an expert at listening to girls, aren't you? I don't get why we're even talking about this. It's got absolutely nothing to do with *This Careless Life*.'

'It has. It reveals a lot about how you cope under pressure,' Cass answered.

You've blown it, you've blown it.

Liv's heart pounded the refrain so loudly she half expected the others to ask what the noise was.

'And the good news is . . .' Cass continued, lightly clapping, 'that was superb.'

Something was wrong with Liv's brain. More specifically its ability to process information.

Superb?

'Rule number one of telly,' Cass continued. 'Entertain, entertain and entertain. And that was electrifying. Edge-of-the-seat star performance. Well done.' A teasing note crept into her voice. 'I told you it was going to be tough, didn't I?'

Liv leaned back in the sofa trying to absorb what was happening.

'The photo on your phone – that's not real, is it?' Hetty said eventually. 'I mean, no one's dead.'

But Cass only shrugged and picked up her pen. 'Evangelina? Does it matter? Liv did brilliantly.'

Duff tipped his head back to look at Cass. 'It's just some shock pic off Google images. Right?'

'I'm saying nothing,' she answered, a half-smile on her lips as she bent over the keyboard. 'Except that was fantastic and we need to get a move on.'

'I don't get it.' Jez clasped his hands behind his head. 'How could you have found all this out?'

'Researchers,' Cass said, not lifting her eyes from the laptop. 'I told you they were thorough. Anyway, we really do need to hurry up.'

'Maybe I should have heard her out,' Liv said quietly.

Hetty smiled sympathetically. 'There's no way your dad would rip people off. Your dad's lovely; he'd never do that.'

'Evangelina's probably back home in Whatever-istan

right this minute, eating cabbage and drinking vodka or whatever they do for fun,' Duff said.

Liv half-smiled. 'Dad's always moaning that the field staff don't stay.'

'Exactly.' Duff grinned. 'Look, you can't beat yourself up over something you never knew about, can you?'

'That's quite philosophical,' Jez said, brushing a speck of fluff from his trousers. 'For you, anyway.'

'What can I say?' Duff replied mildly. 'Maybe Liv's not the only one with hidden depths.'

'Actually,' Cass said, 'she did OK after she left the farm that night. Got another job pretty much straightaway.'

'So that can't be her in the photo then,' Duff said. 'See, I told you.'

The TV screen blackened then flicked to life. Cass peered down at her notebook, tapping the remote control against her thigh.

'OK, let me see, who's next? Ah yes, Hetty. Can we . . .?'

She waggled two fingers, indicating for Liv and Hetty to swap places. As Hetty went to take Liv's place on the sofa, she lightly patted her arm.

'Don't feel bad,' she whispered. 'It was just a mistake. Everyone makes mistakes.'

⊙ 5 Hetty

1 July, 11.22 a.m.

'Good to go?' Cass's finger hovered over the play button.

Give it all you've got, Liv had said three weeks ago. *We need our application to stand out from the crowd.* Hetty's gaze flitted up to the paused image on Liv's enormous TV: goggle-eyed, open-mouthed. And why *the hell* that vest top? A blurry pink wave of underarm flab bulged from the armhole like an uncooked chicken breast. Jesus. She looked like a children's TV presenter on hallucinogenics.

From the rug beneath the screen, Duff threw her a reassuring wink. Jez stretched his legs out, his chest on a level with the sideboard camera, and flicked her a double thumbs up, low down by his hip where no one else could see.

Cool as ever, in the sleeveless silk jumpsuit that

hugged her lithe curves, Liv steadied herself with a perfectly pointed toe as she slid on to a stool under the breakfast bar.

'It's OK, Hets,' she murmured.

OK? God, she must have been insane to agree to this. Balling the cuffs of her sweatshirt into her fists, Hetty choked down a nervous giggle. *Visualise success.* That's what they told them at school. She tried to imagine herself nailing the audition, as laid-back and confident as Liv.

From the corner of her eye, Hetty registered the breeze nudging the door ajar, like an invitation. *Never mind the killer dogs*, it whispered. *Run away! Run away now!*

But in two quick strides, Cass was across the room and the latch clanked down.

'Ready?' she said, waggling the remote control towards the TV.

'Absolutely,' Hetty said, curling her lips in a smile.

Henrietta Barraclough Pretty Vacant Productions #2

[*Waving.*] Hi, my name is Hetty Barraclough and I'm auditioning for *This Careless Life* along with my friends Jez, Liv and Duff. We all go to the same school, St Benedict's [*claps hand to mouth, giggles*]. Except we don't. Wow, still feels weird to think we're never going back. Oh well, onwards and upwards, I suppose [*shrugs*].

I have to describe myself in three words. Gosh. Not sure I can do that – three sentences maybe [*laughs*]. Erm, OK, crazy, obviously [*rolls eyes, waggles tongue*], at least, that's what my friends would say. What else? Oh yeah, good friend. Seriously. You can trust me with your darkest secret and I will never, ever breathe a word. And last one . . . kind-hearted. Or too soft as my boyfriend Duncan would say. I can't even watch the news without crying, which is not the best thing [*exaggeratedly turns down corners of mouth*] when your entire family is in politics. [*Puts on serious expression.*] OK, I may as well get it out there now: my mum is Rose Barraclough MP. That's right, from the Britannia Party [*laughs, flaps at the screen*].

Stop! Before you switch this off, love her or hate her, she certainly gets people talking. And the papers.

So what can I say about being the daughter of Britain's Lady Marmite? [*Wags finger.*] That's the name of her newspaper column, by the way. They can't get enough of her.

Well, we've got quite . . . how can I phrase it? . . . *different* personalities. I'm much more like my dad, but I'm very proud of what she's achieved. And Duncan, my boyfriend, works in her office, which is only weird occasionally [*laughs*].

We've been together for over eight months. We have our ups and downs, sure, like anyone does, but he's the love of my life. [*Claps hand over mouth.*] He'll probably kill me for saying that. But I'd do anything for him. Actually, forget kind-hearted. Hopelessly devoted, that's me.

Cass pressed pause.

'Remember the camera needs to see your face,' she said gently. 'And I know it's easy for me to say, but try to relax.'

Hetty obediently pushed her spine against the sofa; crossed then uncrossed her legs. The effort of relaxing made her muscles spasm.

'Sorry about that,' she said.

Cass waved her notebook dismissively.

'Don't apologise. I love your audition clip. You've got a fantastic down-to-earth likeability. Now, Hetty's . . . questions . . . questions . . .'

Smoothing invisible creases from her white linen dress with one hand, Cass consulted her tatty notebook, flicking back and forth through pages filled with indecipherable writing.

Hetty couldn't imagine wearing linen. Or white for that matter. Sweat tickled under her arms and she immediately clamped her elbows to her side. What had possessed her to wear a sweatshirt? She was melting. Literally.

'Here we go.' Cass ran her thumb down an open page. 'Right, first question. In the clip, we saw you talk about your boyfriend Duncan. He's clearly a huge part of your life, so what are his thoughts about you applying for the show? I mean, hopefully he'll be happy to be part of it too?'

'He thinks it's a brilliant opportunity. For both of us,' she said confidently.

'And your parents? What do they think?'

Without warning, a lump rose in Hetty's throat. 'Parent. Dad passed away last Easter.' Since the accident, she'd lost count of the number of well-meaning relatives and family friends who patted her arm and assured her 'time heals'.

It didn't. If the last year had taught her anything it was that grief hid everywhere. In a TV theme tune, the shape of a stranger's head, a waft of aftershave. Even the letter 's' tacked on the end of an everyday word. One minute, she could be almost nearly fine, the next ambushed by misery.

She exhaled slowly, exactly as the grief counsellor had taught her.

'Of course, that's in my notes. I'm so sorry.'

Hetty watched Cass's features adopt the sympathy position. Tight lips pulled halfway between smile and frown? Check. Forehead crease? Check. Slight head tilt? Check.

Starting with the police officer turning her hat in her hands like a steering wheel, Hetty had watched this expression drop like a mask on many, many faces since the accident. From an anthropological perspective, the universality was quite interesting really. Irrespective of age, gender, ethnicity . . . The look never wavered and neither did Hetty's response.

Don't make eye contact. Don't elaborate.

She focused hard on her stubby nails, bitten so far down they made Duncan wince.

Hetty intensely despised those people who deliberately ignored the subject-closed vibes, disguising their curiosity as concern. *A lorry driver sending a text in the fast lane? Dad died on impact, you say? No chance to say goodbye? Poor you.*

Thank God Cass wasn't one of *them*. Instead she continued briskly, 'I know your mum is the local MP here. Did she mention any worries about you being in the limelight?'

Thank you. Hetty lifted her head.

'My mum's as thrilled as I am,' she said.

Which was kind of true, once Duncan had called a crisis meeting and reframed it as 'a pre-election campaigning opportunity' and 'a stage to mould public perceptions' and 'the ideal platform to reach out to a new demographic'.

'OK, but if this backfires . . .' Mum had said, clearly envisioning scenarios too horrific to verbalise. 'Just don't mess up, Hetty. Please. This is a big year for me.'

And riding high on that vote of maternal confidence, Hetty had recorded her audition clip.

Cass looked down at her notes. 'In your video you described yourself as "hopelessly devoted". I know Duncan works with your mum now, but you met him at the gym last summer, yes?'

'Sort of. It was the Manor Country Club. He worked there for a bit after uni.'

'So would you say it was love at first sight?'

Hetty laughed. 'Well, for me, definitely. Lightning bolts, arrow to the heart, fireworks . . . the works.'

'And for him?'

She peeked over at Liv who was using her tiptoes to rotate herself on the stool. Her jumpsuit made little *swish swish* noises with each turn. On the floor by the lamp, Jez fiddled with the lace on his trainer, looping it around his finger then letting it go.

'I think it took Duncan a little longer,' Hetty said.

Twelve months earlier

Hetty Barraclough didn't do sport.

That was the thing she hated about St Benedict's . . . or rather *one* of the things she hated: the ridiculous insistence that participation in sport was in some way character-building. Apparently, the potentially bearable ones like swimming or Zumba, where the only person you competed against was your own un-sporty self, didn't count. The upshot was St Benedict's infatuation with team games and their numerous attendant humiliations (the kit, the cold, the pain, the perfect blend of patronising and nasty practised by team captains) only managed to nail a big old RIP to Hetty's sporting ambition.

Sitting with a book in one hand and a mug of hot chocolate in the other – she could see how that built character. But waving a stick around a muddy field? All six years of hockey had built in Hetty's character was a deep hatred of sport.

And so last summer, when Liv became totally obsessed by the gym at the Manor Country Club (for reasons that only became clear later on), Liv's suggestion she join her made Hetty want to run far, far away.

Well, maybe not run exactly. Briskly walk.

But with Dad's absence and Mum's sadness haunting every room, Hetty was grateful for any excuse to escape.

'Trust me, you'll love spinning. It's fun,' Liv had insisted in that way that people who are naturally good at things do. Of course, it was easy for Liv to go all shiny-eyed over wearing Lycra. In tight black leggings and a neon vest clinging to her toned legs and flat belly, she could have been off to a casting call for *Women's Fitness* magazine.

'And it keeps you fit,' she'd continued, pulling her glossy hair through the back of her cap as she strode into the studio. Hetty sloped in behind, tugging her Connecting Together T-shirt down over her bottom and silently vowing never to fall for Liv's persuasion/blackmail again.

Watching Liv's balletic leap into the saddle jolted Hetty's already plummeting heart into a tailspin. *This is a breeze*, Liv's feet whispered, slipping light as butterflies' wings into the alarming metal cages while

Hetty, burning with self-consciousness, clambered up on to her seat. A hippo, that's what she felt like. A hippo hanging out down the gym with a particularly gorgeous panther. As Hetty wiped her already sweat-slicked hands down her thighs, Liv leaned over the handlebars.

'You'll love it,' she said. 'Just don't forget to —'

'*To what?*' But the rest of the sentence disappeared under the rapid thud of hi-energy music and the whirr of wheels.

'It's party time!' the instructor yelled, grinning like a horror-film maniac.

'Yay,' Hetty muttered and began to pedal.

But astonishingly, Liv was on to something. Yes, Hetty's thighs burned, her lungs rasped, even her fillings ached by the end and still . . . as she climbed off the bike, flicking droplets of sweat in the air, she felt oddly elated. Endorphins?

Over the next couple of weeks, as much to her own surprise as Liv's, she added Zumba then combat to their weekly schedule. What a revelation that turned out to be: it was entirely possible to enjoy 'exercise' while simultaneously hating 'sport'.

Ironic then that it was the arrival of Duncan Lambert that killed Hetty's newborn gym bunny stone dead.

She'd had crushes before on popular boys at school, friends' brothers, and even friends' dads. And – ugh, acute cringe – she'd spent most of Year 7 compulsively doodling Duff's older brother's name over and over in her planner. *Gus and Hetty. Mrs H Duffy. Hetty and Gus.* Usual kid's stuff.

But in the context of *this* guy, who bounced on the balls of his feet while he talked to you, as though at any second he might drop for a cool hundred or sprint off into a marathon or abseil down a rock face, 'crush' sounded so . . . inadequate.

Was it really love at first sight? It was something anyway. Obsession. Like when you stare at a light and the after-image tattoos itself on everything else you see.

Spinning had just finished. Next to her, Liv towelled off the dewy sheen that passed for sweat in Liv-world, while Hetty, the blood supply to her legs replaced by rapidly hardening concrete, lifted her water bottle to her lips.

And there he was in the doorway, with a smile for everyone. From his wide shoulders to the faultless sculpture of his cheekbones, Duncan was every romcom cliché rolled into one perfect package. And then some.

'Tough class, ladies?' he said when it was their

turn to leave or, in Hetty's case, woodenly stagger from the studio.

Hetty froze, careful to avoid eye contact.

'Yes, but we're tough girls,' Liv said. She tugged her ponytail with both hands, pulling the band tighter.

He pushed his hands together so the muscles in his forearms stood out.

'Well, it'll be even tougher when I take over at the end of the month.'

Liv smoothed wisps of hair back from her temples, a smile playing across her lips. 'Really? We'll look forward to that, won't we, Hets?'

'See you later.'

'Hot, hot, *hot!*' Liv whispered the second they were out of earshot, and Hetty could feel the new instructor's eyes following them through the gym and up to the changing-room door. Or, more likely, following Liv.

And that was the trouble. It wasn't that she envied Liv, more that she was under no illusions: Hetty could flip her heart inside out pining for a guy like Duncan, but guys like Duncan would always go for girls like Liv. It was just the way life was.

The following weekend she watched him chat up the Botox ladies in their velour tracksuits, and discuss handicaps with the golf crowd until he seemed as at home as the shiny trophies under lock and key in the bar.

Back at school that Sunday night she couldn't stop thinking about him.

He was clever: a *politics* graduate, of all things. He was gorgeous. He was popular. He was sporty. And he was so far out of Hetty's reach he may as well have been from another dimension.

'What do you mean you're not coming any more?' Liv's incredulity rattled down the phone when Hetty told her. 'I thought you loved it.'

But what choice did she have? Duncan had wrapped himself round her brain like a python, squeezing the life out of any other thought. She could barely talk to him; would rather die than *sweat* in front of him. Maybe crush was the right word after all. She certainly felt crushed.

When school started again in September, she filed him away alongside Duff's brother and a whole load of random heartbreakers and got on with pretending he didn't exist. Until October when her mum came to pick her up from school for half-term.

It was a blustery Friday, too warm for autumn, but inside the car hot air blasted from every vent. It was like having a hairdryer fired directly on to her eyeballs.

'Can I turn the heat down?'

'What?' Mrs Barraclough gripped the steering wheel, her knuckles pearly through the skin. 'How can you possibly be hot? It's so cold in here.'

Hetty wiped a circle of condensation from the window. Streams of red light blurred through the opaque glass as another car overtook them.

'Doesn't matter,' she said.

Mum's chat filled the car: the garden, the neighbours, redecorating Hetty's room. *I thought blue might work? We could pick up some samples tomorrow in town. Or green. Something restful. Or bright, liven us up a bit. Maybe purple. I don't know. Red?*

Sentence after sentence, stretching like elastic bands to hold their fragile post-Dad lives together. The definition of ironic, wasn't it? The one thing they needed to talk about was the one thing they never mentioned.

Before Dad's accident, Mum's driving had mirrored her attitude to politics: focused, fast, ruthless. Nowadays she crawled. And *talked*. From the minute she put the key in the ignition to the minute she clicked the fob to lock the doors. As if words were some magic talisman because accidents just couldn't happen mid-conversation, right?

Cocooned in stifling warmth, Hetty recalled her mum

confessing to talking to the satnav when she was alone.

Outside, the lights had gone and the darkness of the country lanes enveloped them. One of the rubber bits on the windscreen wipers had come loose and the rhythmic screech up the glass set Hetty's teeth on edge.

And then, suddenly, floating on the stream of inconsequential chatter came: 'I forgot to mention, you might know my new intern, Duncan Lambert? Had a holiday job at the Country Club gym.'

Hetty's mouth dried so quickly it was only capable of producing a *hmmm*?

Mum carried on, oblivious. 'His university tutor sent a glowing reference and I spoke to Bob at the Manor and he'd been an exemplary employee. Never missed a shift, not a single customer complaint, and you know what they're like at the club – they'd find fault with Mother Teresa. And the interview . . .'

Her mum dropped a gear in anticipation of a distant Give Way and raved about Duncan's 'acumen', 'likeability' and his 'bright future'.

Meanwhile, Hetty feverishly gnawed the skin around her thumb. Fate? Coincidence? Karma? What was this?

Mrs Barraclough looked left, right, left again into the inky darkness, then inched the car over the white line in the direction of home. Half a mile to go and it would have been quicker to get out and *crawl*.

'Hetty?'

She realised her mum had asked her a question.

'Sorry, what?'

'I said are you OK to go to the office and help Duncan with the leaflets for the campaign tomorrow?'

And that was how it began.

She'd had a weird out-of-body experience the first time he asked her out. And when he made it clear he meant an actual *date*, her shocked subconscious presented her with a mountain of reasons to refuse.

Look at him, it whispered, *look at this brilliant, gorgeous, fit guy who can run 5K in less time than it takes you to lace up your trainers. Look at his many friends, his fierce intelligence, his sporting ability. He could have anyone. Why on earth would he want you?*

Over the last few months, her memory had edited out the panic and self-doubt and softened the edges, turning that moment into a scene from a film.

But fear still hummed in the background.

You do realise it's because he wants to work for your mum?

She was afraid of so many things. Of his Wednesday playing five-a-side at the Country Club, 'having a few beers' afterwards with 'the lads'. Of the volleyball matches he refereed every other Sunday. Of his appallingly perfect family and their Sunday-supplement-lifestyle existence. Of the parade of blow-dried, made-up exes who sprinkled his social media posts with hearts and kisses; the random girls whose gazes slid over him like possessive hands.

Some nights she lay staring at the ceiling, trapped under the weight of a tremendous silent panic while her imagination spooled through endless break-up scenarios.

Ground down by worry – this wasn't how she'd imagined being in love.

So when Hetty heard her recorded voice saying, 'I'd do anything for Duncan,' she knew it was true.

⊙ 6 HETTY

1 July, 11.35 a.m.

'So,' Cass continued, 'I know you boarded at St Benedict's during term time. That must have been tricky, only seeing your boyfriend at weekends and in the holidays. You must have a strong relationship to survive that.'

'I think we have,' she answered.

'What about politics? Would you say it impacts on your relationship?'

Careful, Hetty. *If they ask you about politics*, Duncan had warned her, *try to steer it on to something else.*

'I totally support Mum and Duncan,' she replied, 'but I'm not really interested in their work, if I'm honest.'

'Some people might interpret your mum's views as quite extreme. For example, her manifesto stated . . .' Cass flipped through the pages in her notepad. 'Here we go. *Free movement of people undermines British society, our values and what we have always represented as a nation.*' How do you feel about that?'

Hetty flashbacked to the previous November. Mum, her hands braced on the kitchen worktop, leaning forward. She looked and sounded completely incredulous.

'What do you mean, "No"?'

'I mean I am not doing it,' Hetty said, picking at a large orange leaf stuck to the side of her shoe. 'I am not carrying *that*.'

She pointed to the table where a sign waited, printed with the Britannia Party's slogan *British Jobs for British Workers*.

Hetty heard the spray of gravel under tyres outside. Duncan had arrived to take them to the rally. The car door slammed.

'For God's sake, Hetty, how can you be so unsupportive?' Her mum lowered her voice. 'Your dad would have been very disappointed.'

'I think he would say I had to do what I thought was right,' Hetty said, slowly tearing a piece of kitchen paper off the roll.

With a noisy sigh, her mum clattered the breakfast

things into the dishwasher and shut it, slightly harder than necessary.

Hetty wiped the mud from her shoe, crumpled the paper and dropped it in the bin. 'I'll come with you if you want me to, but I am *not* carrying that sign. And that's final.'

Brisk footsteps crunched past the kitchen window and around the corner. There was a pause followed by a sharp double knock.

Rose Barraclough MP wiped her hands aggressively on a tea towel and, shaking her head at her defiant daughter, opened the back door.

Back on Liv's sofa, Cass was looking at Hetty, waiting for an answer.

She tapped her hands on her thighs and repeated, 'I'm not really interested in politics.'

Cass nodded, wrote something down. 'OK. Let's talk a bit more about you then. What do you like doing in your free time? Cinema, meals, gym? Hanging out with your friends?'

'Definitely!' Hetty answered over-brightly.

'One of you mentioned the Manor Country Club. I get the impression that's the place to hang out?'

Liv chipped in, 'Yeah. The *only* place. It's so tedious

living round here. There's literally nothing to do.'

'But I think we're all kind of over it now,' Hetty said guardedly. 'I've hardly been this year.'

'I don't go much either,' Liv said. 'Especially now people drive and it's easier to get out of the village.'

'*Some* people drive,' Jez said pointedly in Duff's direction.

'Whatever,' Duff said, putting his hand in the air. 'Cass, you said no secrets? I'm not proud of it, but I lost my licence. For speeding.'

He glowered at Jez. 'Happy?'

Jez shrugged and stared intently at the wall.

Hetty sighed. *Oh Jez, why do you have to wind him up?*

'I appreciate your honesty, Duff,' Cass said. 'What about you two?'

'I've failed six times already,' Liv said sadly. 'And Hetty hasn't even started lessons yet.'

'I don't need to,' Hetty added quickly. 'Duncan's got a car.'

'So tell me a bit more about the Country Club.'

'There's a big café and a bar, and they hold a lot of social events. Two swimming pools, a gym, exercise classes. Tennis courts.'

'And it's a private club, right?'

'That's correct,' Jez said. 'You make an application then you have to get a reference from an existing member.'

'Sounds rather unfair. What if you'd only recently moved to the area?'

Hetty shrugged. 'There's always a massive waiting list. If you'd just moved, you wouldn't get in.'

'That's not totally true,' Duff said. 'My brother met his wife there – her family joined right after moving to the area.'

'How did they get in then?' asked Cass.

A thunderous clanking interrupted his reply. Sliding off the stool, Liv had caught her foot on the metal base and sent it rocking on the wooden floor.

'Sorry, sorry,' Liv said. 'I'm just getting a drink. Does anyone else . . .?'

No one did.

Meanwhile, Cass continued speaking. 'What about people who aren't lucky enough to join?'

'There's nothing round here,' said Jez, waving in the vague direction of the farm gates. 'You have to go into town. Which is an utter dump.'

In the kitchen, Liv gripped a glass and turned on the tap. Her shoulders looked hunched and stiff as she gulped down the water.

'There's the swimming baths, I suppose,' Jez went on. 'And I guess shopping. But that's not very good either. Pound shops mainly.'

'Dodgy nightclubs,' Duff added. 'Me and my brother went to one at the weekend. Full of dodgy girls like –' He mimed doll-sized clothing.

'Sounds grim.' Cass pulled a sympathetic face then turned to Hetty, raising her voice in a question. 'So you and Duncan got together at the Country Club?'

Hetty shook her head. 'Afterwards. Staff and guests don't mix with each other.'

'Except Rafa, remember?' Liv said, neatly perching her ridiculously compact bottom back on the seat.

Hetty grinned. 'Oh yeah. How could I have forgotten about him?'

Liv and Hetty caught each other's eyes and burst into giggles.

'Well, he was about the fittest lifeguard ever. I mean, everyone fancied him. Everyone.'

'Go on,' Cass said, lighting up at the suggestion of gossip.

'So, Rafa was this Brazilian lifeguard and he was just unbelievable. Fit, tanned, all round gorgeous. On his shifts, honestly, I've never seen the swimming pool so busy. Two summers ago, do you remember what the weather was like? It poured down for practically the whole of August and still the pool was packed. Everyone just went to stare at this Brazilian guy.'

'He was so worth staring at,' Liv added, going all dreamy-eyed. 'Those pecs, those arms . . .'

'And the waiting list for private swimming lessons with him! And then, oh God, Liv, when you pretended to drown so he'd jump in and save . . .' The rest of the sentence dissolved into Hetty's giggles.

Liv swivelled round in mock outrage. 'Shut up! I had cramp. I wasn't pretending.'

'Sure you did,' Duff sniggered.

Hetty fanned her hands over her red cheeks. 'Sorry, but it was funny. And then in the end, after all those girls spent the whole summer throwing themselves at him, he went off with —'

Hetty was laughing so much now that she couldn't get the words out.

'Emily Lloyd Martin's dad!' finished Liv.

Jez's eyes widened. 'I heard about that. I didn't know it was the lifeguard though. That's hilarious.'

'I suppose it is funny,' Cass agreed, running a finger along a crease in the notebook, 'unless you happen to be Emily Lloyd Martin's mum of course.'

'Er . . .' A sheepish note damped Hetty's hysteria. 'True. They got divorced in the end. Emily was really cut up about it.'

'And speaking of staff,' Cass said, her eyes trained on where her finger stopped on the page, 'I wonder if you remember any particular one. A member of the waiting team perhaps.'

Hetty stopped giggling completely.

She shrugged her shoulders in what she hoped was a casual gesture and said carefully, 'They have loads of different guys there every week.'

Cass continued. 'This particular waiter was female. And she was there for quite a while.'

Duff propped himself up on one elbow as if he were lying in bed. 'Ask me. I can tell you whatever you want to know about the girls who work at the Country Club.'

Jez snorted. 'I'm sure they'd have plenty to say about you as well.'

For one second, Hetty thought Duff was going to lose it. Instead, he grinned broadly and flexed his biceps. 'I'm sure they could, mate.'

For a split second Hetty caught Cass's eye. Then she

stared down at her hands. 'Honestly, I hardly ever go there any more. Haven't been for ages.'

'Any reason?' Cass asked and there was something slightly ominous about the question.

A sense of foreboding gnawed at Hetty's stomach.

She tried to smile. 'We've kind of moved on. You know, grown out of it. I can tell you about what I'm doing after the summer if you like.'

She heard Cass in her head.

Everyone's got secrets.

Her pulse began to hammer da *dum* da *dum*.

On the floor by her feet, Duff was blowing out his cheeks and releasing tiny puffs of breath. Genuinely bored or acting bored? It was impossible to know with Duff.

Cass picked up the remote control and aimed it at the TV.

'You said you hadn't been for a while. I was wondering if you remembered this?'

A photo appeared. A group shot of predominantly young women dressed for a night out. A statuesque girl at the centre posed with a bottle of champagne. Banners and balloons hung in the background.

Hetty flinched.

'Sorry, I don't,' she lied. Her voice sounded echoey

and distant, as though she were listening to herself from a long way off.

Liv's surprise was evident. 'Course you remember! It was Duncan's sister's eighteenth.' She shuffled her feet in tiny steps, turning in Cass's direction. 'That was New Year's Day.' She twisted her head to look back at Hetty. 'How can you not remember? That was the night your purse got stolen.'

The room was stifling. Wetness seeped under Hetty's arms. Could she take the sweatshirt off? What was under it? Oh God, a vest top. An old one. No. The sweatshirt stayed.

'Where did you get that picture?' Liv asked, peering closely. 'I don't think I've seen it before. Are we in it?'

Before Cass could answer, Jez said, 'The researchers. They must have lifted it from the Country Club website.'

'You're not in it,' Cass said. 'But this person is.'

She tapped a key and the image zoomed in on a figure shadowed in the background. The others scrutinised the pixellated blur wearing the Country Club's distinctive polo shirt.

Hetty didn't need to inspect the screen. That face had lurked in the back of her mind since the night of Anna's party.

Anxiety sloshed in her stomach like acid. Something burned behind her breastbone and for a horrible moment she thought she might be sick. When she tried to breathe she found her lungs had shrivelled up.

Click-clack went Cass's sandals as she walked up to point at the figure.

'This is who I was talking about. She's called Ellie. Does that ring any bells?'

Hetty swallowed past the stone in her throat. 'Yes, I remember her.'

I'd do anything for Duncan . . . Hetty's voice came back to her as clearly as if she'd spoken the words aloud.

'OK then, Hetty,' Cass said. 'Why don't you tell us exactly what you did?

⊙ 7 HETTY

Six months earlier

Duncan's mother climbed down from the stool with the grace of an Olympian descending from the podium.

'Does that look straight to you?'

Marguerite Lambert didn't actually need Hetty's input. She didn't need anyone's confirmation about anything. The *Happy 18th Birthday* banner pinned to the function room wall couldn't have been straighter if it had been hung by someone armed with a spirit level and a degree in banner hanging.

Hetty opened her mouth to respond, but Marguerite was already snapping a lifeless balloon over the nozzle of an air pump.

New Year's Day, and while the rest of the country slobbed on the couch, nurturing its collective hangover with caffeine and carbs, Hetty and Marguerite were at the Country Club surrounded by cardboard boxes with things like 'spare linens' and 'balloon equipment' printed neatly on them in marker pen.

This was not the first time Hetty had met Marguerite. Her initial invitation to Lambert family HQ arrived around two weeks after she and Duncan became an official item, back in November. With Hetty fully immersed in the pinch-me-I'm-dreaming state of dizzying infatuation, Marguerite's greeting brought her back to earth with a bump.

'Welcome. We're thrilled Duncan is working for your mother.'

Marguerite was wearing a dove-grey chiffon blouse with a bow at the neck, military-ironed black trousers and black suede court shoes with a dazzling silver buckle. Her blonde hair was rolled into a complicated twist fastened with a pearl clip and her mouth was a lipsticked slash. She looked like Hollywood's idea of a prison governor. *Terrifying.*

Hetty smiled, unsure how to respond. Marguerite's expression didn't change and Hetty's stomach churned.

Things didn't get much better once she'd followed Marguerite inside.

'And this,' Marguerite had said, waving a marker pen, 'is our communication station.'

For a second Hetty thought she was joking. But no. A deadly serious whiteboard dominated one corner of the large farmhouse kitchen, divided into colour-coded columns headed with the names of the Lambert family. At a glance Hetty noted that Duncan's sister Anna's mani-pedi clashed with Mr Lambert Senior's scale and polish that afternoon. And for today, 4–6 p.m., Marguerite's schedule read 'write non-family Christmas cards'.

She noticed the ironing drop-off, the gardener and the organic vegetables each had a designated colour. And right now so did she. Light blue, under the Duncan column.

'And there we go,' Marguerite said, slashing a red line through 6.30–7.30: *Henrietta (friend) visit.*

'Henrietta' (friend) discovered Marguerite (never Maggie) had fine-tuned her organisational skills during her many years in special forces. Well, allegedly she owned a dental practice, but that had to be a cover, right?

No one made shopping lists, appointment lists, to-do now lists, to-do later lists . . . *and then listed them on a master list of already-written lists.* No one normal anyway.

Until she met Duncan's mum, Hetty had considered

her own mother to be the lean, mean, bullet-point queen, but Marguerite redefined the term anally retentive.

Today's party checklist clearly featured *involve son's girlfriend*. Hetty sighed inside. It was thoughtful yet at the same time really quite depressing.

'Is there anything I can do?' Hetty asked dutifully.

'Let me see . . .'

Marguerite set down the pump and knotted the end of a balloon with deft fingers. A vision of her flipping through a book called *101 Easy Tasks for Inept Girlfriends* popped into Hetty's mind and she stifled a nervy giggle.

She'd almost finished distributing the stack of place cards Marguerite had handed her ('Here's the seating plan!') when the door opened.

'Hello,' called a female voice. 'Would you like any extra help?'

The girl in the doorway somehow transformed the Country Club's polyester shirt-and-trousers combo into haute couture. Her name tag read 'Ellie' and she had the kind of face that Hetty associated with ornate frames hanging on museum walls. Skin with a mother-of-pearl glow, artistically proportioned nose, lips and chin, and beautiful eyes fringed with thick dark lashes. Right on cue a shaft of late winter sun filtered through the blinds, illuminating her in a golden halo.

Of course it did.

'I've just started my shift; I'll be serving at the party,' the girl went on to explain. 'But I thought I'd pop in and say hi, just to see if there was anything you needed.'

'I think we're fine,' Marguerite said, craning her neck towards the seating plan. She frowned. 'Although . . . Oh dear. Maybe you could help with the place cards.'

Laughing possibly the least sincere laugh of all time, she whipped the plan out of Hetty's hand.

'Whoops-a-daisy. We've had it upside down, haven't we?'

Her granite grey eyes practically bored twin *you*-shaped bullet holes into Hetty. *You've* messed it up. *You.*

A wave, no, a tsunami of inadequacy washed over Hetty as perfect Ellie swiftly re-dealt the cards in to their correct positions before looping an elastic band around the emergency blanks Marguerite had so thoughtfully included.

Meanwhile, Hetty twisted her fingers together and desperately wished she were somewhere else. Anywhere else. The effort of maintaining a smile made her cheeks twitch.

'Oh, you marvellous girl,' Marguerite gushed, perusing the tables. She placed the small bundle of cards in a box.

The approval in her voice unleashed a further rush of misery in Hetty. She felt herself fading into the floral riot of the wallpaper. For God's sake, who even *had* a seating plan at their eighteenth?

Not good enough, the mean voice in her head whispered. *You'll never be good enough.*

Ellie checked the water level on a vase of white lilies by the door.

'You're welcome. I'll drop by in a while to make sure everything is running smoothly. In the meantime, if you think of any—'

'Buckets!' Marguerite exclaimed, her hand flying up to her mouth. 'Ice buckets for the champagne. I almost forgot.'

'Don't worry,' Ellie said calmly. 'I've counted them out and set the ice aside. The champagne is chilling and I'll bring it up right before the guests arrive.'

'Thank goodness. And you'll have the wine and beer ready too?'

'Of course. I'll be behind the bar all night,' Ellie answered. 'There's no need to worry, Mrs Lambert. Your daughter's party will be a night to remember.'

'Why don't you go home?' Marguerite said, turning to Hetty. 'I'm sure you want some time to get ready for the party. If you're anything like Anna you'll need hours.'

The last of Hetty's self-confidence shrivelled like a week-old balloon. *Get ready? You think I wear £180 of Ted Baker every day?*

It didn't help that Ellie's immaculate polo shirt and simple black trousers showcased a trim waist and long legs that made Hetty feel dumpy and overdressed in comparison. Or that Marguerite clearly considered Hetty surplus to requirements.

'Here.' The older woman unzipped her elegant cream leather purse and thrust a twenty-pound note into Hetty's hand. 'I'm sure Ellie can call you a taxi, can't you?'

'My pleasure,' the girl said, holding the door open.

To make matters worse, Hetty had just fastened her seat belt when a telltale cramp signalled her period had showed up a few days early. Meaning the cream silk Ted Baker she'd agonised over choosing ended up relegated to the wardrobe anyway.

By the time she'd got back home, showered, dried her hair and tugged a pair of opaque tights over her bloated belly, she was tearing up.

She wriggled in to a little black dress, 'little' being the operative word. The black lace panel stretched across her chest and rippled from hip to hip.

This would never happen to Marguerite or Anna. Undoubtedly, periods in the Lambert household were carefully calibrated to arrive during an allotted time, tidily slotted between Pilates and the veg box. Actually, forget that. They probably just outsourced them entirely to the manicurist or the lady who collected the ironing.

Peering at her face in the mirror, Hetty smoothed on concealer and foundation, fanned blush along her cheekbones and coated her lashes with mascara, top and bottom. Carefully blotting with a tissue first, she smacked her lips together then dabbed a little gloss in the centre.

She was reaching for her make-up bag when her hand stopped midway to the eye-shadow compact. What was the point? Her shoulders sagged. All the make-up in the world couldn't disguise who she was: Hetty Barraclough. Hetty never-good-enough.

The low-level grief inside her instantly sharpened to a painful stab of misery. Dad. The one person who knew the exact words to make her feel better was the one person whose voice she'd never hear again. Her chest, covered now not in expensive cream silk but in slightly faded black polyester, threatened a ragged, sobbing breath. Gripping the edge of the dressing table, she commanded her reflection, 'Do not cry. Do not cry.'

The worst part was hating herself for even caring. After all those hours with the counsellor, the affirmations, the self-esteem exercises, motivational quotes. The motivational quotes! What started life as a couple of scribbled Post-its stuck to her bedroom wall had now grown into a floor-to-ceiling barrage of motivational propaganda. *There's only one you! Be your own best friend! Love yourself!*

And she'd allowed a pretty waitress and an over-Botoxed Nazi to ping her straight back to square one.

'Liv is going,' she reassured her mirrored self. 'You won't be alone.'

Downstairs she stuck her mouth directly under the tap and gulped down a painkiller. She felt the heavy ache below her belly button spread to encompass her lower back. Flicking stray drops from her lips with her thumb, she shook out another pill.

By the time the taxi pulled up outside Liv's house, a reassuring text from Duncan (*Where are you?*) coupled with a motivational one-to-one with herself meant she'd cheered up. OK, so the chances of her shaking her booty on the tables remained slim, but at least she wouldn't be sobbing in the toilets either.

'What happened to Ted Baker?' Liv said, sliding on to the leather seat. Despite her improved mood, Hetty felt a twinge of panic at her friend's endless slim legs, her model-perfect make-up.

'Visit from the red queen,' said Hetty. 'Showed up early.'

'Nightmare.' Liv clicked the seat belt and pulled a sympathetic face. 'Still, I've always liked that dress on you.'

Hetty wrapped her arms around herself, suddenly defensive. 'Are you saying I always wear the same thing?'

'Noooo,' Liv replied patiently, rooting through her sequinned clutch bag. 'I'm saying you look lovely. Now just let me . . .'

In one seamless movement, she took her phone out of its red leather pouch and angled it above them.

'Taxi selfie time.'

'Oh no,' groaned Hetty, trying to shuffle out of range.

'Oh yes.' Liv nudged a strand of hair from her forehead with her thumb and reached an arm around Hetty. 'One, two and —'

Click.

Liv tapped her handbag. 'Gorgeous. Now, stop panicking and tell me exactly how late are we?'

Pushing open the door to the function suite, they were greeted by the brand of R & B Anna favoured . . . and the swift assessing glances of her friends.

Funny, really, how the boys puffed out their gym-built chests, their eyes gliding over Hetty to latch on to Liv in a genetic man ritual ingrained in their DNA. *The slim one with her long shiny hair and neat features*, their stares said. *She would be a match for us.*

But never mind that. Hetty squeezed through the mass. Skinny girls with long hair and long legs. Boys with too much aftershave, untucked shirts and warm alcohol breath. Briefly smiling at familiar faces, but not stopping. Pushing through, feeling anxiety swell in her chest. Where was Duncan?

'Hi.' A voice separated itself from a knot of glowy girls. Anna. A *Happy 18th* balloon jiggled above her head, tethered to the stem of her glass by a shiny string. 'Hey, everyone, this is Duncan's new girlfriend, Hetty.'

'Oh. My. God. That dress is stunning,' gushed one of the girls.

'Love your shoes,' said another.

'And the bag. Is it Hermès?'

Anna giggled, lifting a hand to her mouth. The balloon bobbed above her head.

'No, no, no, *this* is Hetty.'

The girls' eyes pivoted away from Liv and found Hetty. Three pairs of eyes widened in exaggerated shock. Hetty wasn't entirely sure if her imagination added a gasp.

Stripped naked. That's how it felt. Stripped naked and placed under a spotlight like a particularly bizarre lab specimen.

'Happy birthday!' Liv exclaimed, breaking the awkward silence. She stepped between Hetty and the girls and clutched Anna's shoulders, *mwah*-ing the air by both cheeks. 'We're trying to find Duncan actually. Have you seen him?'

'You could try the bar,' said Anna.

Liv linked arms with Hetty. 'Fabulous. See you later.'

The girls didn't even wait until they were out of earshot. Scandalised cries of 'You're kidding me!' pierced through the music and the hubbub of conversation, killing off the last shreds of Hetty's self-esteem.

Liv held her arm firmly and marched them towards the bar.

'Take no notice of them,' she said.

Hetty felt a rush of love for her friend, swiftly followed by a rush of panic.

Behind the wooden counter with its shiny brass rails, Ellie was rattling ice cubes off a plastic scoop as she chatted to a customer hidden by the impatient queue. Hetty watched her throw her head back with laughter, her satin hair catching the light. Then the seated customer leaned forward, his familiar broad back creating a corner of intimacy in this public space.

Ellie dropped coins in the tips jar and wiped her hands on a striped tea towel. She walked to a coat hanging by the bar and took out a phone.

Hetty froze, rooted to the spot by . . . humiliation? Misery? Fear?

Duncan, elbows propped on the bar, recited something. His number! He was giving her his number!

Someone touched her; she jumped.

'There you are! We were all wondering where you'd got to.' Marguerite's spectacularly insincere voice grated in her ear. She touched Hetty's bare arm with a paper-dry hand. 'And this is your friend . . .?'

'Olivia Dawson-Hill,' Liv finished. Marguerite graced her with the replica of a smile and twitched a frozen eyebrow. Her highlighted blonde hair hung stiffly in a solid sheet, as if it had been laminated.

'Ah. I've heard a lot about you.'

Liv switched on her most dazzling smile. 'And I've heard a *lot* about you too, Mrs Lambert.'

The mischievous undercurrent to Liv's words sent Marguerite's pencil-thin eyebrows into orbit and the older woman's smile set like concrete. She stretched her neck like a tortoise, searching the room behind Hetty.

'And is your mother here?'

Hetty shook her head. 'She's sorry, but she had to stay in London. Work.'

A waiter held out a silver platter piled with fiddly snack things. 'Those canapés look lovely,' Liv continued.

She took one and popped it in her mouth.

Hetty helped herself to two of the things. One fell, squidgy side down, and tracked a greasy trail down her dress and dropped on the carpet.

Marguerite's smile became even more strained.

'I have to go,' she said in her finest Ice Queen tones. 'Lovely to meet you.'

'You too,' Liv replied.

As soon as Marguerite's elegant back was turned,

Hetty grabbed a glass of champagne from a passing tray and took a sip. A gulp.

'Jeez. You on a mission or what?' said Liv.

'I am indeed.' And she tipped back the rest with a grimace, like a child forced to swallow medicine, and strode the last few paces to the bar with her head held determinedly high.

'Hey,' Duncan said, throwing his arm around her as she reached the bar. 'I was wondering where you got to.'

Hetty let him kiss her on the lips, willing her heart to slow down, her limbs to loosen up.

Was he looking over her shoulder down the bar?

'Let's get you a drink. Ellie? When you're ready,' he called.

Why had he gone all melty-eyed?

'Hets?' he said, tapping her hand. She turned and swayed, the champagne already fizzing through her bloodstream.

'Sorry, yes,' she answered, vaguely aware he'd asked what she'd like. 'I'll have . . . wine, please.'

Later, when she tried to justify to herself what happened that night, she blamed the alcohol. She blamed the pills she'd popped on an empty stomach. She blamed the girls who sneered at the thought she could be Duncan's girlfriend. She blamed her hormones.

But if Hetty poked down into the true depths of her motivation . . . honestly? None of those played a starring role. Each barely had a cameo.

No, her actions that night, her awful, terrible, shameful actions, sprang from unadulterated jealousy.

Oh God.

Afterwards, she'd hurried out of the manager's office. Shunning the Ladies closest to the function rooms, she lowered her head and quickly ran down to the ground-floor toilets. Quieter. More private. The cold hit her like a slap.

She fumbled for the switch and the fluorescent tube buzzed and flickered to life, illuminating a stranger in the mirror. Shiny skin flecked with mascara. Hair like limp string, wisps clinging to her scarlet cheeks. Crumpled rag of a dress. She could hardly bear to look at her reflection.

Her throat felt scratchy and raw. Undoubtedly a side effect of the poison she'd just coughed up in the manager's office. She pictured a dark green liquid surging up from her soul, some mysterious alchemy converting it to lies the second it hit her tongue. She lowered her head to the tap; gulped down cold water, feeling drops slide over her chin when she straightened up. The bitter taste remained.

Selecting the furthest door, she locked herself in. She closed the toilet lid and sat down, the chill plastic burning her thighs. The cubicle had pink tiles and smelled of synthetic cinnamon. Somewhere close by, drops of water plinked repetitively. *I'm a bitch. Bitch. Bitch. I'm a bitch.*

A sudden image of her dad flashed in her mind. He was shaking his head, the corners of his mouth pulled down in disappointment.

Oh, Hetty, love. What on earth did you do that for?

And that's when the tears finally came, gushing so suddenly it was as though a tap had been turned on behind her eyes.

What have I done?

⊙ 8 HETTY

1 July, 11.55 a.m.

'Please could you explain why you did it,' Cass said.

'No,' she moaned into the crackling tension. 'Please don't make me talk about it.'

'For the cameras, please,' Cass said gently.

Hetty responded on autopilot, bracing herself against the button-backed leather of Liv's sofa. She sniffed and rubbed her nose.

'That's OK, I can do it.' Cass spread out one hand, bending a finger with each number as she counted down Hetty's crimes. 'One. You went to the manager. Two. You told the manager that this girl had taken your purse. Three. When she denied it, you said she was lying. Four. They found it in her coat pocket. Five. She got fired. Does that sound about right?'

Jez clapped his hand on the wood floor. 'You've got it completely wrong. Hetty was the victim. Everyone was talking about it. Hetty saw her steal her purse.'

'He's right, Cass, you really have got it wrong. I mean, I was *there*. I know exactly what happened,' Liv said.

'I told you, we do our research,' Cass said pleasantly. 'I'm not wrong, am I, Hetty?'

Hetty looked up at her friends, aimed for a smile, but couldn't make it. Tears stung in her eyes. These were decent people who, despite the evidence, refused to believe she was capable of such a terrible, terrible thing.

She'd watched a documentary once, about a girl in Cambodia with a cyst in her belly the size of a watermelon. When the doctors cut her open they found the remnants of a twin inside her. Since her first breath, she'd been the oblivious host to a mass of teeth and bones and hair. That's how Hetty pictured her remorse. A tarry ball embedded deep inside, feeding on her unhappiness.

The confession was out before she even realised she was speaking. 'Cass is right. I made it up. That waitress – Ellie – never stole my money. I put my purse in her coat pocket.'

The muscles of their faces contorted in synchronised

shock. If she hadn't felt so distraught, she'd have burst out laughing.

Liv was the first to speak. 'No, I don't believe it.'

Her shocked disbelief, her complete faith in Hetty's inability to do anything so cruel, so wicked, hurt way more than any criticism ever could.

Hetty laced her fingers across the top of her skull.

'Why?' Liv said, nonplussed.

Why?

The truth erupted from her in one, breathless, cathartic rush.

'I wasn't thinking straight. I'd had these super-strong painkillers, right, and then a couple of drinks. My head was all over the place and —'

She took a deep, shuddering breath.

'No, that's not right. I knew exactly what I was doing.'

The truth was she'd wanted to retract the accusation almost as soon as she'd made it. But that was the trouble with words: you didn't get a cooling-off period and there was no option to return them if you changed your mind.

She'd been so insistent on seeing the manager and then — oh God — used her mum's position as leverage. What option did they have? It was Hetty's word against Ellie's.

And they believed Hetty. Or at least they said they did.

'That's not the worst part. There was nothing dodgy in it. She just wanted to try out for his volleyball team. That's why I don't go to the Country Club any more,' she finished up. 'And there hasn't been a day when I haven't regretted what I did.'

'Jesus, Hets. That's just . . .' Duff tailed off, scratching his nose.

The room blurred and wobbled before her eyes.

She leaned forward to rest her forehead on her hands. She heard the stool creak softly, then Liv's sandalled feet padding across the floor before the sofa cushion sank a little and her friend gave her arm a reassuring squeeze.

'You OK, Hets?' she whispered, close to her ear.

Hetty nodded. Her throat ached from holding back the tears, but now Liv was hugging her she felt them spill over.

'Duncan is mad about you,' Liv said. 'He wouldn't cheat on you in a million years. I don't get why you're the only person who doesn't see it.'

And then Cass clapped her hands together. 'Right, that's you off the hook, Hetty. Thanks for your input.'

There was a brief, shocked pause. Someone – Jez maybe? – gave an incredulous snort.

Liv exclaimed, 'You've got to be kidding!'

'No.' Cass sounded faintly amused. 'We all make mistakes, right, Hetty? We've still got the boys to do and time is ticking on and –'

A shrill ringing interrupted her, coming from deep inside the Pandora bag.

'Excuse me one second . . .' she said, unzipping an interior pocket and extracting a rose gold phone. She frowned at the screen, looped a hank of hair behind her ear and held the phone to her head. 'I need to take this, sorry, guys. I'll just . . .'

Cass stood up and walked to the door, which closed behind her; evidently it was a private matter. Tony checking up on them perhaps?

Duff said into the brief silence that followed, 'I don't know what you've got us into, Liv, but this really isn't what I was expecting at all.'

'What? And you think I was?' she hissed. 'Tony pitched this to me as crazy rich kids and their exciting lifestyles thing.' She waggled jazz hands in emphasis. 'I thought it was going to be fun.'

'Fun?' echoed Hetty in a flat voice. 'Well,

I don't know about you, but I'm having a *ball*.'

'Oh, Hets,' Liv sighed, hooking her arm around her friend's shoulder. 'Honestly, Tony told me it'd be a laugh and good exposure.'

'Exposure?' Jez said. 'Considering what just happened, you should be on your knees totally praying you *don't* get any exposure.'

Hetty felt her insides tighten. 'What do you mean?'

'Think about it. I know they say there's no such thing as bad publicity, but do you really want that getting out?'

'It must already be out there though,' Duff said thoughtfully. 'Or else, how would anyone know?'

Hetty's guts twisted and again she had the horrible sensation she was going to be sick.

'Oh God, the papers will totally crucify my mum,' she moaned, clasping her hand to her mouth. 'She parked in a disabled space outside Waitrose once and it made the front pages. What are they going to make of this?'

She didn't mind her friends knowing the truth. Surprising, really, the liberation finally admitting it had brought. At last with the secret out she had nothing left to fear, nothing to run away from.

Although at that exact moment there was nothing

Hetty wanted more than to run out of the house, out of the gates, through the village and to home, where she could shut her bedroom door on the world.

The mean voice in her head: *Oh Hetty, you know you're useless at running.*

The absolute killer though was the idea that her actions could destroy her mum. Hetty thought about her increasing hours in the office. Of her lined face and the way her clothes hung off her. Of the nights Hetty heard her crying in bed, softly, discreetly. And, OK, Hetty might not always see eye-to-eye with Rose Barraclough MP, but she was her mum and Hetty loved her.

Work had given her something to focus on after dad died. And the thought she could have jeopardised that sent fresh waves of misery coursing through her.

'I still can't understand how Cass could have found out,' Liv said. 'The only way anyone could know this is if they've spoken directly to both girls.'

'One of whom may be dead,' Jez reminded them.

'Don't be stupid,' Duff scoffed. He patted Liv lightly on the top of the head. 'Ignore him. Someone photoshopped it to try and get a rise out of you.'

'It worked,' Liv said quietly.

'Yeah, but you said this girl's hood was all . . .'

He mimed peering out of a deep cave. 'And it was dark and raining.'

'And it was a long time ago,' added Hetty.

'And I only saw the back of *a* head in the photo,' Liv admitted. 'There's no proof it was *her* head.'

'Exactly.' Duff squeezed his arm round Liv's shoulder in a brief conspiratorial hug. 'Jesus, I wouldn't recognise a girl I met last weekend, never mind last year. So don't get all wound up over that, seriously. Forget it. No, what we need to focus on is how Pretty Vacant got to these girls. They must have had a rant on Facebook or whatever, and when the researchers Googled your names, it came up. I mean, they'll have known what you were called, right?'

Hetty's voice approached a wail. 'If the research team could find them, that means the papers can too. Oh God, they're going to sell it to the papers.'

Clasping her hands to either side of her reddening face, the room wobbled under a film of tears.

'Nah,' Duff said. 'They'll have made them sign something and paid them off so they don't talk.'

Liv gave a rueful smile. 'I guess we won't find out until we're on the telly.'

'That's if you're chosen, of course.'

Hetty let out a startled gasp. Cass had come back in, silent as a ghost.

'Jesus!' Duff said, clutching his chest dramatically. 'You nearly gave me a heart attack.'

'Sorry about that,' Cass said, dropping her phone back into the bag. She placed it on the worktop. 'Honestly, the quest never ends. OK, Jez, your turn on the couch, please.'

Prickles flared through Hetty's foot, making her gasp again when she stood up, flexing and pointing her toes in an attempt to get the blood supply to return.

'Pins and needles.' She hopped over to the floor lamp and lowered herself into the space vacated by Jez. The floor was still warm.

Jez hesitated by the sofa, his forehead creased, his mouth pulled down in a frown.

'I'm not sure about this,' he said slowly.

'About what?'

'The whole thing. I'm not sure it's a good idea. What Hetty said about the waitress, she could get into a lot of trouble, and I'm not sure any of us should be putting ourselves in that position.'

'Does that mean there's something you need to tell me before we start?'

Duff interrupted with a laugh. 'You're barking up the wrong tree there, Cassandra. Jez is a total saint. He's never put a foot wrong in his whole life.'

Jez waved his hand, dismissing Duff's words. 'What I mean is, you say this won't ever be made public. But how can we be absolutely sure?'

Hetty turned her head at the sound of Liv's murmuring agreement.

'Look,' Cass said. Her sandals clattered in the silence as she moved to the camera on the sideboard. Now she was facing all of them. 'Liv and Hetty, I know you're feeling emotional about this, but let me tell you, you've both been amazing. Think of this as an exam. Pass or fail, that's all anyone will ever know. The process stays in this room. Unless any of you choose to take it outside, of course.'

Hetty glanced at Liv; both girls nodded. But Jez was still shaking his head.

'Sorry, Cass, but I'm still concerned for the girls.'

'Which girls?' Cass snapped.

Jez's brow knitted in confusion. He raised his voice in an *Isn't-it-obvious?* question. 'Liv and Hetty?'

Cass nodded slowly.

Hetty picked her phone up and began turning it over and over in her hands.

'I think if you're really worried,' Cass said, 'the only option is to withdraw completely from the casting. I can let Tony know that you've decided to drop out.'

'No!' exclaimed Liv, wide-eyed. 'We don't want to drop out, do we, guys?'

She looked round the room, but before anyone could reply, Cass continued. 'Of course that's not what I want. Both the performances we've seen so far have been very strong, exactly what we're looking for. It would be a pity to stop now.'

Performances? It certainly hadn't felt like a performance to Hetty.

A sudden draught chilled her. Weird. She'd been burning up all morning and now her skin felt clammy with a slick of cold sweat. She shivered and tugged her sweatshirt down. But the movement dislodged her phone and it skittered on the polished floorboards.

'Sorry,' she said meekly, shuffling to retrieve it.

'Actually . . .' Jez stared as Hetty picked it up. 'Actually, Cass, I've had an idea that might just work. We could film you explicitly confirming complete confidentiality.'

'Yes!' Hetty leapt to her feet, clutching the phone. 'That's a great idea. Right?'

Duff and Liv murmured their assent.

Cass gave an amused smile. 'That is in black and white in the contracts, you know. And recorded on these cameras. But if you'd like your own copy for reassurance, I'm happy to be on the other side of the lens for a moment.'

'I'll do it,' Hetty said quickly. She stood in front of Cass, turned the phone sideways and looked back over her shoulder at Jez.

'You going to ask the questions?'

He nodded.

She framed Cass's face. 'Ready?'

Cass gave a thumbs up.

Hetty tapped record.

Jez spoke slowly, enunciating each word. 'Cass, please can you confirm that the casting auditions today for Pretty Vacant Productions will never be made public.'

Cass nodded and stared straight into Hetty's phone.

'I can indeed. I, Cassandra Verity, confirm that none of the footage recorded here will be broadcast or discussed outside this room by me or anyone I know.'

Pause. She crinkled her forehead and went on. 'The only way any of this will leave this room is if one of you chooses to take it outside. Maybe you should film each other agreeing, in case there are any misunderstandings later on.'

Hetty nodded and turned the camera on herself. 'I won't tell anyone.'

She panned the room, alighting on each of them in turn.

'I won't tell,' said Liv.

'Me neither,' said Duff.

'I won't speak of what has happened in this room,' Jez said solemnly.

'Happy?' said Cass, while Hetty tucked her phone away.

'Yes,' she said, feeling her shoulders relax just a little.

'Great, glad that's all cleared up,' Cass said. 'Jez, are you ready? We really need to get on with things if we're going to wrap up on time.'

Hetty watched his tongue protrude slightly between his lips, the way he always did in tests at school.

'Come on,' Duff said impatiently. 'There's no dirt to dig on St Jeremiah. This is going to take, like, five minutes. Then she can spend hours on me.'

He gave a suggestive wink.

Jez sank on to the couch, his body and head upright.

He crossed his leg over his knee and jiggled his foot. 'Ready.'

Cass stepped back behind the laptop. 'Don't worry. What can a saint possibly have to worry about?'

⊙ 9 JEZ

1 July, 12.15 p.m.

Jeremiah Livingston's earliest memory involved looking down on people.

Bald heads, curly heads, blonde heads, one spectacular rainbow hat topped with a fat orange pompom. Perched high on his dad's shoulders, he was a giant towering above the roar and crush of the crowd. Hat Lady shook her cardboard sign at one of the shouty men and the pompom wobbled irresistibly.

Could he reach it? He stretched out his arm, felt Daddy's strong grip on his legs.

He rocked as he leaned over further. The woolly ball bobbed tantalisingly close. So close. So close. So . . . Rough wool brushed his fingertips.

Then little Jerry felt himself whirled away, one hand

blindly digging into his father's dense hair as he watched the lady, then the hat and finally the pompom swallowed up by the crowd.

He'd filled his lungs ready to unleash a wail when a shape caught his eye.

Who was this man? And why was he holding a big black box in front of his daddy?

Mr Livingston Senior's deep voice vibrated through his chest. Big words came out of his mouth – 'integration' and 'compassionate' – complicated words little Jerry had heard plenty of times before, but still didn't understand. What he *did* understand was that his dad's soft beard tickled his legs deliciously. He promptly forgot all about the pompom and squirmed so suddenly his dad had to take a step to regain his balance. Instead of a scream of frustration, he let out a gleeful giggle.

Immediately, the stranger in front of them swung the black box up towards Jez until huge brown eyes, curly black hair, round cheeks reflected back from the almost-mirror. He giggled again. That's me! Smiles bloomed across upturned faces.

'Wave at the camera,' his dad had whispered. Jez obediently flexed his chubby fingers to the crowd's further delight and everyone cheered.

144

That was Jez's first taste of the limelight. And it tasted good.

It wasn't Jez's first protest march though. As far back as he could remember, a photo of his mum waving a placard inscribed with 'No tuition fee increase!' in firm black capitals had been tacked to the wall of her study. And closer inspection, under the broad length of cloth wrapped around her chest, revealed the outline of a baby. But it was the day of the pompom that had woven itself into Livingston family lore. No matter how many times he'd been photographed or filmed since then, the experience never lost its gloss.

And today was no exception. OK, so Hetty and Liv had had a hard time, but with the tangible promise of privacy safely stored on Hetty's phone there was no reason why they should drop out.

Yes, lucky for Hetty and Liv he was here. They could all relax now.

He set his shoulders, rested his elbow on the arm of the sofa and, casting a satisfied downward glance at the firm curve of his bicep, angled his head very slightly to the left.

'Ready,' he said.

The screen flickered. Jez appraised the stilled image:

the bookcase with the oh-so-casually arranged spines of political biographies and tales of ordinary struggle. *Good.* The awards carefully pushed back until they were just visible. *Good.* A curated selection of newspaper clippings pinned to the wall. *Good.* Only when his eyes caught the framed picture of Prince Harry shaking his hand did he feel a niggle of doubt. Overkill?

No. Everything was good.

Cass looked happy, and if Cass was happy, then so was he.

She pressed play. Jez sat back and watched himself.

Jeremiah Livingston Pretty Vacant Productions #3

Hi, my name is Jeremiah Livingston and I'm eighteen years old and this is my application audition for Pretty Vacant Productions. So I need to tell you some facts about myself. I am, or rather I was, a student at St Benedict's school and I'll be going to Cambridge in October to study law.

[*Camera keeps a steady focus on him.*] Ever since I can remember, it's been my ambition to become a lawyer. The law is in my blood, you could say. My parents are both in the legal profession: my mum's a judge and my dad's a barrister.

You might be wondering why a man like me would be interested in auditioning for a reality TV show, so let me tell you. Firstly, because my friend Liv asked me and I'm too much of a gentleman to turn her down. Reason two is I have an ulterior motive: to demonstrate to the nation that not all young adults are the selfish and self-obsessed narcissists endemic to the press. Some of us do think about others. [*Leans closer to the lens.*]

My parents helped me to set up a charitable organisation called Connecting Together and I am passionate about it. We work in partnership with local groups and businesses to provide opportunities for disadvantaged children and young people. My main role is to promote the fundraising. I've organised balls, auctions, celebrity sports matches and, wow, too many to mention. I was honoured to receive an award at Buckingham Palace last year, along with a couple of other team members.

What else? [*Leans back, folds arms.*] Studying and my charity work take up most of my time, but my one indulgence is my car. My parents bought it for me for my eighteenth and it's a beautiful silver BMW 4 x 4. Top of the range. [*Grins.*] I know, I get it. You're thinking self-indulgent hypocrite, he could have used that money for noble causes, but hear me out. We live somewhere really rural and the roads aren't great. The only way my parents would let me drive around here is if I had a car that could cope with the atrocious driving conditions we have in winter. So, although I'd have preferred something less ostentatious, it genuinely was the most practical option. [*Nods rapidly.*]

The last thing I guess is to say thank you for the opportunity to speak on behalf of the young people who are determined to change this world for the better. Please choose our team.

Was it twenty or twenty-five? Jez lost count of the number of times he'd rerecorded the footage. What was certain was he'd spent forty-eight hours virtually doing nothing else. Rewinding, checking, scrutinising every minute detail to get to this point. He'd nailed it early on, the elusive blend of entertaining and serious that was surprisingly tricky to pull off. But then, playing it back, he'd realised his eyes were too fixed on the script he'd pinned to the wall, giving them a weird glassy look.

So he'd memorised the whole thing. After all the speeches he'd delivered, reeling off a monologue was the proverbial walk in the park. But bizarrely, sounding genuine had proved almost impossible, at least at first.

Too try-hard, like a dodgy car dealer. Stilted. Insincere.

Cass gave him a reassuring smile. 'Well, we can all see why Jez's audition stood out. And to have set up a successful charity at your age – that's just phenomenal.'

Those self-deprecating shrugs, the timely serious expressions . . . in the end it'd all come together brilliantly. He glanced up at the screen where his face, caught between two frames, fluttered oddly. Was it necessary to pause it at such an unflattering moment?

'Oh, I'm just part of an amazing team,' Jez said, straightening his trousers and leaning back. 'Really. It's a group effort. I'm just a cog in the machine.'

From his place on the floor, Duff made a sound midway between a laugh and a cough. *Pfff.* He waved his hand. 'Don't mind me. Something caught in my throat there.'

Why the hell had Liv asked Duff?

Right through school, Duff had been an embarrassment. Not only to himself, but to the St Benedict's community as a whole. Admittedly, the football and cricket teams relied on him and he was far from academically stupid. But in every other way Duff was a fool.

And now, here, jeopardising their chance to be on the

show. Jez's big opportunity to make a name for himself before he started uni.

And promote Connecting Together, of course. That went without saying.

He'd never admit it to anyone, but he'd started having dreams about uni. Ridiculous dreams, of the walking-through-the-hall-and-realising-he-wasn't-wearing-trousers variety. Or sweating helplessly in an exam hall, watching the clock speed to the end before he'd even put pen to paper.

Frankly speaking, at St Benedict's he'd been a big fish in a little pond. Head boy, cricket captain, school ambassador . . . And now it had finished. Not just the sports teams and prefect stuff, but every last bit of it. Uniform, dormitories, never-ending assemblies, hours of prep. Taking with it a huge chunk of Jez's identity.

St Benedict's had been his life since he was four years old. Without it, he felt adrift, thrown in at the deep end with only a leaflet on how to swim.

Hence the bad dreams.

And then there was home.

The media said no teenager got on with their parents. Writers told you how dumb they were, how unfair. You learned to roll your eyes at the sight of them, slam the door when they asked you to do something, and disagree with every word from their dumb mouths. That was the law, right?

Wrong. At least in the Livingston household. He'd googled the drive to Cambridge. It was a long, long way.

He realised Cass was watching him.

'Sorry, what was the question?'

'I asked you about Connecting Together. About the slogan.'

'We're all in it together?'

'Can you talk us through what that means?'

He nodded, uncrossing his legs and placing his elbow on the arm of the sofa. The last traces of nervousness dissolved. This was going to be a piece of cake. 'Basically, exactly what it says. You know, the chances of any of us being alive are a billion to one as it is, so the odds of us existing at the same time . . .' He swept out his arm, taking in everyone. 'Us four, and you, Cass, are just incredible. So that's got to mean something. There has to be a purpose to it.'

Even as he was speaking, he was conscious of Cass making notes, the steady red dot of the cameras trained on him. Maybe if he shifted slightly further down the sofa his new trainers would be in shot.

'So the "it" in the slogan refers to life,' she said, glancing up.

'Exactly.' He leaned forward, making sure to hold his head upright. He'd had one too many desserts at one too many functions lately and, although he'd kept a shadow of fuzz deliberately, experience had taught him double chins lurked in camera lenses. 'And the together part is that we should all look out for each other. That's what the trust is for, to offer a helping hand to people who need it.'

'What a brilliant way to give back to your community.'

After the total roasting Cass had given Liv and Hetty, he allowed himself to bask just a little in the compliment. 'Thanks, it's really rewarding.'

'And this helping hand, would you say that extends to your personal life?' She laughed a little. 'When you're not wearing your Connecting Together hat, could you say you walk the walk?'

'Absolutely.' He nodded forcefully, the motion dislodging his glasses a little.

Cass sucked in her cheeks as she wrote in her notebook.

'That's certainly wonderful to hear. I take it that beautiful silver beast parked outside is the car you mentioned in the application. Certainly looks a prized possession.'

Jez puffed up with pride. 'My parents bought it for my eighteenth.'

'Lucky you,' Cass said. 'When was that?'

'Near the beginning of the year.'

'You'll have to forgive me, I'm not great with car makes and models, but doesn't it cost a small fortune to run?'

Jez nodded. 'And insure. But Mum and Dad take care of that and, like I said in the application, I need something reliable because the roads around here can be terrible.'

'I can imagine what it must be like in winter. Winding roads and snow. Black ice. No street lighting.' Cass mock-shivered. 'Of course you need a car you can rely on. I completely sympathise with you there. Anyway, back to the events you organise. You mentioned something about a charity auction and, Liv, I think you touched on it too. That sounds like a glamorous event. Where was it?'

'The Manor Country Club.'

'Ah. There,' said Cass meaningfully. 'The same place where the thing with the waitress . . .' She gestured at Hetty.

'Er, yeah. There.'

Jez deliberately avoided looking in Hetty's direction. At some point, when this was over, he'd have to consider Hetty's colossal bombshell and his reactions to it. But right now it was up to him to repair the damage.

'Brilliant night,' Liv chipped in. 'I donated a Hermès bag.'

'Yeah,' said Duff. 'My brother and his wife won a weekend in Venice. I didn't get anything though.'

'Sounds like a roaring success,' Cass said. 'Why don't you tell me all about it, Jez? Start from the beginning.'

He settled back in the sofa, pulling his ankle across his knee.

'Well, because of the snow we were a little late getting started —'

'No, earlier,' Cass interrupted. 'Start earlier in the day.'

'Well, I spent most of the day there supervising the set-up and about four o'clock headed home and, yeah, had a quick shower, got dressed and drove back in time to meet the camera crew at 6 p.m. Like I said, the weather had turned and the roads aren't great around here, so I was maybe fifteen minutes late. The interview went fantastic—'

Cass interrupted again. 'So you left your house

at about quarter past five and set off for the Country Club.'

'Ye-es.' Jez felt a flicker of unease creep up his spine.

'And was there anything unusual about that journey?'

The atmosphere in the room shifted very slightly.

The flicker fanned into a flame. He twisted one cufflink round, then the other.

'Nothing in particular.'

'What about the weather?' Cass insisted. 'Was it cold?'

'Naturally. It was February.'

'Icy?'

'Yes. Why?'

'I'll get to that in a moment,' she said with a clear edge to her voice.

He heard Hetty draw in a breath. 'Jez, if anything happened,' she said softly, 'you should probably just admit it.'

He interlaced his fingers and steepled his thumbs. 'Nothing happened.'

Cass raised an eyebrow. 'Are you sure?'

'Not that I recall.'

'OK, maybe you'll let me remind you,' Cass said. She pushed the ribbon to one side and turned a page in the red notebook. 'It must have been quite a rush, getting home and back in time for the start. It was icy and cold. As you pointed out, the roads can get treacherous around here and . . . how long had you had your licence?'

'Two weeks,' he said abruptly.

In two strides he had his finger on the camera facing the sofa.

'Actually, I'd like to switch this off. I've changed my mind.'

'Whoa!'

Duff pulled himself to his full height and manoeuvred between Jez and the camera.

'Liv and Hetty both took one for the team. If they can do it, I think you should too, mate.'

'Get off me,' Jez said, embarrassed to hear his voice rise to a near-squeaky pitch.

Duff held his hands above his head. 'Not touching you.' He turned to the other cameras, arms still high. 'Not touching. Not touching.'

'Please sit down, both of you,' said Cass mildly.

'What did you do, Jez?' Duff said, sliding his back down the wall. He tucked his right foot under his right knee and placed his palms flat on his thighs.

Jez felt his shoulders go rigid. 'Nothing,' he said. 'Like I told you.'

'Maybe you could just talk us through the drive to the Country Club,' Cass said.

'Remember the video,' Hetty said, waggling her phone at him. 'It won't get out.'

Jez pointed at Duff. 'What about him? He won't keep his mouth shut, will he? No, sorry, Liv. I'm out of here.'

'You can't leave now!' Liv whirled round on the stool. 'If you drop out we're all out.'

He shrugged and put his hand on the latch.

'That's OK,' Cass called out. 'I can tell them what happened. I know exactly what you did. Lena told me.'

He turned slowly.

'We don't have much time left,' Cass said, gesturing at the sofa. 'You tell them or I tell them. You didn't know she was called Lena, did you? Because you drove off before you could find out.'

Liv's hands flew to her mouth. 'Oh my God, did you run someone over?'

'Don't be stupid.' He sat heavily on the sofa, his mouth set in a sullen line. 'OK. But if a word of this leaks out I will be suing Pretty Vacant Productions.'

'Understood. For the cameras, please. What happened on the day of the Valentine Auction?'

⊙ 10 JEZ

Five months earlier

Every year, St Benedict's invited a guest speaker in for prize-giving and, during the course of Jez's secondary education, he'd politely tried not to fidget through a range of addresses from 'fascinating' to 'excruciatingly tedious'.

The boring physics man. Hetty's mum. The police inspector whose short skirt caused a tremor of disapproval from some of the stuffier members of staff.

And now, in his final year, they'd got around to asking Justice Patience Livingston, his mum.

In every court case, his mum explained, there is a moment, usually when one side is summing up for the jury, on which the case depends. The tension is palpable. You can see it on every face, you can feel it in the air like a change in the wind direction, and you can almost taste it in the back of your throat. Everything that has happened to this point, everything that happens after, all of it hinges on this crucial moment.

A decision.

If Jez rewound the series of events that led up to him freezing his testicles off on a frosty verge by the B1376 on Valentine's Day, physically convulsing with panic, then pinpointing the first hinge moment was easy.

It was when he discovered that Pat the housekeeper had failed to iron his shirt properly.

He was running ridiculously late. Temporary lights in the centre of the village plus the threat of freezing fog slowed the traffic to a snail's pace. An arthritic snail's pace.

Shower. Shave. Quick tooth brushing and his phone beeped. One hand clutching the towel round his waist, he snatched the phone off the top of his dresser. A text from Hetty.

Sorry, Mum won't drive in this. Can't make the auction. xxx

For a second he contemplated making a detour to collect her. No. No time. Hinge moment number two. With hindsight, making the trip to Hetty's would have been a smart move.

The coat hangers rattled as he walked his fingers over the row of neatly pressed shirts before unhooking the one he'd bought specifically for tonight.

No Hetty. And no Rose Barraclough MP, which meant a potential drop in media interest. He frowned, peering into the darkness. How many other people would let themselves be put off by a bit of frost? He did not need empty tables. He did not need unsold auction lots. If tonight was a flop . . . God, it didn't bear thinking about.

He held the shirt to his nose, inhaling the new linen scent before slipping it off the hanger and on the bed.

What?

He couldn't believe it. The housekeeper had ironed the front, the back, the collar and only one sleeve! Creases criss-crossed from the shoulder to the cuff! How could he turn up like that?

Fastening his dressing-gown sash, Jez thundered down the stairs, holding the shirt.

Thank God. Pat was in the kitchen, illuminated by the spotlights over the granite island, one arm in her coat sleeve.

He tried to keep the irritation out of his voice.

'Glad I've caught you, Pat. Thanks for washing my new shirt. I was just wondering whether you could run the iron over it really quickly?'

Pat's swollen hand, veins like blue string, emerged through the cuff of her grey anorak and threaded through the handle of a plastic bag. She smiled apologetically. 'If I miss this bus I won't get home until late.'

'I wouldn't ask except . . .' He lifted the shirt up by the shoulder seams for inspection. 'I really need it,' he finished firmly.

The unspoken 'you should have done it properly in the first place so it's your fault' hung in the air.

She sighed, took her coat off.

'Thanks,' he said. 'I'll be back down in a minute.'

Jez had dismissed the faint twinge of guilt before he was even up the stairs. What would the TV crew think if he turned up crumpled like a homeless person? Everything came down to professionalism: you had to look the part otherwise how could you expect people to hand over their money?

Hurriedly fastening his (passably ironed) trousers with one hand, he yanked out the drawer in which his ties nestled, coiled in individual compartments. Selecting this one had been easy: blue silk damask with a subtle peacock feather print that could only be seen if you looked closely enough.

He heard the front door slam shut and automatically glanced out of the window. The sky had turned inky blue, the first stars visible as tiny pinpricks of silver. A crescent moon shone brilliant white in the darkness Pat was hurrying into, white breaths billowing from the hood of her anorak.

Orange flashes lit the distance and a faint rumbling grew louder. At least the gritters were out.

Jez frowned. Could she really have ironed it that quickly?

Back in the kitchen he found the shirt back on the hanger and hooked on to one of the bar stools by the counter. Looked OK. But then —

He spotted a single knife-edge crease between the cuff buttons and the elbow. The old cow!

Well, there was no time now. He pulled it on, his already heavy mood darkening further when he looked at the clock. He buttoned the shirt rapidly, put a neat Windsor knot in his tie and grabbed his suit jacket. From the cloakroom he selected a thick overcoat with a faint herringbone pattern, and ran to his car.

The temperature had dropped so quickly that

clusters of ice were forming on the windscreen. In. Turn ignition. Headlights came on automatically, icing the driveway in white light. Sparkles of frost gleamed on the pavement. He shivered, turned the heating to full.

People must expect the cold in February, right?

He drove past Pat, hunkered into her coat and illuminated by the weak light above the bus shelter. He kept his eyes fixed ahead, floored the accelerator and drove right past in a spray of loose salt from the freshly-gritted road. If he'd had time, of course he'd have stopped and offered her a lift. Mind you, he had no idea where she lived. It was probably miles out of his way. And she could always ring for a cab, couldn't she?

About half a mile up the road he caught up with then overtook the bus. The windows were completely fogged with condensation. One bus an hour! How did he manage before he got the car?

Between his house and the Country Club lay several miles of unlit, twisty-turny roads. The gritters hadn't made it this far and permanent triangular signs warned of the dangers of ice.

The ice that had thawed during the day hadn't disappeared completely and, with the sudden drop in temperature at sunset, the slushy puddles were refreezing. Twice he felt the steering wheel go suddenly light in his hands. Felt the heart-stopping thud as the anti-lock brakes kicked in.

Very occasionally a car would creep down the other side of the carriageway, turning out from some lit cube of a farm or secluded home. But otherwise nothing was on the road besides Jez and his car.

And his growing fear of being late.

His phone rang, the sound unusually loud in the eerie darkness.

Dad's voice sounded tinny through the speakers.

'Jez, are you on your way?'

'Yep.'

'The TV crew is waiting.'

Jez smacked the flat of his hand on the steering wheel.

'Won't be long,' he said, keeping his tone light.

'OK,' Dad said. 'But drive carefully. It's icy out there.'

No kidding. The signal cut out at the exact second Jez felt the car slide away from him again. This time though he couldn't catch it. Panicking, he wrenched the steering wheel back, sending the car into a graceful sideways skid across the other side of the road.

Headlights!

He had half a second's glimpse of a red bonnet, silver

doors and the terrified pale face of the driver before he swerved back hard left, missing the ancient hatchback by inches. There was a loud thud and he slammed on the brakes and – thankfully! – this time the BMW came to a halt. He killed the engine and the lights faded.

His heart was pounding so hard his ribs vibrated. He put his head down over the steering wheel and took a deep, deep breath, willing his heart back to a steady rhythm.

With trembling fingers, he clicked the door open and climbed out into the cold to quickly check for damage. He rubbed a mark with his thumb. It faded to nothing. No scrapes, dents, only a few spatters of mud. Thank God the airbag hadn't gone off.

Emergency stop. That's all it was, an emergency stop.

He turned. He could make out the silver panels of the other car, lights angled into the field. The mismatched bonnet had crumpled and, above it, a fence post was apparently floating in mid-air. The driver's door opened. A figure emerged.

He couldn't make out any details, but the figure could walk and that was all the information he required.

At that exact moment, the phone began ringing in

his car and a collection of thoughts whirled through Jez's mind.

I'll lose my licence. I'll lose the car. Mum and Dad are going to kill me. I'm going to get caught up here for hours. The police will get involved. The police!

Adrenaline kicked under his ribs, triggering panic mode. Jez's brain liked to weigh things up in his own time, spending a while figuring out the pros and cons of each option before making a best-case decision.

This was the third hinge moment.

He did not have time to weigh up anything.

He did not make the best-case decision.

He hurried back towards the car.

'Hey!' shouted the figure, waving at him. Definitely female. 'Over here. My car is in the ditch. Can you help?'

He peered down the dark road in the direction he'd just come, cocked his head, listened. No cars. And more importantly, no CCTV on roads like this.

Ramming the gear stick into reverse, he floored the accelerator, praying the wheels wouldn't find another patch of ice, and turned. First gear. Sped off away in a squeal of tyres and heavy breathing. Almost light-headed with relief, he flicked his eyes up to the right and caught the static tail lights of the other car dwindle to red pinpricks before disappearing in the darkness. Only then did he slow down a little.

The illuminated digits on the clock told him he was late, but not catastrophically, and the glow in the distance gradually took the familiar outline of the Country Club.

The guard at the gate knew him by sight and waved him through, and as Jez drove up the long driveway he let out a sigh of relief. He'd made it.

Now he just needed a minute or so to calm down.

What did they call it in war movies? Collateral damage, that was it. Not stopping to help might seem like the wrong thing to do, but honestly, what use would he have been? He couldn't fix cars and he was hardly dressed to go pushing them out of verges. Anyway, she'd have a phone, everyone did, and she'd be better off calling the AA or a friend who knew about cars.

Plus, replaying the events in his mind, he actually hadn't done anything wrong. OK, he'd wandered across the central line, but that was the ice, not his fault at all. She could have pulled on to the verge to avoid him. He didn't barge her off the road; there was no real danger of a collision.

And, most importantly, if he'd stopped he wouldn't have arrived in time to do the TV interview. The more people who knew about Connecting Together, the more successful the charity would be. So collateral damage it was. If someone, who drove a car that frankly did not look roadworthy to start with, had to hang about for a bit waiting to get towed away, that wasn't nearly as important as the hundreds of young people's lives that could potentially be improved if tonight were a success.

Inconvenience for someone who probably shouldn't be driving anyway versus helping hundreds? It was a no-brainer.

By the time Jez pulled up in the car park, he'd managed to convince himself he'd done totally the right thing.

He hooked his finger behind the sun visor and tiny LEDs lit up the concealed mirror. He patted his hair, checked his face from both sides. And said out loud to his reflection, 'You got this.'

His dad was waiting in the foyer. 'Hurry up,' he said, shooting his arm out to examine the gold Rolex on his wrist.

'Sorry,' Jez replied just as the runner from local news walked around the corner.

He strode towards her, holding out his hand. All the professionalism you could ever dream of, and the memory of what had happened by the side of the road faded completely.

1 July, 12.25 p.m.
'We raised almost ten thousand pounds that night,' Jez said, folding his arms.

Cass had the notebook open in front of her. 'That's amazing,' she said. 'Was there any damage to your car?'

'Look.' He pinched the bridge of his nose. Was he coming down with something? 'I agreed to explain what happened that night, which I have, but that's me done now. I hold my hands up. I hit some ice, I swerved. End of story.'

'Don't you want to know what happened to Lena, the girl you ran off the road?'

He shuffled in the seat. 'I am not comfortable with the term "ran off the road". It was an accident due to the icy conditions. Completely out of my control.'

'But it was within your control to stop and help, wasn't it?' Her cheerful tone was at odds with her words. She flicked her eyes at the others. 'I mean, you guys, in the same situation, would you have left a lone driver stranded on a country lane on a night when the temperatures were expected to stay below freezing?

170

Hetty, Declan, would you abandon the scene of an accident?'

'I don't drive,' Hetty said quickly.

Jez found he couldn't meet her eye. He couldn't meet anyone's eye. The unspoken fact of Hetty's dad's accident hovered over him like a malignant thought bubble.

'She got out of the car so I knew she wasn't injured,' he said.

Cass continued flipping through the notebook. 'Well, let me just tell you quickly what happened to Lena.'

Jez shrugged.

'OK.' Cass located a page. 'Lena. Here we go. Our research team says Lena had worked very hard to save up for her car. Admittedly, it wasn't a top-of-the-range BMW, but it meant a lot to her. She had a job working for a florist, cash in hand. She'd had a recent change in circumstances that made it difficult to work any other way.'

'Working illegally?' Jez said.

Cass ignored the interruption. 'Lena was on her way to deliver the last Valentine's Day bouquet. She'd been flat out since 5 a.m. and was already dreaming of a hot bath and her bed.'

Jez opened his mouth, but Cass held up a finger

for silence. 'And before you say the accident must have been her fault because she was tired, remember she was driving within the speed limit and on the correct side of the carriageway. Unlike you. So she was driving safely and legally along a winding country road when you appeared out of nowhere, forcing her to swerve into a fence post, writing off her car in the process. Apart from the fact she didn't have breakdown cover, there was no signal on her mobile phone. She had to leave the car and walk almost six miles in sub-zero temperatures, shaken up from an accident.'

Jez stared at his cufflinks. Solid gold. They'd been a present from his grandparents for his eighteenth birthday.

'Anyway, before I bore you too much, let me finish. Because you didn't stop, Lena wasn't covered for the cost of her car on the insurance policy. She also had a loan for the repayments on the car. Payments she couldn't keep up with because she could no longer work for the florist as she no longer had a car. All of which could have been avoided if you'd made different decisions that night.'

Duff whistled a long, low sound.

Jez processed this information. How could Cass —?

'So that's what happened to Lena after —'

'Hang on a minute.' He placed both his feet firmly on the ground and got back up, his brow furrowed with suspicion.

'How can you possibly know this? There's no way that girl identified me. She didn't get close enough.'

Cass waved dismissively. 'I have no idea. I'm just going on what the researchers say.'

'No.' He put his chin in his hand. 'I get how you could have found out about what Liv and Hetty did, because that farm girl and the waitress at the club could have easily found out who they were. But not me. This was a thirty-second encounter. Max. There's just no way. It was dark and it happened so quickly. Absolutely no chance.'

'She must have got your registration number,' Liv said.

Jez shook his head, trying to align his thoughts. 'No. It was just too fast. And too dark. She can't have seen it.'

'Have you ever told anyone about it?' Hetty asked, nibbling the skin around her thumbnail.

'Why would I?' He spread his arms wide. 'I'd completely forgotten about it myself until Cass brought it up.'

'Dashcam,' Hetty said suddenly. 'My dad had one. The insurance company and the police used it as evidence against the lorry driver. I bet she had a dashcam.'

Jez bent down to wipe an imaginary smudge off his trainers. 'Not in a crappy car like that. Probably didn't even *have* insurance'

'What happened to her afterwards?' said Liv. 'Was she OK? I mean, I know you've not spoken to her, but did she get things sorted?'

'I get sent these notes from the research team and that's as much as I know,' Cass said, tapping the book with the flat of her hand. 'Anyway, time is not on our side, guys. So, Duff, maybe if you and Jez swap places now.'

There was a pause.

'Duff?'

'Oh, sorry,' Duff said, feigning surprise. 'Miles away trying to figure out what you're going to pin on me. Was it that girl who . . .? Or maybe the one who . . .?'

'What were you expecting from today?'

'Oh, I don't know.' He cracked his knuckles. 'Not this gloom-fest. Liv kind of made out it was going to be a laugh.'

'I thought it was!' she protested.

He smiled at her. 'I know you did. I'm not having a go at *you*.'

'As I explained earlier,' Cass said, 'we have a tough

⊙ 11 DUFF

1 July, 12.41 p.m.

Declan Duffy liked girls. *Lots* of girls.

Very often, before he'd even finished with one girl, he'd be craving the next. Like a food addict devouring a twelve-inch pepperoni while he fantasised about a kebab. Some days he fancied chips and sometimes he fancied a steak. Or sashimi. Or a Sunday roast. Or an entire lobster.

The one thing he *never* fancied was going on a diet. And with a brother who owned a promotions company, he never had to. Once he'd spent weeks chasing after a particularly gorgeous girl who worked for his brother. When she finally gave in, he lost interest almost straightaway. It was the thrill of the chase with Duff, and he couldn't stand to lose.

'So,' Cass said, 'I understand you've been helping out

your brother over the holidays. Can you tell me a bit about that line of work?'

'*Riiiight*,' said Duff, then sat up straight suddenly. 'Oh, oh, *oh*. My big bad is going to be some mischief I've got into with Crawlers, isn't it? Can I guess?'

Cass inclined her head a little. 'Crawlers?'

'Yeah. You'll know this from your "researchers".' He mimed air quotes. 'But for the cameras . . . Crawlers is my brother Gus's business, which in the last five years has gone boom.' He mimed exploding hands. 'Basically, it's pub crawls in fancy dress. Mainly students, but hens and stags and eighteenths, that kind of thing. So he themes it –'

'Themes it?' echoed Cass.

'Yeah, Walk of Shame or School Disco or whatever. They turn up in fancy dress and there are, like, games and competitions and drink. Lots and lots of drink. Basically, you pay for the ticket and it's free booze all night.'

'Would you say you were close to your brother?'

'Very.' Duff grinned. 'He's an absolute legend.'

'Tell us about him.'

'Does he have to?' Liv interrupted.

'Is there any reason in particular that you don't want Duff to talk about his brother?'

Liv shrugged. 'I just thought if we're short of time maybe there were more relevant topics to discuss.'

'Your brother,' prompted Cass. 'He got married recently, didn't he?'

'Yep,' Duff said, 'to Claudia.'

'And do you like her?'

'Yeah,' he answered, his expression softening. 'I don't know how anyone couldn't like her, to be honest. I've never seen Gus so happy. She's completely changed him, calmed him down.' He stiffened. 'Hang on. This isn't going to be about Claudia, is it? She doesn't deserve to get dragged into anything. Seriously, she's way too nice. I'm not going to talk about her.'

'Don't worry,' Cass said. 'This isn't about your brother's wife.'

He heard someone breathe in sharply. Liv? He looked across to where she was sitting rigid on the stool. Her head was bent, her hair shielding her face.

Poor Liv. She looked gutted. That thing with the girl on the farm must have really got to her. More than he would have expected.

'Go on then,' Duff said, leaning back into the sofa, legs sprawled under the coffee table.

'Go on then what?' Cass said.

She glanced up from her notebook and threw him a pleasant smile.

'Whatever you've found out. Must be something to do with Gus, right? That's why you're asking all the questions.'

He tapped his foot under the coffee table almost without being aware of it.

But Cass was shaking her head. 'You applied for this, not your brother. It's you I'm interested in.'

'OK,' he said, still smiling. 'You're in a hurry, right? Let's get down to it. Tell me how naughty I've been and we can move on, yeah?'

If Duff were the kind of pushover who allowed himself to get nervous, then the long silence that followed, during which Cass stared at him levelly with an unreadable expression on her fairly pretty face, might have made him feel a little uncomfortable. Maybe fidget a bit. Maybe stare fixedly at the floor. Maybe even allow a few doubts to creep in.

But he wasn't that kind of guy. So he met her eyes and coolly stared back, waiting for her to speak.

In the end it was Liv who broke the silence. Duff twisted his head instinctively at the sound of her sharp intake of breath. She opened her mouth, about to say something.

'What?' he said.

'Nothing.'

'*What?*' He sat up suddenly and crossed his arms.

'OK,' Cass said, with another abrupt change of tone. 'You're right. We don't have time for games, unfortunately.'

'Let me guess. You've found some girl shooting her mouth off about me on the internet.'

Cass inclined her head, but didn't reply.

He blew out his cheeks and exhaled slowly through pursed lips. 'Could be literally anyone. You're going to have to give me a clue here, Cassandra. Like when I did the bad thing.'

'Very recently,' Cass said, picking up her notebook. She ran her finger down the page. 'Saturday, in fact.'

Duff frowned. Last Saturday? That didn't sound right. He distinctly remembered the girl in the club, Jodie or Jade, something like that. He'd bought her two drinks, chatted to her and then, when he was about to make his move, her mate had been sick down her dress and she'd taken her home. Selfish cow.

'Really not ringing any bells,' he said. He picked up the cushion and plumped it up a little before wedging it into the small of his back.

'OK,' said Cass. 'Maybe this will help. You went to a bar called Sugar Sugar . . .'

He looked up and caught her eye. Her mouth was fixed in a smile, but her eyes weren't. He felt a shiver of – what? Not anxiety exactly; Duff didn't do anxious. But something not altogether comfortable.

'Oh God, seriously?' Liv was saying. 'That's low even for you, Duff. What the hell were you doing there?'

'I wasn't there by choice. My brother's thinking of buying it and so we went on, like, a research trip.'

'And what happened at Sugar Sugar?' Cass said. 'For the cameras, please.'

Duff stared at the ceiling. 'If that girl has been saying I made her do anything she didn't want to do, then she's lying.'

'OK,' Cass said. 'Why don't you tell us what happened?'

Six days earlier
'You have got to be kidding me.'

Duff stared in disbelief at the soot-covered building on the edge of town. The taxi's departing headlights illuminated patches of weeds straggling through cracked tarmac before sweeping across a boarded-up office block to one side. Even the cars parked by the entrance seemed depressed.

'Nope,' Gus said, slapping him on the back. 'This is the one, bro. Come on.'

It was the first time the two brothers had been out together since Gus got back from his honeymoon. Duff followed Gus up the steps and past the grunting doorman, Jase stumbling a few paces behind.

Oh dear. Jase.

He and Gus had zero in common, as far as Duff could make out, but it was one of those weird things, where they'd known each other for so long that their shared history kept them together.

Jase was getting married. What was even more astonishing was that Jess was his high-school sweetheart. They'd started going out in Year 9 and somehow forgotten to ever split up and move on. Incredible.

All the years while Gus was playing the field, usually playing several fields simultaneously, Jase was going to university with Jess, moving in with Jess and now about to become Mr Jess.

But the most shocking thing of all – and this was something Duff could never get his head round – was that Jase had never had another girlfriend. He had not so much as kissed anyone other than Jess since he was fourteen years old.

And that was behind Gus's plan to take Jase to Amsterdam for his stag do in three weeks' time.

'What kind of best man would I be,' he'd said to Duff in the taxi on their way over to Jase and Jess's. 'No, what kind of *friend* would I be if I let him get married without knowing what else was out there?'

And that made sense to Duff.

Of course, Jase knew nothing about that. He thought they were just out tonight to have a bit of best man/groom bonding before the stag do proper. And what better place to prepare for a weekend of debauchery than a night in Sugar Sugar?

Sugar Sugar did not smell sweet.

The waves of old sweat and damp that hit him in the lobby intensified in the bar. More gross odours joined in. Blocked drains, dirty urinals and a whiff of fermented school dinners. Duff didn't consider himself to have particularly high standards; in fact he'd spent some top-quality nights in places with similarly sticky carpets and scarred tables. Even so . . .

'Tell me you are not investing in this dump,' he whispered in his brother's ear.

The three of them were snaking round clusters of tables, sparsely populated by seedy men holding pint glasses.

One of the men made a noise like a revving Formula One engine, flushed beetroot red and began to cough so alarmingly Duff fully expected to witness one of his lungs flop out on to the table. It didn't. No one else seemed to have noticed. The man himself continued staring at the girl listlessly dancing on the podium.

They'd reached a booth now, cracked leatherette seats darkly stained with . . . well, no point speculating. He sat down hesitantly, perching as close to the edge as possible while making a mental note to bin his trousers the second he got home.

'Why d'you bring us here?'

Jase, who'd been tanking up since lunchtime, enunciated each word precisely the way very drunk people do.

'Mate, I told you in the taxi,' Gus said, patting Jase's hand. 'This is going to auction next week and I'm putting in an offer. Expanding my empire.'

'What for? It's a dump,' Jase said. 'Can we go?'

'Potential. I'm not interested in this.' He waved his hand dismissively at the punters and the dilapidated stage. 'I'm interested in what it could become.

Crawlers is great, I'm not knocking it. But I want my own venue, somewhere where I'm not giving half the profits back to the bars in commission.'

Duff slurped on his pint, trying not to worry about the smeared glass it came in. 'Yeah, I get that.'

'So,' Gus said, leaning forward as though to put his elbows on the table and then thinking better of it. 'Hen and stags in the bar, birthdays or whatever upstairs. Maybe karaoke. Limo taxi service door to door, the full package. But class, you know. Not —'

'Hello, gentlemen.' The waitress gave them a tired smile. 'Can I get you anything else?'

Tall. Tiny waist, but a bit thin for Duff's taste. Coat-hanger bones across the top of her chest; rack of ribs visible under her crop top with its faded Sugar Sugar logo. Long legs dotted with bruises in minuscule shorts. Very black hair, ironed flat with a square-cut fringe. Under it, dead eyes. Bad skin.

'Bottle of . . . whatever champagne you've got,' Gus said, extracting a couple of crisp notes from the pile in his wallet. 'Let me know if that's not enough.'

'Of course,' she said. 'Champagne. Three glasses.'

Gus looked her up and down, smiled. 'Four,' he said. 'You can come and sit with us.'

They watched her leave. 'Pretty,' Gus said. 'In a skanky, wouldn't-touch-it-with-yours kind of way.'

The champagne (or what passed for it in here) slipped down surprisingly well. And the girl turned out to be surprisingly intelligent, answering Gus's questions about the building and the business, even though she'd only been there a few months. She lived nearby, knew the area and could talk a bit about the regeneration that was going on.

Meanwhile, Jase had given up any effort to enunciate his words clearly and was sitting slack-mouthed with his eyes on the girl's chest.

'Angel,' he said, batting at the name badge on her chest. 'You're called Angel.'

Out of nowhere, a security guard with fists like breeze blocks materialised in front of them.

'No touching,' he growled, in heavily accented English.

Jase slumped back as though he'd been tasered and raised both hands. 'Sorry, sorry.' He sounded like Hugh Grant.

He turned to the girl. 'Sorry, it's just you're so . . .'

Angel twitched her eyebrows slightly at the doorman in what was clearly a prearranged *I'm OK* signal. She smiled kindly at Jase and for a moment Duff had a fleeting glimpse of how, in a different life, she could be pretty. Really pretty.

'No worries,' she said, with a slight roll to the 'r'.

'Are you Welsh?' Jase said suddenly. 'My nan was Welsh.'

The girl laughed, not a dirty laugh; a pure, light sound completely at odds with the surroundings. 'No, I'm not Welsh.'

Duff felt someone grab his wrist. Gus.

'Bro, come with me a minute,' he said, pulling him to his feet.

'Where?' Duff said.

But Gus peered at the girl's name tag. 'Angel? Could you keep an eye on Jason here for two minutes while we . . .' He waggled a packet of cigarettes in one hand and extracted his wallet from his back pocket with the other, thumbing out a fifty-pound note. 'Here, in case he's thirsty. Declan, you come with me.'

'Where?' Duff said again, squeezing past Angel. God, even from here the stink of urinals made his eyes water. He put his sleeve to his nose. The whole place smelled like a zoo.

Jason's voice drifted behind him, choked with emotion. 'I loved my nan. I really miss her.'

A different girl was on the stage now, contorting herself round a pole. Midway through a turn her hand slipped and she scrabbled inelegantly to regain her hold. The watching men broke into a ragged cheer. She spun round, stood upright and scratched her butt cheek with a pointed, glittery nail.

Seriously, this had to be one of the most profoundly depressing places Duff had ever set foot in.

He followed Gus outside. The night air tasted sweet and he treated his body to a couple of mouthfuls.

'So listen,' said Gus, tapping the bottom of the cigarette packet. French, unfiltered and specially imported. Stinky, yes, but compared with Sugar Sugar, the smoke smelled of roses. 'I've got a plan.'

Duff listened.

'Man, that is a bad idea,' he said when Gus had finished. 'On every level, that is bad.'

Gus put his arm round Duff's shoulder, pulling him in close. 'I can't be best man and let Jase promise to forsake all others just like that.'

'*You* did,' Duff replied, pulling away. 'I was there. Claudia. Big white dress. Ring any bells?'

'Not the same. I'd been playing the field since I was fourteen.' He stuck a cigarette in the corner of his mouth and patted his coat pockets. 'Every man should try before he buys and Jase never has. Look, the girl takes Jase back to hers for an hour. What they do is up to them. I am doing Jess a favour. Providing the means to remove any lingering doubt that she is Jase's one and only love.'

Duff's scepticism must have been clear on his face. Gus grinned, cupping his hands around his gold lighter. He lifted his cigarette to the flame.

'Think about it. Nothing happens: it proves he loves Jess and is ready to take the big step. Something does happen: at least it's out of his system before the big day. Then he can play happily-ever-after with Jess and not spend the next forty years wondering *what if*. Every man deserves one for the road, right?'

Duff shook his head. 'What's in it for the girl?'

Gus blew out perfectly formed smoke rings. 'What's always in it for girls?' He rubbed his fingers against his thumb.

When they got back inside, Gus sat by the girl and began whispering in her ear. Duff watched while her smile drooped then hardened like setting plaster. She looked down at the tabletop while Gus talked and Jase flopped his head against the back of the seat and closed his eyes, oblivious to the negotiations taking place on his behalf.

Gus raised his eyebrows and she sighed and nodded once, just a dip of the head.

As arranged, Angel disappeared to tell the manager she wasn't feeling too good. Gus had her address keyed into the maps app on his phone and five minutes later they got up to leave.

'Wha' you doin'?' burbled Jase, staggering to his feet. He bumped the table, knocking the glasses together, the sound drawing the bouncer's attention to their table.

He moved surprisingly swiftly for a man built like a minibus.

'It's all right, boss,' Gus said, laying a twenty-pound note on the table. 'We're taking him home.'

Palming the note without a word, the bouncer followed them along the greasy carpet to the exit. He held the door open and the three of them staggered out like a gross, many-limbed monster.

'Gonna puke,' Jase said in bleary-eyed horror. His face went grey, clammy-looking and a string of saliva dangled from his chin.

Gus stepped back sharply. Duff jumped, but not

before the sour contents of Jase's stomach hit the tarmac and splashed his shoes.

Duff yelled in disgust and tried to get back inside, but the bouncer blocked the door.

'No re-entry.'

He opened his mouth to protest, but the bouncer flexed his meaty hands and Duff thought better of it, going back to where Jase stood doubled over while huge belches tore out of him.

'Need to go sleep,' he mumbled.

'Soon, soon,' murmured Gus in the soothing tones a mother would use on a highly strung toddler. 'First of all, we're going to go for a little walk. Can't send you back to Jess in this state, can we?'

He thumbed down his phone screen. 'Left here. Behind the kebab shop.'

By the time they climbed the narrow steps to her door the girl had changed out of her work clothes into jogging trousers and a grey vest top, her hair dragged back into a ponytail.

'Nice to see you made an effort,' Gus said with a sneer, although Duff thought that without all the make-up and the skank clothes she looked far prettier than she had in the club.

She didn't smile.

She glanced down the fire escape. 'Come in.'

They stepped inside. Well, Duff and Gus stepped. Jase kind of collapsed into the narrow hallway. Damp. Music from the shop underneath pulsed through the floor.

'Would you like a drink?' the girl asked. 'Cup of tea?'

Gus shook his head. 'We need to be quick, love.'

The girl shrugged and gestured to a greasy-looking sofa part covered with a red throw.

Jase was trying to focus.

'Where am I?' he said, gazing round him.

There wasn't much to see. The lounge was claustrophobic with the four of them in it. Smaller than the utility room at his parents' place. The sash window was bisected by a chipboard partition that didn't quite reach the ceiling. Presumably, the bedroom was on the other side. Equally dark, equally depressing.

Duff looked at the room and looked at the girl and suddenly really, really wanted to go home.

The girl looked at Gus, her face a blank mask.

'Oh, right,' he said, reaching for his wallet. He counted out five twenty-pound notes, but as the girl went to take them he snatched them back.

'Here.' He peeled off a single purple note. 'You get the rest afterwards.'

She looked at the two men on her sofa and then at Jase who had slumped down on the only other seat, a plastic garden chair tucked under a small table.

Without a word, she shrugged again and rubbed her palms across her face.

'Jason,' Gus sing-songed, 'time for a lie-down. Now you get up, mate, and go with the lovely . . .'

'Angel,' supplied the girl in a flat voice.

'Angel, beautiful Angel. She's going to take you for a little lie-down, all right?'

Jase's face was a picture. He could barely open his eyes and his Paul Smith shirt was splattered with puke. Ruined, he'd have to chuck it out.

Gus was helping Jase to his feet and gesturing to Angel, who came over to take his hand. As she turned to leave the room, Jase in tow, she caught Duff's eye and for a moment held his gaze.

He looked away first.

Then she shut the door.

'Ssshh,' hissed Gus, barely holding in his own laughter. He leaned against the door, put his ear to it. But that was just for show really: the walls in the flat were so thin they may as well have been made of paper. Duff heard the bed creak and muffled voices. Then there was silence.

'He'd better not have gone to sleep, not for a hundred quid,' Gus said after a minute. He pressed his ear to the door, frowning. Then came the unmistakable sound of a snore.

'Cheating cow!' he exclaimed and banged on the door. 'Oy, what's going on?'

What happened next occurred so suddenly that Duff didn't quite work out what was going on. He heard Jase yell; presumably, in his confused state, the noise had freaked him out. From what Duff pieced together, Jase woke up, still drunk, saw he was in a strange room, then saw an unfamiliar face next to him and jumped off the bed. Then he either caught the girl accidentally or deliberately hit her in the face. Whichever it was, the next thing Duff knew, the girl cried out and Jase, still fully clothed, came barging through the door, almost sending Gus flying.

'What's happening?' Jase panted, his expression a masterpiece of panic and confusion. 'Where am I?'

Behind him, the girl came out, clutching her nose. Blood was dripping through her fingers.

'Jase, what did you do?' Duff shouted. Some part

of him knew he should be running to the bathroom to fetch a towel for the girl. If he'd been anywhere else, he would have done. But here, in this damp bedsit with the waitress from Sugar Sugar, his feet were rooted to the spot and he watched in horrified fascination as drops of scarlet blood hit the grubby threadbare carpet.

The girl snatched up a tea towel from the kitchen. Not a real kitchen, just a sink, a kettle and a microwave on a shelf. She held the tea towel to her nose, wiping away the blood. Even in the gloomy half-light, Duff could tell she was going to be waking up with a black eye tomorrow.

'God, Jase,' said Gus. 'Can't you do anything right?'

'I wanna go home,' Jase said.

'Yep,' Gus sighed. 'Let's go. We can flag down a cab on the high street.'

No one spoke to the girl. Gus had his fingers on the front-door handle when the girl placed her hand over his, stopping him from leaving.

'Please give me the money,' she said, muffled by the bloodstained towel. Very bloodstained. Duff wondered if her nose was actually broken.

'Do you think we should take her to hospital?' he said under his breath to Gus.

Gus ignored him, instead narrowing his eyes at the girl, 'What money?'

'The hundred pounds, please.'

He laughed then, right in her face. 'You're kidding, right? You want a hundred quid for letting him —' he jabbed his finger towards Jase '— fall asleep in your bed? You're lucky you got twenty.'

He pushed past her.

'But my face,' she said desperately. 'I won't be able to go to work tomorrow.'

'Not my problem,' Gus said. He pushed the handle down and opened the door, herding Jase out on to the rickety staircase. Jase tripped, sending a plant pot crashing over the edge to smash on the flags below.

Gus swore loudly.

'Declan, will you come and help me before he kills both of us?'

The last glimpse Duff had of the girl, she was standing in the doorway holding out a hand covered in blood.

⊙ 12 DUFF

1 July, 12.53 p.m.

'What?' said Duff. '*What?*'

Even though no one had uttered a sound, he knew the others were accusing him of something. A sensation nagged at his stomach, one he wasn't very familiar with. Guilt? He'd done nothing wrong. Fair enough, if the girl had said no then that would have been a totally different story. But she took them back to her flat. She led Jase into her bedroom. No one made her do any of that. For girls like that it was a job. Like working in a clothes shop or whatever would be for a normal girl. She chose that life, right? She didn't have to do it.

'I haven't done anything wrong,' he said, folding his arms and staring into the camera. 'Got that?'

'Of course you haven't,' said Cass, directing her

words into the notebook. 'She could have stopped it at any time. She didn't have to say yes.'

Duff looked sharply at her. Something about the tone of her voice didn't quite match her words. No. She was just writing in the notebook, a pleasant expression on her face.

He'd done nothing wrong.

But there was Jez, tugging at a thread out of the rug and pulling an *Oh yeah?* expression.

Gazing down at him from the moral high ground, acting so superior. The hypocrite. Duff felt anger bubble up inside him.

'I *didn't* do anything wrong. Not a thing.'

Still with his focus on the rug, Jez said mildly, 'No, of course not. You and your brother, total Boy Scouts.'

'Yeah,' Duff agreed sarcastically. 'Not like maybe someone who'd crash into a girl's car and leave her by the side of the road.'

That got Jez's undiluted attention. 'That was an accident and nobody got hurt. What you did is totally different.'

'Leave it, Jez,' said Hetty. 'Please.'

Duff felt his hands begin to clench as Jez sat up straight.

'No. He sexually exploited a girl and then left her bleeding.'

Duff hadn't spent the last few days haunted by guilt. Hand on heart, he hadn't given the girl much thought since Saturday. He'd been more concerned about Gus moaning that Jase wasn't talking to him. And now he could feel serious stirrings of guilt and he didn't like it.

The anger manifested itself as cold fury.

'Get out of my face or I swear you are going to regret it.'

'He's threatening me,' Jez said, swinging round towards Cass. 'You saw that, right? I'd like a copy of that recording, please.'

Cass pointed at the clock. 'We don't have much time left. Let's just move on, please. Duff, you say you did nothing wrong?'

What option was there but to brazen it out?

'Nothing illegal.'

'Illegal,' Cass repeated slowly, writing in the book.

Jez sucked his teeth and shook his head and it took all of Duff's self-restraint not to walk over and punch him.

'And we gave her twenty quid,' he added.

Liv, who for some unfathomable reason looked

as though she were about to burst into tears, laughed sarcastically. 'Oh yeah, twenty quid.'

He wriggled in discomfort. It was one thing for Jez to have a go at him, quite another when it was Liv. He opened his mouth, ready to placate her, but she hadn't finished.

'Can you even hear yourself?' She was literally shaking. Hetty put a hand on her arm, but Liv whipped it away. 'Your brother. The way he treats girls is disgusting. Using them and just throwing them away like they don't count. Like they don't have any *feelings*. He makes me sick.'

Something started to softly turn in the back of Duff's mind, like a key.

'He should be locked up,' she continued, spitting out each syllable. 'Somewhere miles away from the nearest woman. In a cage because he's an animal.'

The words seem to tear out from somewhere deep inside Liv. The muscles on her neck were straining and her usually pretty features contorted. Not a good look.

'Why are you so bothered about Gus?' Duff said, genuinely puzzled now.

Then suddenly *click*.

A heap of disparate images shuffled then straightened, forming a neat line like a neon arrow. Pointing directly at —

Oh God. No!

He dragged his hands across the top of his head and then down over his face, and let out a deep groan.

'Tell me I am imagining this.'

He opened his hands, cupping them round his temples and peered at her from in the shadow.

'Olivia, tell me I am wrong.'

Liv didn't reply. She didn't need to.

He groaned loudly again then spun round to slam both hands against the wall above the sofa. The crack echoed like a thunderclap. Tension thickened the air, crackling like electricity.

When he turned back to face her, he spoke quietly. His shoulders slumped.

'Please tell me your mystery man was not my brother.'

Liv hung her head.

⊙ 13 LIV

The last year

Who said love had to be neat and tidy? Dangerous, exciting, unpredictable love, that was what Liv craved.

The first time they met, *deliberately* met, it was the end of May and he was sitting in one of the tiny rooms tucked off the main bar in the Church Inn with his phone against his ear. When he saw Liv, he quickly finished up his conversation and switched the screen off, laying his phone carefully on the table.

'Hello, gorgeous,' he said, breaking into a smile. 'Glad you could make it.'

Act cool. Do not fall over. Stop shaking.

She concentrated on putting one foot in front of the other to override the self-consciousness that flooded her cheeks with heat. Between the ages of eleven and fourteen her bedroom wall had been papered with faces similar to the one in front of her now. Symmetrical angles; peacock-blue eyes; stubbled jaw.

Surreal, as though one of the poster boys had sprung down off the wall and materialised in 3D before her.

Breathe.

He was on his feet now, skirting round the edge of the table. His pale blue shirt with the sleeves rolled up revealed tanned forearms covered in light golden hairs. A heavy watch, silver with a brown leather strap, circled his wrist. Indigo blue jeans. Soft denim. Designer cut, of course.

When his lips brushed softly against her cheek, the hairs on the back of her neck stood on end.

Return the greeting.

'Hi.'

He pulled out a chair for her and she sat down with a bump.

Breathe. Breathe. Breathe.

His phone vibrated on the table; he glanced at the screen before shutting it down with one final angry buzz. His blue, blue eyes stared into hers, his expression unreadable. He rotated the dead phone in his hand.

'You are going to get me into so much trouble,' he said finally.

Gus Duffy.

After that first time, she felt guilty. Grubby even. He was Duff's *brother*. He had a *fiancée*.

All through the summer, she tried to distract herself. Really, she did.

She recorded video after video for her blog.

She went out with her friends.

She read.

She shopped.

She dragged Hetty to the gym. A lot.

She even offered to help Mum with the horses.

And still, whenever he texted *are you free?* somehow, she was. Every time.

She hadn't planned for this to happen. What had started life as a few flirty texts had grown into an obsession that dominated her existence. She floated in a cloud of denial. Whenever Duff mentioned Gus or Claudia, she blanked it.

By the time the new term began, she'd gone well past the stage where she could kid herself this was just a bit of harmless summer fun.

All that had come screeching to a halt when he came back from five days in Venice with Claudia. His stupid auction prize from Jez's stupid Valentine's Ball.

He'd really enjoyed spending time with her, he said. She was loads of fun. But it had always just been temporary. He was getting married in June; it was time for him to progress to the next level, which unfortunately meant he couldn't see her any more.

Her ears couldn't take it in.

Progress to the next level? She was being dumped because he needed to *progress to the next level*?

What did he think he was . . . a game on the Xbox?

Liv felt her cheeks blaze as if he'd slapped her in the face. *Do not cry. Do not cry.*

He stroked her cheek with his thumb. 'Come on. You knew it was only ever going to be a bit of fun.'

The night before she'd dreamed of Claudia's tear-stained face. Months of building their future in her head and he'd sent it tumbling with 'bit of fun'.

She swallowed, tried to look nonchalant, as though, yes, this did happen all the time and, no, it was no big deal.

But it was impossible.

She started to cry. Screwed up her eyes and moaned. Covered her face with her hands. Salt tears spilled off her chin.

He pulled a packet of paper tissues out of his pocket, opened it and slid it across the table.

She stared at them. Paper tissues? She'd never seen him with paper tissues.

Her already heavy heart plummeted. He'd planned this!

She heard herself gabbling about how much she loved him, how he couldn't leave her, how she couldn't live without him.

He murmured, making soothing, calming noises. For a while at least. Then he stopped listening and looked at his watch and her already frayed control snapped.

She didn't know what she was going to say before she opened her mouth. When she spoke, her voice was husky.

'I'll tell Claudia.'

At first, he laughed, as if she'd said something funny! Then, when she began to describe how she'd kept a diary and told Hetty everything, he looked angry. More than angry; furious. She could feel his shock vibrating through the air like a trapped bluebottle and it thrilled inside her. Her lip curled into a twisted smile.

Then he held his finger up, signalling for silence.

'Remember these?' he said, his blue eyes suddenly as cold as steel.

He clicked through two password-protected screens

and there was a folder. Of photos of Liv. Ones she'd sent him (at his request). Photos he swore he'd deleted immediately. He scrolled through them quickly.

The smile froze on her lips.

'Now listen carefully,' he said quietly. 'One word, and I mean *one word* of this gets back to Claudia, and these go everywhere. Your parents first, obviously. But then all the guys at the Country Club. Your little mates from school. Everyone.'

'You wouldn't,' she said, eyes wide. 'You haven't shown them to anyone else, have you?'

He stashed the phone in his pocket. Pushed back the chair.

'Not yet,' he said, turning to leave. 'But watch this space.'

And he winked at her.

She hadn't seen him since.

He hadn't totally destroyed her heart, but he'd definitely left it a different shape.

1 July, 1.00 p.m.

Duff was staring at her, his face pulled into an expression of utter horror that under any other circumstances would have been hilarious.

'Liv,' Cass prompted. Liv wiped at her eyes quickly with a finger. 'Is there anything you'd like to tell us?'

Liv shook her head, gently at first and then more decisively. 'No, absolutely not. That's everything.'

Hetty wrapped an arm around her and pulled her into a shoulder-to-shoulder hug.

Duff turned on her. 'You knew?'

'I kind of thought you must have worked it out,' Liv said softly. 'That hidden depths stuff you were saying before?

Duff sighed and put his hands on the top of his head.

'How could you let her do something so stupid? I can't believe he'd do that, with one of my friends.' He patted his back pockets and dragged out his phone. 'I am going to call that –'

'No!' Liv shouted, stricken with panic. 'Don't. Please don't. It's over and done. I'm sorry.'

Duff ignored her. A ping signalled he'd switched his phone back on.

'No,' Cass said firmly. 'Remember: no phones, no communication. It's part of the deal.'

Liv could see Duff shaking, battling to control his emotions.

'Please don't be mad at me,' she said.

He whipped his head round, laugh-snorted. 'I'm not mad at *you*. I know exactly what he's like. It's him. I can't believe he'd . . . when he knows how I feel about –'

He swallowed hard.

Liv tried to get her head round what Duff appeared to be implying.

He liked her? And Gus knew?

'And photos of you. I can't believe he'd do that.'

'He did,' she said quietly, and he nodded.

'Don't worry about the photos.' He went as though to put his arm round her and then clearly thought better of it, dropping it back to his side. 'I'll make sure he deletes them.'

The photos have already been taken care of.

Liv knew no one else could hear the whisper. Cass was smiling at her. No one spoke.

'OK,' Cass continued into the silence, 'we're almost done here and before we finish I want to say a huge thanks to you all for being so amazing. And by amazing I mean you've shown how willing you are to open up about yourselves. It would make mesmerising TV.'

She held up her hand before Jez could protest.

'Not that any of today will be shown on TV of course. These four walls, remember?'

She closed her notebook and tied the ribbon as she spoke. 'Really great insight into your world. Now . . .' She looked up at the clock. 'That's me just about done.'

Amazing. Mesmerising. Liv knew she should be ecstatic to hear those words from Cass. But the atmosphere in the room was grey and flat; the words fell like stones.

'Last thing,' Cass said, tapping away on her keyboard. 'Just a couple of final images the researchers found. Tying up the loose ends.'

The TV screen flickered and then came to life, the screen divided into two.

The picture showed a girl smiling and proudly pointing at a beaten-up old car, and Liv realised several things simultaneously.

First, the girl was wearing the Country Club staff uniform.

Second, the car's bonnet was red, but the door and roof were silver.

And finally, with a sick thud, Liv realised that over her uniform the girl was wearing a red parka, with a folded-back furry hood.

Cass zoomed in on the image and *click*, the final piece slotted into place. Close up, the girl in the photo had beautiful, unusual eyes. Fringed with dark lashes, they were as green and slanted as a cat's.

No one spoke.

Even the clock hesitated. It seemed to Liv that minutes passed, but the white digits remained static, while the silence stretched on and on.

⊙ 14 LIV

1 July, 1.18 p.m.

'You honestly expect us to believe this has all been about *one* girl?' Jez's incredulity gave a voice to Liv's thoughts. 'That the farm, the Country Club, that knackered car, Duff's dodgy nightclub hostess – everything we've talked about today happened to the same person?'

'That's exactly right,' Cass said with a beaming grin.

Liv choked down a hysterical urge to shout *Ta daaa!* like a magician's assistant. It was so utterly . . . surreal.

'Up on the screen we have Evangelina. Also known as Ellie. And Lena. And Angel. Depending on where you met her.'

'Nope,' Duff said, ''fraid not. No idea who she is.'

'Yes, you do,' Cass said. 'You only saw her on Saturday, though she looked very different then.'

Duff squinted up at the photo. Cass continued. 'Her hair's different, of course. No fringe. But it's more than that. This photo shows someone who's content with their life.' She shook her head sadly. 'No, she certainly didn't look like that last time you saw her. With her bloody nose and her twenty pounds.'

'It's not the same girl,' Duff said, sounding less confident than before.

'It is,' Cass insisted. 'Living a life very unlike the one she'd envisaged when she first got off the bus almost two years ago. All of you — whether she called herself Evangelina, Ellie, Lena or Angel — you all met the same girl and, without even realising it, you changed her life. And today she is going to change yours.'

'What can she change about us?' Duffy said.

'Everything,' Cass said simply.

'Who are you?' Hetty said suddenly. She'd been quiet for so long, her voice made Liv jump.

'Me?' Cass said, surprise on her face. 'I told you, I'm Cassandra.'

'Why are you here though?' Hetty said. She waved at the cameras. 'I don't mean all this about the show. Why are you really here? How do you know so much?'

'This.' Cass rifled through the document wallet,

214

plucked out the contract. 'Hetty. Here we are. *I give Pretty Vacant Productions permission to continue to obtain and research my online presence including accessing my social-media profiles.*'

'I'm just saying it doesn't feel like an audition for a reality TV show.'

'Really? What does it feel like?'

'Court,' Jez said slowly. 'It feels like we're on trial.'

'Rest assured, I'm not judging any of you,' Cass answered. 'All that interests me is getting the best out of you.'

Still reading from Hetty's contract: '*I understand that participation could have a profound impact on my life and those around me.*' She darted her eyes at the cameras and pinched her finger and thumb together. 'Seriously, guys, we are *this* far from the end. You've blown me away today. Just hold on for a few more minutes and I'll be on my way.'

'I still don't think it's the same girl,' Duff said. 'All we've got is your word for it. You could have Photoshopped it or whatever. All you've found are some rants off the internet or dashcam footage and you've cobbled together this story for God knows what reason. I don't believe it. It's not even one girl and not one of us has done anything illegal. Except for him maybe.'

He jerked his head at Jez.

'It was not illegal. Skidding on ice is not illegal.'

'You keep telling yourself that, mate,' Duff said.

'What's that supposed to mean? And what about what you did? Your brother, you, that's disgusting.'

Duff wasn't going to let that pass. In one fluid movement he sprang off the sofa, fists clenched by his sides. Only dimly aware of the shock on Hetty and Liv's faces, he heard one of them let out a gasp. He leaned forward, so close that his nose was almost touching Jez's.

'Do not talk about my brother,' he said slowly and with unmistakable menace, 'or you will regret it.'

'Will I?' Jez had got to his feet now. What he lacked in width he made up for in height.

'Hey,' Liv said, sliding down from the stool. She stood between them, palms almost touching each chest. 'This isn't the right thing to do.'

'He shouldn't talk about my brother like that,' Duff said.

'He's got a point,' Liv said. 'How could you leave someone with a broken nose?'

Duff stood rigid and narrowed his eyes at her. 'Don't you give me a hard time. Remember, you started this.'

'*Me?*' Liv said, gasping. 'No one twisted your arm to come here today, Duff.'

'I don't mean today, do I? I mean the whole thing. If this is the same girl, which I still don't believe, but just for a minute let's say totally coincidentally this Evangelina is all these other people. Then it all started with you. If you hadn't threatened to set the dogs on her, then she wouldn't have been working in the club last weekend.'

'That was a mistake!' she said, louder. 'I couldn't know what was going to happen. I admit what I did was wrong, but what you did was far worse.'

'Would it have made any difference?' Cass interrupted suddenly. 'If you knew what was going to happen? Would you have missed your big night to help that girl?'

'Yes,' Liv said quickly. 'Of course.'

But even as the words were coming out of her mouth, Cass's voice was in her head. *Would you really?*

'Interesting,' Cass continued. 'Imagine if we could suddenly see all our connections, every interaction we ever made with another human being. Like a Tube map maybe. Every line a different colour. So, yellow for when we helped someone. Red when we hurt someone. Black

217

when we turned our back on someone who desperately needed our help. If the consequences were made visible, would it make a difference?'

Liv squirmed and tried to imagine it. A vast shimmering map, and in one tiny, tiny corner of it the four thick black lines linking them to Evangelina.

'You can't help wondering,' Cass said, 'if any one of you had stopped to think, maybe Evangelina would have been OK.'

In the silence, broken only by the retro clock flipping over to 13.25, the steady red eye of the camera ahead of her stared silently back. Liv bit the inside of her cheek. She thought about Hetty, so devastated by her dad's death that she refused to see the truth about her boyfriend. Hetty, one of the kindest, sweetest people you could ever meet, let jealousy and insecurity cause an innocent girl to lose her job. She thought about Jez and how he was so concerned with other people's opinions that he ruined that same girl's chances just when she'd got back on her feet. She thought about both the Duffy brothers and how they got away with treating women as commodities, used then discarded, and everyone just chuckled and let them get away with it because they were rich and charming.

And then she thought about herself and the night of

the awards when Gus had turned up and she believed every lie he told her. The night she carelessly set in motion the events that would destroy the stranger who smiled at her now.

And then she realised Cass was standing next to her. It was weird like a dream. The others were frozen, as if someone had pressed pause. And she felt Cass's hand on her shoulder and suddenly she was thinking, not of Hetty or Duff or Gus or Jez or even herself, but of Evangelina. Without realising, she closed her eyes. Cass's hand weighed heavy on her shoulder. Hot. Not burning exactly, more like she'd rested her back against a radiator. And she had the strangest desire to just let the warmth from Cass's hand flood her entire body until she felt certain that if she opened her eyes, her body would be glowing like one of those infrared images from police helicopters.

And with the weight and the heat and everyone around her still, the room shifted and dissolved, re-forming as a different room with shabby wallpaper for exposed brickwork; stained carpet tiles for reclaimed church floors.

Her reflection, or rather someone else's, stared back at her from a narrow blackened rectangle where the sagging curtains didn't quite meet. There were grey hollows under her cheekbones. She blinked, and the reflection blinked back with swollen, bruised eyes.

Instead of the silk jumpsuit, she was wearing a pale blue dressing gown with white stars on it. Despite the fleecy material and the chunky socks scratching at her shins, she couldn't stop shivering.

Liv felt Cass's presence through the weight on her shoulder and cold burned like ice through her veins until it reached her heart. She felt a despair like nothing she'd ever experienced deep inside her; a dark, welling misery that radiated out to engulf her whole body. She felt as though a huge black hole had opened inside her, sucking in the memory of every smile, every kiss, every sunny day until the black hole was all there was.

Then Cass lifted her hand, the room shimmered and Liv was back in her own comfortable, warm home. And she was crying. No, not crying. That didn't come close. Through a blur of tears she saw Hetty rush towards her as she crumpled on the sofa, drawing her knees up to her chest and howling for this girl, with a bitter taste in her mouth, like melting paracetamol.

'I started this,' she coughed out. 'I did this. Cass, please tell me what I can do.'

'The only thing you can do is understand,' Cass

replied. 'That's all any of you can do. Try to understand.'

She pressed another button and the screen split in two. On the left-hand side, the image of a girl appeared.

'Oh God,' said Liv, turning her head away. The image was similar to the one Cass had shown her earlier on the phone, but the TV screen sharpened the details.

The girl still had her back to the camera, wearing the same tatty dressing gown; the room was still dim, but the shapes revealed themselves into a bedside table with a digital clock on it. The red digits glowed in the gloom: 14.33. The empty packets were now clearly identifiable as assorted painkillers.

'She couldn't see any other way out.'

Liv rubbed her palms over her eyes. The pain lingered, not physical but intensely sharp like the worst imaginable toothache. The tears had dropped on to her jumpsuit, marking the silk with dark spots that Liv didn't even notice.

Hetty's eyes were bright with tears. 'I wish I could go back.' She looked imploringly at Cass. 'You know I'd do it differently. There's no way I'd have done it if I'd known this was going to happen.'

'That's kind of the point though, isn't it?' Duff said. 'You can't possibly know. Like Liv, yeah? She didn't let this girl in – if it's even the same one. She didn't fetch her dad to speak to her. But so what? Because of that, she actually went off and got another much better job. I mean the Country Club – got to be better than picking broccoli.'

'True,' Cass said.

'But then I lost her that job anyway,' Hetty added miserably. 'So it cancels it out.'

'The point I'm making,' Duff said, 'is that you have to look out for yourself. Everyone has good and bad thrown at them: it's how you deal with it that counts. Don't get me wrong, I do feel sorry about this girl, but we didn't break the world. We didn't invent the system that says you'll do all right; you won't. This is what we've been given. Maybe it's not fair that life's not fair, but what can you do?'

Cass nodded. 'I guess you could look at it like that. But maybe if you knew the full story about Evangelina, you might not see it as so black and white.'

Duff folded his arms. 'Right. But what has this got to do with auditioning for reality TV?'

'OK.' Cass pointed at the two images of Evangelina. One happy and the other . . . not. 'Reality is how you can go from this . . .' she said, 'to this . . .'

Evangelina's silent, smiling face watched them from the screen as Cass began to speak.

Evangelina was born in her parents' bed in a place very few people have heard of. Her father died in an industrial accident when she was ten and Evangelina, her mama and her little sister, Elena, moved out of their large, comfortable home into a much smaller, much less comfortable flat. In the years before Evangelina came along, her mama had been an English teacher and now she found a job in a school. But teachers aren't paid much and Elena was ill all the time.

One Monday morning, when Evangelina was sixteen, she answered the door to two men who pushed past her and then left, taking most of the family's valuables with them.

Her mama said, 'Next time they come, it will be to put us on the streets.'

She said, 'I really don't know how we're going to pay.'

One day, walking back from a failed interview, Evangelina passed a man in dirty overalls pasting flyers on a derelict warehouse. *Jobs in UK. Guaranteed work. Good pay. Come to the square tomorrow to hear more.*

She'd heard rumours — there were always rumours — of the kinds of jobs girls ended up doing. She asked the man if he knew what type of work it was. Something in the man's stare made Evangelina pull her cardigan tight at the throat. He shrugged, dipping his brush into the paste. 'I just put up the posters,' he said.

Two weeks later, Evangelina found herself choking on diesel fumes in the back of a stinking bus driven by the man who put up the posters. *Pavel.*

And on the same bus were twenty-seven other young people, each of whom now owed an arrangement fee, a travel fee, accommodation fees and a document fee – to Pavel.

'But don't worry,' he told them. 'You can pay me back out of your wages every month. You'll still have plenty to send back home.'

Then he lit a cigarette and carried on driving through countries until they boarded the ferry that would take them across the North Sea to their new life.

At first life in the bunkhouse was good. Evangelina spoke excellent English, thanks to her parents, and for that reason she became a kind of unofficial translator. The work was back-breaking, sunrise to sunset out in the fields, but she gritted her teeth and got on with it, thinking about her mama and Elena and how much the money would mean to them. Everyone warmed to her and she was happy. Until she got her first pay packet.

'Ten pounds?' she said, pulling the creased note out

of the envelope. 'There's got to be some mistake.'

Pavel sucked hard on his cigarette, didn't even look at her. 'No mistake,' he said out of the corner of his mouth.

'But how can I only have earned ten pounds?'

There was a long pause. When Pavel looked up he wasn't smiling. 'Arrangement fee, travel fee, accommodation fee, food fee, bills fee. You think I'm running a charity? When you pay me back, then you get more money.'

Evangelina went away, not exactly satisfied, but resigned to it. The next four months were the same until finally she asked Pavel when she'd have paid off the fees.

'One year,' he said.

One year. To earn £520 working sixty hours a week.

Liv shook her head. 'There's no way my dad could know about this. He'd never let it happen.'

Hetty added, 'Seriously, Liv's dad is like the nicest man ever. He just . . . wouldn't.'

Cass shrugged. 'That's a conversation you need to have with him.'

Shoes . . . £200 . . . Never been worn . . . The voice was so fleeting Liv thought she'd imagined it. Or at least she must have imagined it, right? Because it was her own voice and Cass hadn't moved. Liv put her hands over her ears.

'Not much more,' said Cass. 'After four months, Evangelina had had enough. She planned a time when she knew Pavel wouldn't be around and walked all the way to speak to Mr Dawson-Hill, the boss. She knew he was her only hope.'

'Don't say any more,' Liv said miserably. 'We know this part.'

Cass studied her for a second. 'She knew Pavel wouldn't change his demands, so she ran away. Part scared and part relieved.'

'And that was when she started at the Country Club?' Hetty said.

'She even saved up and managed to buy what Jez referred to as her "crappy car". It was no BMW, but it meant a lot to her. Of course, she had to buy it from a fairly dodgy garage, but she wasn't in a position to get finance through official channels, like going to a bank, so she had no option but to pay the extortionate interest rates. It's illegal of course, but garages like that know there's a market of desperate people like Evangelina. Anyway, everything was going fantastically well: she had a nice flat, good job, car and was sending regular payments home. Things were really looking up for her. And then she lost her job.'

Hetty put her hands over her eyes. 'If I could go back in time, I would never have done it.'

Cass's eyes flicked from Jez, then to Duff, to Hetty before finally pausing on Liv.

'You do know most of what happened, but what you don't know is that Pavel tracked her down, threatened to hurt her mum and sister if the debt wasn't repaid. If your dad had spoken to her, maybe he could have stopped it.'

Liv thought Hetty might cry, but she didn't. Her ponytail had come loose and as she straightened her head, she looped an errant strand behind her ear.

From the TV screen, Evangelina's silent, smiling face beamed down on them.

Cass continued. 'Obviously, she was devastated to lose her job and she knew she had to find another one really quickly to keep up the rent, the car and sending money home. There wasn't much around, being New Year, so it was harder this time to find something else, but she did. Delivering flowers. But of course, she met you, Jez. Late for your date with the TV crew and that was that. As I told you, she took the last two bunches of flowers and walked almost six miles to make sure they were delivered. In all that freezing cold.'

Jez held his palm out in protest. 'For the last time, I did not run her —'

Liv whirled round, 'Don't you dare try to worm your way out of this or I will come over and slap you myself.'

Jez pinched his lips together until they almost disappeared. He walked up to the TV screen and peered closely at Evangelina and the car.

'Two things, right. First, there's no way she could have known it was me. It's obvious she doesn't have a dashcam.' He tapped the screen. 'It was dark and she was further down the road. But that doesn't matter anyway, because even if she could read the number plate, how would she know who the car belonged to?'

'Jez, there's no point arguing,' Liv said. 'It was her; it's obvious it was her. So just . . . shut up.'

'My second point is, I don't think the girl Duff met on Saturday was the same,' he said. 'He'd know, wouldn't he? He went to the Country Club and saw this Ellie. It's not like you'd forget her. But he said the one from Saturday looked nothing like her. Anyway, how could your researchers have found all that out in just a couple of days?'

He folded his arms and shook his head. 'I understand

what you're trying to do, Cass, but this has got set-up written all over it.'

Cass didn't answer for a few seconds. Instead she looked at them, a broad smile on her face.

'OK, thanks, guys. I think that just about wraps things up. Thanks for your co-operation. You've been awesome. We'll be in touch.'

'What?' Liv said. 'Is that it?'

Cass picked up a camera and unscrewed the tripod. 'I've got a plane to catch.' She opened the case and stashed it inside.

'What do you mean?' Hetty said. 'That wasn't the whole interview . . . was it?'

Jez clicked his tongue disdainfully. 'Well, what a total waste of time that turned out to be. Cheers, Liv.'

'Don't blame me!' she said. 'You think I was expecting this?'

'If it was true,' he said, 'then I might be. But it's some fantasy. There's no way all that stuff was just one girl. It's too . . .'

'Awful to imagine?' Cass helpfully supplied, zipping the case.

'Coincidental,' he said.

'Yes, I suppose it is. She came here full of hope,

looking for a happy life. Then she met the four of you and now she's sitting in her flat across town wondering what to do with her life.'

'I thought you said she was dead,' Jez said.

Cass slung the Pandora over her shoulder and lifted up the tech case.

'Not yet.'

Liv froze. 'What do you mean, "not yet"?

'I mean there's still enough time to do something, if you hurry. Jez, could you just get the door for me?'

'She's alive?' Liv said.

Cass nodded. 'For now at least.'

Jez gave an explosive snort of laughter. 'What? Are you seriously suggesting you can see into the future?' He shook his head, chuckling. 'Mystical powers. Crystal balls. I've heard it all now.'

Cass shouldered the Pandora. 'Like I said, you've still got time.'

'Why didn't you say some–'

The door closed behind her.

⊙ 15 LIV

1 July, 1.37 p.m.

The four friends stared at each other.

'What,' Duff said eventually, 'was all that about?'

Something nagged at Liv even as the others were talking. She opened the door and peered into the hall. Then she turned the handle of the door to next door. Coat cupboard. Nothing. Ditto the downstairs loo. She ran up the stairs two at a time and checked every room. Nothing.

'What's up?' asked Hetty.

'Where's she gone?' Liv asked, running back down.

She pulled the heavy front door open and stared. Jez's BMW stood impassively in the courtyard and behind it the gates were closed. She fumbled for the panel and pressed the mechanism and the dogs instantly came running out, howling.

'I didn't hear the front door shut,' she said. 'It weighs a ton, it's impossible to shut the thing quietly. And look, the gates are closed . There's no other way out.'

'Over the wall?' Hetty offered.

Liv shook her head. 'Too high. Too quick. And she had that big case.'

'Why weren't the dogs barking?' Hetty asked slowly.

'They weren't barking this morning either,' Liv said slowly. 'Or at least, they didn't until I shut the gates. Jez left them open.'

'No he didn't,' Hetty said.

'No I didn't,' Jez said simultaneously.

'That's true,' Duff added. 'Hetty made a big deal of me getting out to press the button. Even though the dogs were on chains and couldn't reach. Gates definitely closed.'

'But Cass can't have known the code,' Liv said.

Duff was gazing around the hall. His face broke into a broad grin as he shouted, 'I get it! I know what this is. It *is* one of those prank things. Liv.' He clapped his hand on her shoulder. 'You set us up, right? Cass and some guy are going to come out and go "Surprise!" yeah?'

'Don't be such an idiot,' Liv said, pushing past him.

'It's probably out there already.' Duff tilted his head back and looked up at the beams. 'There's probably hidden cameras everywhere.'

'Oh God.' Hetty's voice came out part moan, part wail. 'Everyone is going to know what I did. My mum is going to kill me.'

'No,' said Liv. 'How could Cass get in to place the cameras?'

Liv was standing in front of the framed portraits Cass had admired earlier. It felt like a lifetime ago. Her eye caught on the vivid blue of the Greek sky and that fragrance filled her head again. That's what it was. Herbs, sunshine, the sea, an earthy undertone. The smell of Cass.

'Don't you have staff she could bribe?' Jez was saying impatiently. 'They managed to get those girls –'

'Girl,' corrected Hetty.

Jez shrugged. 'Girl, girls. My point is they sold their stories easy enough. Do you trust the people who come in your house, Liv? Can you say for definite they wouldn't take a kickback to plant a hidden camera?'

Liv thought back to Martina/Marina . . . no, *Martha*. That was it. Definitely. *Martha's* sullen expression.

Except . . . the voices. Cass's perfume. The horrible soul-sucking emptiness she'd felt when Cass placed a hand on her shoulder. The way she'd apparently vanished into thin air. How could anyone fake that?

Jez put his hands over his face. 'I'm going to lose my uni place for sure.'

Duff said, 'Look on the bright side: if you're lucky, maybe you could do your degree in prison. You've confessed to dangerous driving on film, remember?'

Jez leaned against the wall as though his knees couldn't hold him any longer.

'My parents' trust is going to be destroyed by this. They're going to hate me. They probably do already! I bet when she came out to take that phone call she was really setting it all up.'

Liv jumped as Duff suddenly slapped his forehead and shouted, 'We are so *dumb*! *Phone*. Hetty's phone. We've got her on camera saying this won't be broadcast.'

Jez jabbed his finger in the air, words tumbling out. 'He's right! Get your phone out quick. Play the footage. We need to send it to whatshisname, Tony, tell him if they even put one second on air or online we're going to sue their arses off.'

Hetty already had her phone out and was fumbling with the buttons.

'Come on!' Jez said.

'I can't believe you lot,' Liv said, grabbing the phone. 'Were we even in the same room? Did you hear what I heard? Don't you even care the tiniest bit about that girl and what we did?'

'*What* girl?' Jez said, almost shouting. 'Get real. There was no girl. It was a set-up. Don't be so stupid, just give me the phone.'

'Am I the only person who gets this?' Liv said. 'Hetty, come on. You have to. For God's sake. We. Did. Those. Things. I could've helped that girl stop that Pavel guy. Hets, you got a girl sacked from the Country Club. Jez, you drove someone off the road and left them. Duff, you let . . . your brother . . .' she couldn't bring herself to say his name '. . . leave her with a broken nose and act like it's a big joke. We *did* those things. Whether it was all the same girl or this is some massive set-up, we still *did* that. Don't you realise? Maybe it *will* get broadcast and if it does we won't have a leg to stand on because it's all true.'

The others blurred as her eyes brimmed with tears. She could hardly catch her breath.

'Here,' she said, thrusting the phone at Jez. 'Take it. Watch it. It doesn't make a single bit of difference.'

Jez took the phone silently. The four of them stood in

the hallway in a circle, no one speaking. Hetty stepped forward to hug Liv, but Liv shook her off.

Jez tapped the buttons and frowned. He tapped again. 'What the —' he muttered. Shaking the phone, he handed it to Hetty.

'Where's the video button?' She peered down at the screen.

Liv moved in closer. 'Can you put the sound up?'

'Sorry, I'm still getting used to it.' Hetty said. 'OK, that's definitely the one.'

Liv stared at the coach-house brick wall on the screen. The image swung round to the granite worktop and a view of the kitchen. No Pandora bag; no laptop; no tatty red notebook.

No Cass.

'Well, that's not it,' she said. 'Look down the —'

'Shhh,' Duff said.

Jez's voice was coming out of the phone. Close.

'Cass, please can you confirm that the casting auditions today for Pretty Vacant Productions will never be made public.'

Then there was a pause. A long pause broken only by faint rustling, like someone shifting position and the gentle sound of Jez's breathing.

Hetty's face appeared. 'I won't tell anyone.'

Liv. 'I won't tell.'

Duff. 'Me neither.'

Jez. 'I won't speak of what has happened in this room.'

Four friends.

No Cass.

Even the air felt heavier, as though someone had turned up gravity. It pressed down on the top of Liv's head, through her skull, and made her eye sockets ache.

Duff was the first to speak. 'Can someone please explain what the hell is going on?'

Liv looked at the shock on each of her friends' faces, reflecting the expression on her own.

'I don't understand it,' Hetty was saying. 'I had my phone the whole time.'

'It's some tech thing,' Jez said. 'She must have sent some bug to your phone that wiped her off the video but kept us.'

'Does that even exist?' Hetty said.

Jez shrugged, gesturing for Hetty to give him the phone. 'Cass works in TV. So I suppose she'd know about that kind of stuff.'

But Hetty continued staring at the screen. She opened her mouth, her cheeks suddenly pale.

'That's impossible. This is a new number. Cass doesn't know my number. No one knows it except for Duncan. And my mum.'

Buzz.

Liv stifled a yelp. Her phone buzzed in her pocket.

The screen said 'sender unknown.' 'I can't do it,' she said, passing the phone to Hetty.

Hetty clicked and a picture message unfolded, top downwards. It was the close-up of the girl lying on the bed. So close the image only caught the black roots of her hair in a straggly ponytail and her bedside cabinet. The photo on it showed two girls and an older woman. Evangelina's family? Next to it, red digits glowed: 14.33.

'Look,' she said, holding the screen up for the others. 'It must be Cass. Who else would have the picture?'

It pinged and she nearly dropped it. A text, sender unknown again.

'Open it,' Duff said.

There is still time.

'Oh man,' Duff said. 'This is freaking me out now.'

Liv scrolled back to the photo of the girl.

'Duff, you've been to her flat. You should be able to find it again, right?'

'Yeah . . .' he said slowly. He narrowed his eyes at her, drawing the single syllable out. 'Why?'

'You are not seriously suggesting that we go round there?' Jez laughed. 'It's a set-up. There won't be anyone there.'

'Or there'll be a camera crew like one of those TV programmes where they catch dodgy builders. They'll come running down the road after us, trying to get us all to confess again,' Duff added.

'Tell me the address and I'll go on my own,' said Liv. 'I'll get a taxi and you can stay here if you want. Hets?'

Hetty blew out her breath in a deep sigh. 'I don't know, Liv. It's so weird. I don't understand what's going on. I just want to go home.'

'You got that girl sacked. Maybe this girl doesn't even exist.' She jabbed the phone screen. 'Maybe they are four different girls and that photo outside the Country Club *was* faked.' She realised she was suddenly shouting. 'Maybe it *is* all a set-up. But we all did what we did, and if there's the tiniest chance we can do something to make it better, then I think we should take it. Otherwise . . .'

'I still don't actually think I did anything wrong,' Jez said, folding his arms. 'That car was a hazard. I probably did everyone a favour taking it off the road before —'

Liv jabbed her hands in the air, making him flinch.

'Shut up! Just get in the car now, and Duff, tell us where it is. You said above a kebab shop, right? Near Sugar Sugar.'

'I don't think any of us should go,' Jez said.

'I'll come,' Hetty said. 'You're right. But it's nearly ten to two already.'

Liv put her hands over her eyes, groaned in frustration. 'We won't get a taxi to come in time. Jez, you have *got* to do this.'

He shook his head, 'Sorry, Liv. All this was your idea. I wish I'd never agreed to it. I'm going home.'

Liv took a deep breath.

'If you go home, I'm going to tell everyone about you running her off the road.'

Jez stood stock still. 'You wouldn't.'

'Try me.'

'Then I'll tell everyone what you did.'

'Do it,' she said. 'I take total responsibility for my actions.'

'You're mad,' he said. 'We've got absolutely no proof any of this is true.'

How could Liv explain what happened when Cass

touched her without sounding like she'd gone mad?

'Jez, what does it matter if you don't think you did anything wrong?' said Hetty softly. 'We may as well go.'

He rattled his car keys. Sighed. 'Whatever. Come on then.'

In the car, through the gates and on to the road that led to the village. The four of them were silent, gazing out of the window at the fields and the dull, grey sky. Next to Liv in the back seat, Hetty compulsively refreshed the time on her phone. 14.01 . . . 14.02 . . . 14.03 . . .

Liv zoned out until she became aware that the car had slowed to a halt.

'Why have we stopped?' she asked.

Jez didn't even bother to answer, just jerked his thumb first at the long line of cars snaking out of sight then out of the window.

She peered out at the cones lined up by the side of the road. The plastic fencing.

'Is there another way?' she said.

'Not unless you know one,' Jez said tersely. 'They're digging up the whole centre of town.'

'God!' she said through gritted teeth. 'We're not going to make it in time.'

'There's no such thing as "in time",' Jez snapped. 'Because this is all just a trap to make us look stupid and I cannot believe I am letting you lead me in to it.'

The car inched forward then stopped again.

'God!' Liv wailed.

'It'll be OK,' Duff said, leaning across to take her hand. 'It's not much further.'

In the corners of Liv's mind, another girl was fiddling with the seat belt, tapping her feet in the foot well and cracking open the window to let cool air breeze in. Another girl who, apart from not having the good fortune to be born a Dawson-Hill, probably was much like herself. Except better.

Liv closed her eyes. She visualised the map charting her interactions with friends and strangers. Like a colourful web. What had Cass said? Yellow for good. Red for hurt. Black when she'd turned her back on someone who desperately needed help. In one tiny, tiny corner, a thick red and black line linked her to Evangelina.

Decisions never came labelled with a final destination. No action, good or bad, came with a guarantee. Life had a habit of changing the signposts, sending you off on a random path. You could cross the road. You could get hit by a bus. You could even get flattened by a frozen stowaway falling from an aeroplane. Or was that an urban myth?

'But you can at least set off in the right direction.' she said quietly.

'What?' said Jez, looking in the rear-view mirror.

'Nothing,' Liv said, switching her phone on. The team selfie she'd posted earlier had attracted the usual likes and *ooh exciting!* comments. Without too much surprise, she noticed that the image had changed. The four of them grinning, their arms across each other's shoulders, remained identical. But where Cass had stood? Only the sofa, wall and a corner of the coffee table.

She showed the image to Hetty, who raised her eyebrows and resumed gnawing at her fingernails. Tiny beads of blood welled from the tattered skin.

'That's it,' said Duff suddenly, pointing to a faded sign on a takeaway. The wind had blown a drift of paper and empty plastic bottles up against the closed metal shutters. The shop next door was boarded up, but one of the plywood panels had been wrenched off and lay splintered on the pavement. It was another world.

'I'm not leaving my car round here,' said Jez, tapping the steering wheel. 'No way.'

'Park outside,' Duff said. 'You can keep an eye on it.'

Jez glanced in the rear-view mirror. 'It's double yellows. The traffic wardens have targets to hit, you know.'

Liv bit back her first retort: *You wrote off someone's car and you're stressing over a parking fine?* 'If you get a ticket, I'll pay. All right?' she said tightly, her voice pierced with scorn.

Jez sighed, flicked his hazard lights on and pulled up in front of the kebab shop.

Liv had slid out of the seat before the others even unclicked their seat belts. She turned around and said in Duff's direction, 'How do we get in?'

'There was, like, an alley round the back, up a fire escape. White door.' He frowned. 'I think.'

'Well, think harder,' she said. She'd opened the passenger door and slapped him on the arm. 'Come *on*. It's nearly twenty past two already.'

'Er, it was six days ago and I was wasted?' he said, joining her on the pavement.

Jez pressed the button and the car beeped. He pursed his lips, looked around the deserted street. 'You know, I think I might stay with the car.'

Liv threw her hands up in the air. 'Don't you get it? We all need to be there.'

'In the highly unlikely event that this girl even exists, which I very much doubt, she's just going to ask for money.'

Liv gaped at him. 'How can you not see this isn't about money? We can't pay our way out of what we did. She had a job, she had a car, she had a life she made for herself and then we came along and took it away, without even a second thought. That's what we're here for. Not to pay her off and forget about her, but to pay our debts to her. To own what we did. To say sorry.'

'What the hell are you going to say to her?' Jez said. 'Sorry I sent your life into freefall. By the way, would you like some beauty tips?'

Liv had her mouth open to respond, but Hetty got in there first. Redness flashed from her sweatshirt collar to the roots of her hair.

'For God's sake. You make this big deal about Connecting Together and saying everyone needs to do their bit, and you can't even man up and admit this one thing. What we did was just as bad, but at least we're trying to do something about it. I will not let you get away with being a hypocrite. So shut up and come with us now or I will never speak to you again for as long as I live. As far as I'm concerned, it'll be you who doesn't exist. Duff, which way are we going?'

Liv's eyes widened in amazement. Hetty's voice may have been shaking, but there was no mistaking the determination behind her words. Or the purposeful way she lifted her chin as she followed Duff towards a narrow opening between the takeaway and the pound shop next door.

Liv raised her eyebrows at Jez, but he refused to meet her gaze; instead he stalked ahead and disappeared into the gloom of the passage.

Finally!

Liv hurried behind him. It was immediately apparent every man who ended his night at the kebab shop viewed the alley as an open-air urinal.

Hetty pulled her sweatshirt up to cover her mouth and the four of them slalomed their way over the cracked tarmac littered with broken glass.

The alleyway led them into a weed-filled courtyard overlooked by tall buildings. The walls were mottled brown where the plaster had flaked away to reveal crumbly brickwork beneath.

A gritty eddy of wind lifted a plastic bag past ground-floor windows, not a single one without a cracked pane or a nailed-on board, and past the exoskeleton of fat silver pipes gusting out warm grease from the kitchens. Liv's eye followed it up, the ancient green stain from a leaky pipe swirling higher, until it wrapped itself around the railings.

A humid cloud of some rank meat belched from the extractor pipes, clinging to Liv's hair, her skin, her nostrils. She wrinkled her nose and stared at the metal staircase. Safe to say Evangelina's home would not get a two-page spread in *Amazing Interiors* magazine.

Duff pointed. 'I'm pretty sure it was that one. Second floor, white door.'

'*Pretty* sure?' Jez said. He lifted his hand to cover the lower half of his face against the fumes, muffling the words.

'Yeah.' Duff skirted round the terracotta shards of a smashed plant pot near the foot of the stairs. 'This is definitely it.'

'What do we do now?' said Hetty.

Liv shrugged. 'If Duff thinks it's that one then maybe we should –'

Bzzzz.

Her phone.

For a second, the four of them froze, then Liv fished in her pocket.

She flipped open the case.

What had Cass said on the doorstep?

He's going to call you later.

She held the phone out; the others peered down at the screen.

Tony.

'I knew it!' Jez slapped the wall with the flat of his hand. 'He's going to tell us this is all a set-up. I told you.' He lifted his head to the grey sky above the courtyard. 'Hello!' His voice reverberated off the concrete walls while the windows stared down, blankly. 'I know you're watching us.'

Duff clenched his fists. 'What do you think you're doing?'

Jez stepped back against the wall. Duff followed, leaning in until the two were almost nose to nose.

'Cameras, idiot,' Jez said, pushing against Duff's chest. 'Rigged up everywhere. Live streaming us.'

'Stop it,' Liv said urgently. She grabbed Duff's arm and yanked him back, lifting the phone to her ear with the other hand.

'Tony, hi.'

She heard a sort of rumbling, then Tony's voice, faintly. In her ear, the slam of a car door. More noise.

'Sorry, I can't hear you very well.'

Was that barking?

'Hello, Liv. Your dogs are quite something, aren't they?'

His words went in her ear, but her brain struggled to process them.

Was he —?

What was —?

'My dogs?'

'We're at the gates now. I think it's the right place. High Acres Farm? Sorry we're a bit late. The traffic was a nightmare.'

'Yeah,' Liv said. 'But I thought . . .'

She tailed off. What did she think?

'We've been pressing the intercom. Didn't you hear it? The dogs certainly did.'

'I . . .'

'Can you let us in?' Tony said. 'We'd love to get started.'

Murmurs in the background. The crew? Of course. What sort of TV company would only send one person?

'Erm, I'm not there right now. Sorry.'

Duff tugged at her sleeve. 'What's going on?'

I have no idea.

She shook him off; held her index finger up and mouthed, *Hang on.*

It began to rain. Heavy drops staining her silk jumpsuit.

'Did you forget we were coming?' Tony was saying.

'Let me talk to him,' Jez said, making a grab for the phone.

Liv turned, hugging it to her ear. Above her, the white door remained closed. Was Evangelina on the other side of it *right now*, thinking she had no way out?

'Look, Tony,' she said briskly. 'I'm sorry about the mix up. We did the interview this morning with the other producer. She said there'd been a change of plan; that you sent her to take your place.'

Pause.

'Sorry, I didn't quite catch that,' Tony said.

'Cass. She came to my house this morning. She had a business card. Cassandra Verity from Pretty Vacant Productions. Assistant or associate something. Wait a sec . . .'

She fished in her pocket, drew out the thick rectangle of card.

Blank.

She flipped it over, her pulse racing.

Blank.

The border was there, the embossed flower was there.

The words weren't.

She blinked just as Tony spoke. 'There's no one called Cass at Pretty Vacant.'

Time slowed down, seconds swelling to fill the space of minutes, and Liv could suddenly see everything so clearly. Every detail sharpened. Microscopic particles of cooking oil entering her lungs with each breath. The scratch of insect colonies in the walls. Spores of damp and mould. The individual spongy filaments on the moss-furred brick.

Tony's voice broke the spell.

'Liv, are you still there?'

'Yes.'

Brisk. 'I don't know what's been going on but if you can get here, we can sort it out. Have you got transport?'

'Yes.'

'Could you get here in fifteen minutes?'

She turned the card between her fingers, tuning out the shouts of Duff and Jez, Hetty playing peacemaker between them. She looked up at the closed white door at the top of the ugly fire escape and sifted through the clutter of her thoughts.

The horror of repeating a year at school.

The desire for revenge on Gus.

The glitter of TV fame.

Another closed door a year earlier, and a girl called Evangelina, who may or may not be alive.

She gripped the handrail and felt the staircase sway under her feet.

'Sorry, Tony. I don't think we can,' she said, and hung up.

When she reached the top, she looked at the others lined up behind her. Then she turned back to the door and knocked.

And waited.

She knocked again slightly louder.

She knocked again.

Nothing.

Knock. Knock. Knock.

Nothing.

⊙ 16 Evangelina

The day before

She was asleep when Pavel came knocking. She was asleep and dreaming. In her dream, she was back home. She did not want to wake up.

The three of them — Mama, Elena, Evangelina — were by the lake. She was lying down, watching wispy cotton clouds float across the brilliant blue sky. Next to her, Elena threaded a daisy chain with skilful fingers.

She'd almost forgotten peace. No men shouting. No shop doors slamming. No stink of damp and rancid meat. No sirens screaming through the night. No constant roar of city life. Just the lazy hum of insects playing over drifts of wild rosemary; the sweet water lapping at the wooden hulls of boats.

Drowsy from the sun, lulled by the waves, her eyelids

were growing heavy when a shout startled her upright.

Papa!

She followed Elena's rapt gaze and there he was, walking towards them through the trees. Not the Papa of the final few months, grey and broken, but before-Papa. Tall, powerful enough to pick her up under his arm and swing her around while she laughed and laughed.

But even then, even as Elena jumped up and ran towards his outstretched arm, the sadness seeped in because she knew it was only a dream. Firstly, Elena was only two when he died and secondly, because Elena's legs never had the strength to run.

Still, she did not want to wake up.

The sun filtered through the leaves, throwing light and dark over the four of them, laughing, hugging, kissing, crying together. Every sensation felt magnified. The smell of wild herbs; the sound of the waves; the breeze rippling through her hair. As though her skin had been stripped away, exposing the nerves beneath. For the first time in months, she felt *alive*.

Over Papa's shoulder she glimpsed a flicker of white through the trees. She tracked it with her gaze, closer, closer, until a woman emerged into the clearing. Her white dress glowed in the sunlight, her black hair shone.

She waved and shouted something, but overhead the sky darkened and thunder pealed a hundred discordant bells. Mama, Papa and Elena vanished. She looked for the woman, but she wasn't there. Evangelina was alone. And cold, so cold.

Bang, bang, bang. The noise woke her and it carried on until her ears rattled.

She opened her eyes. Not at the lake, not with her family, but in England. Alone. Pavel was hammering on the door.

Bang, bang, bang. 'I know you're in there, Evangelina!' he shouted.

What choice did she have? All her choices were gone. She opened the door.

'Hello, *Angel.*' He bared his teeth in a smile like a fairy-tale wolf. 'Aren't you going to invite me in?'

Without a word, she pulled her bathrobe tight and stepped aside. He brushed past, more closely than necessary, and dropped on the couch as though this were his home.

'So.' He crossed one foot over his knee, exposing a stretch of dead white shin, matted with black hair. 'They tell me you've left Sugar Sugar. How do you expect to pay me your debts if you have no job?'

'I couldn't go to work because of *this*.' She pointed at the bruises. 'They fired me.'

He leaned back even further, placed his finger over his lips. 'Now that is interesting. Because my friend Vitaly told me you said you were sick and then took three men home.'

She felt a hot sinking in the pit of her stomach. Of course Pavel knew Vitaly the doorman. Pavel knew everyone.

'It wasn't like that.'

Pavel raised his eyebrows, then patted the couch. 'Come and sit here. Tell me how it was.'

She sat, careful no part of her body came into contact with his. The weight of the sofa caused the ancient seat to sag almost to the floor.

'That's better, isn't it?'

He grinned. His teeth were yellow and his breath smelled of meat. White flecks gathered in the corners of his lips.

'Now, Evangelina, my Angel. We need to talk about how you're going to repay your debt as you've lost your job. You've been such a good girl for so long. Well, apart from running away from the farm.'

'I'll get another job.'

He laughed softly. 'I wish it were that simple. But the interest, you see, all this time it's been going up and up, and now I don't think any job you get is going to meet the payments.'

'How much do I owe?'

'I don't have the exact figures with me right, but off the top of my head, I'm afraid . . .' He sucked his teeth. 'Maybe twice what you owed when you left the farm?'

Inside she was shaking, but she kept her voice steady. 'You can't do that. I have rights.'

He moved so quickly she didn't have time to jerk away. He grabbed her arm so tightly she cried out.

'Don't be stupid. People like you don't have rights.' His hot breath felt clammy on her cheek. 'If you don't pay, I will send my friends to visit your mama and little Elena and see if they can pay.'

'Don't,' she said quietly. 'Please don't.'

'You girls,' he said, letting go of her so suddenly she almost fell off the sofa. 'You're so ungrateful. But you're a lucky girl, Evangelina. I have got a job for you. I will send someone to collect you tomorrow at four o'clock. You need to bring a bag, a few things. Not much. And don't even think about running away.'

'What if I pay you the money?' she said. 'All of it in one go.'

He grabbed her chin in one powerful hand, pulled her towards him, blasted her with rank breath. 'Don't be stupid. Where are you going to get money from?'

He slammed the door so hard the picture of Mama and Elena fell, cracking the glass. She listened to his footsteps clatter down the staircase then took the one chair and wedged it up against the door. It wouldn't stop anyone, but it made her feel better.

Then she lay down on the bed and wrapped the thin quilt around herself. She picked up the cracked photograph and kissed it, not caring if the glass cut her lip. There'd been a few girls at the club who'd told her she could make much more money doing the job Pavel meant, but after Saturday night, she knew she could never do that, no matter how bad things got.

Pavel thought she would do what he wanted. Pavel thought she had no choice.

Pavel was wrong.

And then she closed her eyes and calmly willed the dream to return.

⏭

⊙ 17 LIV

1 July, 2.28 p.m.

Liv knocked again. Her knuckles stung, but she barely noticed.

The staircase creaked behind her as Duff turned.

'Told you it was a waste of time,' he called over his shoulder.

Then Liv heard shuffling footsteps. The staircase creaked and swayed as Hetty climbed up behind her.

She knocked again.

Hetty spoke quietly. 'I told Duff to wait down there. If she is here, we don't want to frighten –'

The door opened a crack.

First, a musty smell, like badly dried bedsheets, drifted out from the gloom. Then one piercing green eye appeared, slanted like a cat's, but smudged underneath with a yellowing bruise. A sliver of face, paper-pale skin and a healing scab on the bridge of an otherwise neat nose. Fingertips with skin gnawed to tattered strips gripped the door frame. Liv saw a thin white wrist, delicate bones, disappearing into a fleecy sleeve, pale blue with white stars.

Liv's whole body sagged with relief. 'Evangelina?'

Suspicion clouded the girl's tired features; she pulled the door closer.

'Please,' said Liv. 'Cass sent us.' She reached out and, as she did, her hand brushed the girl's arm. And in that second such an intense wave of misery washed over that she cried out; that same sensation when Cass touched her arm. An overwhelming loneliness rushed into every cell in her body until she was wracked with a wretchedness beyond anything she could have imagined. As though her body had been hollowed out and refilled with despair.

A thousand tiny insects buzzed in Liv's ears. Somewhere, far off, Hetty gave a panicked shout and the staircase rattled beneath her feet.

As she swayed, a gentle hand held her elbow.

'Are you OK?' a kind voice said.

'Liv, what's wrong?' She felt Hetty squeeze herself on to the platform.

'Bring her inside,' Evangelina said, her voice echoey and distant.

Both girls helped her in and she sank down on a low sofa.

'Would you like a glass of water?'

Liv shook her head and opened her eyes as the room swam into focus. The room she'd seen when Cass touched her shoulder. White wallpaper, that cheap bumpy stuff; a long strip torn off around the door revealing skin-coloured backing. Someone had carefully outlined the jagged edges in blue biro. A wavy yellow patch stained the ceiling. Net curtains sagged across the window. No kitchen to speak of, just a narrow wooden shelf housing a small sink, microwave and a kettle.

And there was Evangelina. Unmistakably her. Tired but alive.

'I'm OK. I just need to –' Liv swallowed. She moved up the sofa, 'Please sit down.'

Evangelina did. Hetty pulled out the plastic chair and perched on it.

Liv took both of Evangelina's hands in hers. The girl looked bemused, but at least she didn't pull away. Liv blinked away tears, steadied herself.

'This sounds mad,' she said with a laugh. 'But we've been sent here to help you.'

'Who are you?' Evangelina said, frowning.

'I can explain everything in a minute,' Liv said. 'But first of all, I need to know . . .' She searched for the right words. Nothing in her life had prepared her for a conversation like this one. 'I need to know – are you OK or should we take you to hospital?'

Evangelina looked at her suspiciously. 'Did Pavel send you?'

'No!' Hetty said emphatically. 'No.'

Liv gripped Evangelina's hands. 'You don't need to worry about him. We're going to sort him out. Somehow. My dad is anyway. But I need to know – did you take the tablets?'

Evangelina slid her hands from Liv's and covered her face for a second. 'How do you know . . .?'

'It doesn't matter. Please tell me – have you?'

'No,' she said. 'I haven't.'

Liv breathed out heavily. 'Thank goodness. There are so many things we need to talk about, so many things we're ashamed of. But it was Cassandra who sent us to you.'

'Cassandra?' Evangelina sounded puzzled. 'Who is Cassandra?'

⊕ 18 Cassandra

1st July, 7.46 p.m.

A shaft of early evening sun slanted through the cabin window.

'The captain has switched on the fasten seat belts sign in preparation for landing. Please could you all return to your seats and ensure your seat backs are upright and your tray tables stowed.'

While the speakers continued to pipe the familiar instructions, Cass placed the red leather book in her lap and slotted her tray table neatly away.

'Have you been to the island before?' the man in the adjacent seat asked.

'Oh yes,' she replied with a smile, closing her eyes to signal the conversation had ended.

In a little over half an hour, she would collect her

car from the airport car park and set off for home. The mountain road's treacherous twists and turns deterred many of the island's tourists and the very few who made it to the top seldom recalled what they'd seen. No one believed those who spoke of it anyway.

The islanders knew not to try.

Of course, the road never daunted Cass. She'd travelled it many, many times before. First, on foot, then by donkey or horseback; later by bicycle and then by car for the last century, give or take a decade.

Home. The scent of rosemary and the sea. The warm sun. Yes, she thought. It would be good to be home, even for a short while.

She sensed the man turn to look out of the window and she opened the book.

If instead of looking out of the window the man in the next seat had happened to glance down at Cass's notebook, he would have been astonished to see spidery black writing materialising on the page. Not that he would have been able to read it. The words were from a dialect of Ancient Greek so obscure even the most accomplished modern scholars were unaware of its existence.

This was the language Cass spoke, although again the

man on her left would not have realised. Wherever Cass went in the world, whoever she spoke to, the listener heard the words in their own language.

The writing stopped. The pages crackled as she peeled them apart, taking care not to smudge the fresh ink. Cass studied the message. Noon. Four days' time. The furthest edge of Europe. She would drink mint tea with the owner of a textile company whose business was booming thanks to the influx of refugees from war-torn regions butting against his city's borders. Refugees whose small fingers moved quickly and who never complained.

Cass waited, but the remainder of the page stayed blank. Sometimes the message appeared instantly, fully fleshed; sometimes it remained vague until moments before.

Over the years, the book had sent her to palaces and slums, in times of war and peace. She'd visited every continent, every country and every generation and gender. There were Evangelinas in every town in every country on every continent. People whose lives crossed and intertwined.

She couldn't force change; she could only speak the truth and hope to be believed. Experience had taught her that from what she'd seen today, they had believed her. That they would change.

Closing the book and retying the ribbon she rested her head against the seat and waited to land.

Somewhere out there, in the darkness, a thirteen-year-old sewing machinist named Fatima was plummeting into sleep, despite her throbbing fingers and aching back. A girl whose life was destined to end abruptly in four days' time when she would step in front of a delivery truck on her way to work.

Usually, Fatima was too exhausted to dream. When she did, she dreamed of returning home. She dreamed of the friends she'd left behind. Her school. Her bedroom. Her grandmother who refused to leave, even when the bombs began to fall. She dreamed of her past.

But tonight she would dream of a woman in white.

And the future.

She nodded.

'I'm sure I don't need to remind you that I promised this was a closed interview. Hetty's got the proof in her hand, so none of you need to worry about anything getting out. Unless you choose to, of course. And, Jez, I do not have time to go off on tangents so please can you move so we can begin.'

The two crossed by the coffee table, careful to keep a clear distance.

'No way is this the interview for a reality TV show,' Jez said, dropping to the floor by Hetty. 'I don't buy all that tough questions stuff; there's something else happening here.'

Duff plonked himself on the sofa, flopped his knees wide apart. 'Whatever. Cass, if you can catch Saint Jez out, I can't imagine what you've dug up on me. So no need with all that fakey let's have a nice chat about things first. Go for it. Tell me what you've got. I can take it.'

Cass laughed. 'I'm sure you can.'

He laced both hands round the back of his neck.

'Give it to me?'

selection process in order to make sure we choose candidates capable of making the right decisions. And so far it's going incredibly well.' She flashed a rueful smile. 'Although I realise it might not seem that way to you.'

'You don't say,' Jez muttered.

Then he slapped his palm to his forehead and exclaimed, 'God, I am so dumb.'

'You don't say,' Duff echoed, raising his eyebrows.

'No. This is it now! This is the show right this minute.' Gabbling now, he continued. 'There is no *This Careless Life*, it's a set up. We're being recorded or live streamed somewhere. That's why she keeps telling us to look in the camera. People are watching us. Right. Now.' He jabbed the air to punctuate the words. 'It's a prank. We're being pranked.'

Cass laughed softly. 'I can assure you, this is no prank.'

Hetty covered her face with her hands. 'Oh God. My mum is going to go mental.'

'You'll be fine,' Jez said.

'Well, you won't,' Duff retorted. 'Running away from the scene of an accident. That's a crime.'

'Enough,' Cass said. 'Now, listen carefully. I can assure you, this is no prank. You still have your phone, Hetty, yes?'